Bing Wang and Ning Zhang (Eds.)

Selected Papers from 6th Annual Conference of Energy Economics and Management

MDPI

This book is a reprint of the Special Issue that appeared in the online, open access journal, *Sustainability* (ISSN 2071-1050) in 2016, available at:

http://www.mdpi.com/journal/sustainability/special_issues/6th-annual-conf-EEM

Guest Editors
Bing Wang
School of Economics, Jinan University
China

Ning Zhang
School of Economics, Jinan University
China

Editorial Office
MDPI AG
St. Alban-Anlage 66
Basel, Switzerland

Publisher
Shu-Kun Lin

Managing Editor
Guoshui Liu

1. Edition 2016

MDPI • Basel • Beijing • Wuhan • Barcelona • Belgrade

ISBN 978-3-03842-274-7 (Hbk)
ISBN 978-3-03842-275-4 (electronic)

Table of Contents

List of Contributors

Xiaolang Chen Key Laboratory of Advanced Materials Technology Ministry of Education, School of Materials Science and Engineering, Southwest Jiaotong University, Chengdu 610031, China.

Ziyue Chen Institute of Industrial Economics, Jinan University, Guangzhou 510632, China.

Leihua Dong Beijing Engineering Corporation Limited, Power China, Beijing 100024, China.

Minzhe Du Department of Economics, School of Economics, Jinan University, Guangzhou 510632, China.

Ye Duan Key Laboratory of Ocean Energy Utilization and Energy Conservation of Ministry of Education, Dalian University of Technology, Dalian 116024, China.

Wuyang Hu College of Economics and Management, Huazhong Agricultural University, No. 1 Shizishan Rd., Wuhan 430070, China; Department of Agricultural Economics, University of Kentucky, 313 Charles E. Barnhart Bldg., Lexington, KY 40546-0276, USA.

Jianhuan Huang School of Economics and Trade, Hunan University, Changsha 410079, China.

Ruting Huang China Center for Energy Economics Research, School of Economics, Xiamen University, Xiamen 361005, China.

Wenchao Huang Research Associate, Fishery Machinery and Instrument Research Institute, Chinese Academy of Fishery Sciences, No. 63 Chifeng Avenue, Yangpu District, Shanghai 200092, China.

Tong Jiang School of Economics and Business Administration, Chongqing University, Chongqing 400044, China.

Peihao Lai Department of Mathematics, College of Information Science and Technology, Jinan University, Guangzhou 510632, China.

Ke Li College of Mathematics & Computer Science, Hunan Normal University, Changsha 410081, China.

Nan Li Key Laboratory of Ocean Energy Utilization and Energy Conservation of Ministry of Education, Dalian University of Technology, Dalian 116024, China.

Boqiang Lin Collaborative Innovation Center for Energy Economics and Energy Policy, China Institute for Studies in Energy Policy, Xiamen University, Xiamen 361005, China.

Jing Lin College of Energy, School of Energy Research, Xiamen University, Xiamen 361005, China.

Qing Liu Faculty of Management and Economics, Dalian University of Technology, Dalian 116023, China.

Yu Liu Institute of Policy and Management, Chinese Academy of Sciences, Beijing 100190, China.

Rundong Luo Business School, Shandong University,Weihai, No. 180 West Culture Road,Weihai 264209, China.

Hailin Mu Key Laboratory of Ocean Energy Utilization and Energy Conservation of Ministry of Education, Dalian University of Technology, Dalian 116024, China.

Peng Ou Electric Power Research Institute of China Southern Power Grid, Guangzhou 510060, China.

Xiongfeng Pan Faculty of Management and Economics, Dalian University of Technology, Dalian 116023, China.

Xiaoxue Peng Faculty of Management and Economics, Dalian University of Technology, Dalian 116023, China.

Haijuan Qiao National Research Institute for Rural Electrification, Hangzhou 310012, China.

Malin Song Collaborative Innovation Center for Ecological Economics and Management, Anhui University of Finance and Economics, Bengbu 233030, China.

Han Su State-Province Joint Engineering Research Lab in Geospatial Information Technology for High Speed Railway Safety, Southwest Jiaotong University, Chengdu 611756, China.

Zuoren Sun Business School, Shandong University,Weihai, No. 180 West Culture Road,Weihai 264209, China.

Bing Wang Institute of Resource, Environment and Sustainable Development, and Department of Economics, Jinan University, No. 601, West Huangpu Avenue, Tianhe District, Guangzhou 510632, China.

Ping Wang School of Management, Jinan University, Guangdong 510632, China.

Ran Wang Resources and Environment Economic Research Center, China University of Geosciences, Wuhan 430074, China.

Wei Wang Co-innovation center for institutional construction for Jiangxi eco-civilization, Jiangxi University of Finance and Economics, Nanchang 330013, China.

Xiaolin Wang School of Economics and Management, China University of Geosciences, Wuhan 430074, China.

Yuelong Wang Center for Regulation and Competition, Jiangxi University of Finance and Economics, 169 Shuanggang East Road, Nanchang 330013, China.

Tao Wu School of Economics, Jiangxi University of Finance and Economics, 169 Shuanggang East Road, Nanchang 330013, China.

Jiejin Xia School of Economics and Trade, Hunan University,.Changsha 410079, China.

Jianzhong Xiao Economic Forecasting Department, State Information Center, Beijing 100045, China.

Jianzhong Xiao Resources and Environment Economic Research Center, and School of Economics and Management, China University of Geosciences, Wuhan 430074, China.

Hualin Xie Co-innovation center for institutional construction for Jiangxi eco-civilization, Jiangxi University of Finance and Economics, Nanchang 330013, China.

Xue Xie Co-innovation center for institutional construction for Jiangxi eco-civilization, Jiangxi University of Finance and Economics, Nanchang 330013, China.

Yaobo Yan Faculty of Management and Economics, Dalian University of Technology, Dalian 116023, China.

Xin Yao Collaborative Innovation Center for Energy Economics and Energy Policy, and China Center for Energy Economics Research, School of Economics, Xiamen University, Xiamen 361005, China.

Yanni Yu Institute of Resource, Environment and Sustainable Development (IRESD), Jinan University, Guangzhou 510632, China.

Conglin Zhang Institute of Policy and Management, Chinese Academy of Science, Beijing 100190, China.

Daobei Zhang Co-innovation center for institutional construction for Jiangxi eco-civilization, Jiangxi University of Finance and Economics, Nanchang 330013, China.

Ning Zhang Institute of Poyang Lake Eco-economics, Jiangxi University of Finance and Economics, Nanchang 330013, China; Institute of Resource, Environment and Sustainable Development, and Department of Economics, Jinan University, No. 601, West Huangpu Avenue, Tianhe District, Guangzhou 510632, China.

Rui Zhao State-Province Joint Engineering Research Lab in Geospatial Information Technology for High Speed Railway Safety, Southwest Jiaotong University, Chengdu 611756, China.

Dequn Zhou College of Economics and Management, Nanjing University of Aeronautics and Astronautics, 29 Jiangjun Avenue, Nanjing 211106, China.

Guzhen Zhou School of Economics, Jiangxi University of Finance and Economics, North Office Tower Rm. 824, No. 169 Shuanggang East Rd., Nanchang 330013, China.

Bangzhu Zhu School of Management, Jinan University, Guangdong 510632, China.

About the Guest Editors

Bing Wang is the Director of the Human Resource Department, and a Professor at the Department of Economics, Jinan University. He received his PhD in Economics from Wuhan University. His current research interests include ecological economics and productivity analysis. His current works have been published in international journals such as *Energy Policy, Energy, Sustainability, Social Science Journal, Technological Forecasting and Social Change* and the top Chinese journal: *Jingji yanjiu*. He is the Pearl River Scholar of Guangdong Province.

Ning Zhang is a Professor at the Department of Economics, Jinan University. He received his PhD in Environmental Economics from Inha University. His current research interests include environmental and energy economics. He has published over 40 papers in Web of Science (SCI and SSCI) journals such as *Technological Forecasting and Social Change, Applied Energy, Computational Economics, Energy Economics, Energy Policy, Resource and Energy Economics, Journal of Cleaner Production, Resources, Conservation and Recycling, Ecological Indicators* and the *Social Science Journal*, among others. Ten of his papers have been selected as the Essential Science Indicators (ESI) top 1% highly cited papers, and one of his papers was selected as being in the ESI top 0.1% of hot papers. He is also serving as an editorial member of *Technological Forecasting and Social Change* and is an associate editor of the *Social Science Journal*, and a guest editor of *Sustainability*.

Preface to "Selected Papers from 6th Annual Conference of Energy Economics and Management"

This book is the Proceedings of the 6th Annual Conference of Energy Economics and Management which consists of selected papers published in a Special Issue in *Sustainability*. This conference was held in Guangzhou, China, 11–13 November 2015. The conference was co-hosted by the National Natural Science Foundation of China (NSFC), the Project Management Research Committee of China, and Jinan University (Guangzhou). Nearly 400 experts and scholars from more than 100 colleges and scientific research institutes in the field of energy economics and management took part in the conference. The theme of the conference was "Toward a Sustainable Low Carbon China: New Global Carbon Reduction Agreements and Energy Policy". In addition to carbon emission reduction, resources and sustainability management, renewable and sustainable energy, energy and environmental modeling, and many other topics in sustainable energy research were tied to the conference theme. This conference provides opportunities for delegates to exchange new working papers and application experiences, in order to establish research or collaboration relations. Our deepest thanks go to *Sustainability*, which provided a Special Issue for this conference. The scope of this Special Issue encompasses topics related to energy economics and management in Asia, at both the macro- and micro-levels, as presented at the 6th Annual Conference of Energy Economics and Management. This book includes 18 papers selected and published in this Special Issue, after the standard peer review process. The first paper is an editorial that introduces all the papers briefly covering both the topic and methodology. These selected papers concentrate on issues of "Energy Economic and Management" in order to control China's excessive energy consumption and resulting CO^2 emissions problem, thus finally improving environmental quality and accelerating the development of an environmentally-friendly society.

Bing Wang and Ning Zhang
Guest Editors

Toward a Sustainable Low-Carbon China: A Review of the Special Issue of "Energy Economics and Management"

Ning Zhang and Bing Wang

Abstract: Severe environmental quality deterioration, along with predatory exploitation of energy resources, are generally associated with economic growth, especially in China. Against this background, the 6th Annual Conference of Energy Economics and Management provides a platform for examining outperforming governance factors and mechanisms of energy economics and policy. Thanks to *Sustainability* for providing this special issue. This editorial highlights the contents and methodologies of the special issue for this conference, presenting diverse issues in energy economics and management. We also suggest guidelines for future study in energy economics and management.

Reprinted from *Sustainability*. Cite as: Zhang, N.; Wang, B. Toward a Sustainable Low-Carbon China: A Review of the Special Issue of "Energy Economics and Management". *Sustainability* **2016**, *8*, 823.

1. Background of the Special Issue

For the past two decades, with the continuous progress of reform and opening-up, China has been one of the world's fastest growing economies. However, serious environmental quality deterioration and predatory exploitation of resources have occurred as the economy has expanded. For instance, China has seen a dramatic increase in energy consumption and CO_2 emissions. It is now the world's largest CO_2 emitter, accounting for 25.9% of the world's total, creating severe environmental problems, such as toxic haze. These environmental effects have stimulated a growing public consciousness of the need for environmental protection. Furthermore, ignoring the effective control of environmental quality is likely to cause further contradictions between economic growth and environmental quality, which is not conducive to social stability and sustainable development [1]. Therefore, lowering the environmental costs of economic growth to strike a balance between environmental protection and economic growth is an urgent issue at present.

Against this background, the 6th Annual Conference of Energy Economics and Management was held in Guangzhou, China, 11–13 November 2015. The conference was co-hosted by the National Natural Science Foundation of China (NSFC), the Project Management Research Committee of China, and Jinan University (Guangzhou). Nearly 400 experts and scholars from more than 100 colleges and scientific research institutes in the field of energy economics and management took part in the conference.

The theme of the conference was "Toward a sustainable low carbon China: New global carbon Reduction Agreements and Energy Policy". In addition to carbon emission reduction, resources and sustainability management, renewable and sustainable energy, energy and environmental modeling, and many other topics in sustainable energy research were tied to the conference theme. This conference provides opportunities for delegates to exchange new working papers and application experiences, in order to establish research or collaboration relations. My deepest thanks go to *Sustainability*, which provided a special issue for this conference. The scope of this Special Issue encompasses topics related to energy economics and management in Asia, at both the macro- and micro-levels, as presented at the 6th Annual Conference of Energy Economics and Management.

A total of 17 journal papers were selected and published in this special issue, through the standard peer review process (http://www.mdpi.com/journal/sustainability/special_issues/6th-annual-conf-EEM). These selected papers concentrate on issues of "Energy Economic and Management" in order to control China's excessive energy consumption and resulting CO_2 emissions problem, thus finally improving environmental quality and accelerating the development of an environmentally-friendly society. The 6th Annual Conference of Energy Economics and Management was the cornerstone of a new platform for academic networking to present ongoing research activities and exchange research ideas in the area of energy economics and management. This conference has made a positive contribution to the overall level of scholarship in the field and inspired young researchers in energy economics and management.

2. Important Issues for China's Energy Economics and Management

Energy economics and management is an important topic that attracts attention from both academia and government policy-makers. As a sustainability science, it is composed of the three pillars of the environment, economics, and social studies. Therefore, most papers on this issue are based on the harmonization of these perspectives and encompass topics related to energy economics and management, at both the macro-levels and micro-levels, such as low carbon and sustainable development, resources and sustainability management, and energy and environmental modeling. Most of the papers address issues of energy economics and management in China. China's economy has made remarkable achievements since the continuous progress of reform and opening-up, but the development of the economy has inevitably generated severe environmental quality deterioration and energy shortages. For example, electricity is a driving force of economic development and the electricity industry is a basic industry of national economy, closely related to economic development. However, China has been experiencing large-scale power shortages, including blackouts and brownouts, which seriously affect the development of the economy and cause large economic losses [2]. To address this problem, quantitative assessment of the economic impacts of power shortages is required. One study presented in this issue uses a static computable general equilibrium (CGE) model to simulate the economic impacts of blackouts and brownouts and compiles a social accounting matrix (SAM) of eight sectors of China in 2007. The simulation results show that power shortages have a significant adverse impact on economic development and that the industrial sector suffers most from blackouts. The main reason for power shortages is the specific administrative pricing system in China. Low electricity prices lead to insufficient construction and blackouts in the long term, with brownouts in the short run. Therefore, the paper proposes power system reforms to address the problem of power shortages [2].

China's economic growth is usually described as "unstable, unbalanced, uncoordinated and ultimately unsustainable" [3]. Based on the concept of sustainable development, achieving green growth has been an important target for the Chinese government. Green growth can foster economic growth and development while ensuring that natural assets continue to provide the resources and environmental services the nation requires. However, accurate measurement of green development performance remains an issue. In proposing a green development growth index (GDGI) for measuring the changes in sustainable development over time, Li et al. [4] adopted a new approach which allows for the incorporation of heterogeneity across provinces and non-radial slack in the conventional green development index. Li et al. measure green development performance based on the four main types of environmental pollution in 30 provinces of China over the period 2000–2012. Their main results include: (1) the average score of GDPI in China during the sample period is 0.398, indicating a low level of green development, but it is improving 2.58% per year driven by innovation effects in recent years; and (2) green development is primarily led by the eastern region. Nevertheless, the average score of GDPI for the eastern region is 0.525, still a low level of green development. This implies that the eastern region's performance is better than the central and the western regions [4]. For the individual provinces, the GDPI scores show huge differences, and the eastern provinces performed better than the central and western provinces.

Interprovincial competition is believed to be one of the most essential driving forces for China's economic growth, as proposed by Li and Zhou [5], based on the promotion tournament model. However, competition among provinces may lead to serious pollution problems and high environmental costs because provincial officials are evaluated based on economic performance. This causes local governments to compete with each other to ease environmental regulation in order to attract investment for economic growth [6]. Thus, whether competition among local governments leads to a race to the bottom, or green development, remains uncertain. A paper in this issue thus explores the strategic interactions in provincial development using an environment-economic indicator, pollution intensity in China. It tests the strategic interactions among two groups of regions according to whether the pollutant is regulated based on the spatial lag term by employing the spatial Durbin model. The results show that the "race to the bottom" is not the case at the national level, and provincial governments do implement strategies for environmental protection. It also shows that the "race to green development" hypothesis is supported in the east for pollutants with obligatory emission reduction targets, such as SO_2 and Chemical Oxygen Demand (COD). Moreover, increasing environmental legislation has been effective in reducing PI, especially for pollutants with obligatory reduction targets [6].

The rapid development of the national economy generates enormous energy consumption and CO_2 emissions. With the increasing prevalence of climate disasters, scholars have begun to focus on CO_2 emissions and their relationship to economic growth, especially for industrial economies. One of the papers in the special issue analyzes China's five major IPPU department CO_2 emissions levels from 1991–2012 and examines the relationship between IPPU CO_2 emissions and the economic level. It shows that the overall emissions level is indeed rising yearly, and that steel and iron alloy manufacturing and nonmetal manufacturing account for roughly 80% of the total emissions. The IPPU CO_2 emissions and the corresponding gross industrial output value do not present a classic Kuznets curve in most industrial sectors because of the increasing size of the population employed in industry [1].

Most previous studies empirically test the environmental Kuznets (EKC) hypothesis at the regional or industry level. Nevertheless, with the development of spatial econometrics, researchers have the opportunity to incorporate spatial dynamics into their analyses of the trends and drivers of China's carbon emissions [7]. Liu et al. proposes the stochastic impact regression on population, affluence, and technology (STIRPAT) model to investigate the factors driving industrial carbon emissions based on panel data from 2006–2010. The determinants of regional industrial carbon emissions are investigated by a spatial Durbin model and a geographically- and temporally-weighted regression is applied to investigate temporal and spatial variations in the impacts of these driving factors on the scale and intensity of regional industrial carbon emissions. The results indicate that spatial correlation, heterogeneity, and spillover effects should be taken into account when formulating policies aiming at reducing industrial carbon emissions [7]. One study in this special issue estimates the contribution of industry structure adjustment to the carbon intensity target of Guangdong Province, which had ranked first in terms of economic performance for over 20 successive years. By using the Markov chain model, least squares support vector machine, and scenario prediction, the obtained results show, in the ideal scenario, that carbon intensity in 2015 will be 25.53% less than in 2010, making a 130.94% contribution to the carbon intensity target. Meanwhile, in the conservative scenario, carbon intensity in 2015 will be 23.89% less than in 2010, making a 122.50% contribution to the carbon intensity target [8].

In order to control the country's excessive energy consumption and resultant CO_2 emission problem, China has committed to reducing its CO_2 emissions by 40%–45% from 2005 levels per unit of GDP by 2020 [9]. Therefore, Sun et al. [9] developed the traditional data envelopment analysis (DEA) model based on improved Kuosmanen environmental DEA technology, which avoids the effect of positive shadow prices on undesirable outputs and designs a dual model to analyze shadow prices on CO_2 emissions for each province and its industrial sectors in China [9]. Likewise, by adopting an energy-input based directional function, Lai et al. [10] have calculated the shadow price of energy (i.e., coal, oil, gas, and electricity) across 30 areas in China. The study estimates the macro-energy efficiency index, which is divided into intra-provincial technical efficiency, the allocation efficiency of energy input structure, and the inter-provincial energy allocation efficiency [10]. This is because industrial production is the primary contributor to carbon emissions in China, and industrial land is an important input to industrial production. Therefore, modeling the dynamic changes in carbon emission performance and industrial land use in recent years and forwarding constructive policy implications is urgently needed [11]. Wang et al. [11] use a global directional function (DDF) approach to compute the total-factor carbon emission performance of industrial land use (TCPIL) and find that the TCPILs for China and its three regions rise over the study period, and the eastern region performed much better in TCPIL than the central and western regions. This study uses a non-radial Luenberger carbon emission performance of industrial land use (NLCPILs) index to model the dynamic changes in the TCPIL, with results showing that the NLCPILs for China were greater than zero in most years of the study period, and their growth was primarily driven by ECs before 2009 and by TC subsequently. Last, the study analyzes factors which significantly affect carbon emission reduction, including more investment in industrial research and development (R and D), the implementation of carbon emission reduction policies, reduction in the use of fossil energy, especially coal, in the process of industrial production, actively learning about foreign advanced technology, solving the problem of surplus labor in industry, and the expansion of industrial development [11].

If we focus on a micro approach to CO_2 emission regulatory issues, the heating industry is a key industry, since China's heating industry is necessary for its tough winters. Further, more residents of Southern China are demanding central heating. China's heating industry, thus, accounts for 89.4% of the total energy consumption during 1985–2010 [12]. CO_2 emissions from the heating industry accounted for an average 6.1% of China's carbon emissions during 1985–2010 and there is a long-run equilibrium relationship between CO_2 emissions and factors such as energy intensity, industrial scale, labor productivity, and energy productivity by the co-integration method. The Monte Carlo technique is adopted for the risk analysis. The results indicate that there is considerable potential for CO_2 emissions mitigation in China's heating industry [12]. In addition to the above research, the threshold effect of the relationship between financial development and carbon emission intensity from the perspectives of financial scale and financial efficiency were tested. The results indicate that both raising the level of openness and improving the industrial structure will have significant positive effects on carbon intensity [13].

In addition to the issues surrounding CO_2 emissions, other resource and management issues have captured the attention of researchers. Water resources are a precondition of improving the environment. From the ecological perspective of water resources management and sustainable development, one study examines water consumption and the determinants of water utilization of Tianjin, which is located in the northeast of North China Plain and in the center of the Bohai economic circle. Using the grey correlation analysis method, the survey shows that the main factors affecting the water utilization structure are industrial structure, per capita green area, cultivated area, effective irrigation area, rural electricity consumption, animal husbandry output, resident population, and per capita domestic water use [14]. Natural gas resources [15] and the crude oil market [16] were also studied. One paper develops an optimization model for the long-term exploitation of limited natural gas reserves in China and analyzes its exploitation scale and related factors, including recovery rate, discount rate, and the gas well exhaustion cycle, that affect the optimal exploration path of this gas field [15]. Further, another paper proposes an improved method, the multivariate empirical model decomposition (EMD)-based model, to take advantage of the heterogeneous characteristics of the price movement and model them in the crude oil markets. The proposed model demonstrates performance superior to that of benchmark models [16]. In addition, agricultural labor productivity was evaluated by using the Difference-in-Differences (DID) method to examine the establishment of the Poyang Lake Eco-Economic Zone, an area surrounding Poyang Lake, the largest lake in China [17]. The results show that the establishment of the zone reduced agricultural GDP by 3.6% and this negative effect has increased each year since 2009. It was found that the lack of agricultural R and D activities and the abuse of chemical fertilizers may be the main reasons behind the negative influence of the policy [17].

With mounting pressure for a green, low carbon economy, materials selection plays a crucial role in product design. Some studies have focused on materials selection by market demand, while others have found that materials selection is also motivated by cost considerations. However, most studies have not treated environmental management as a key aspect of materials selection in product design [18]. Thus, a paper in this special issue presents an integrated multi-attribute decision-making (MADM) approach to aid selection of commercially-available materials for sustainable design. In the study, grey relation analysis (GRA) and analytic hierarchy process (AHP) are used to rank alternative materials in terms of their economic, environmental, and social performance. In addition, three alternative polymer materials, i.e., poly (vinyl chloride) (PVC), polypropylene (PP), and polyethylene (PE), are examined to determine their sustainability for plastic pipe design [18]. Naturally, as in much recent research, consumer willingness to pay (WTP) for sustainable products is analyzed. One paper presents household-level tuna steak consumption, based on an investigation of consumer preference for the attributes of tuna and the impact of eco-labels, particularly Certified Turtle Safe (CTS) labels, on consumer demand, and examines individuals' perceptions and attitudes toward farm-raised and wild-caught tuna species based on an Internet survey [19]. The empirical study shows that respondents on average preferred turtle-safe-labeled tuna steak and were likely to pay more for it and they were less likely to purchase wild-caught species. Moreover, significant heterogeneities were found across individuals regarding tuna steak purchases. The findings indicate evidence of public support for environmental friendliness, particularly with regard to eco-labeling [19].

3. Methodologies of Energy Economics and Management

The field of energy economics and management includes resource and sustainability management, the innovation of methodology, and pluralism; major concerns of the international community, and one of the main challenges of our time. Since economic theory measuring a "sustainable green" economic growth model was developed, it has been argued that sustainable economic growth only depends on technological progress in the long term. Many studies analyze the sustainability of China's economic growth based on total factor productivity (TFP) via growth accounting, or non-parametric methods. However, conventional TFP studies ignore the environmental costs of economic growth (such as CO_2 emissions). Therefore, these studies may overestimate the true contribution of TFP to output growth, thus yielding a false picture of China's green development [4]. Perhaps it is difficult for researchers to put forth a robust theory of energy economics and management, but it is certainly meaningful to propose diverse methodological approaches and raise awareness and emphasize of the importance of sustainable performance.

Most papers in this Special Issue show great interest in the CO_2 emissions [1,7,9,11,12] and the carbon intensity [8,13]. Some of these studies use the conventional regression to analyze CO_2 emissions [1,8,13], but the downside of this method is that conventional regressions cannot solve the problems of multiple inputs/outputs in these models. Therefore, a centralized DEA approach was used to determine the optimal path for controlling CO_2 emissions at the sector level for each province in China, which can avoid positive shadow prices on undesirable outputs [9]. The global directional function (DDF) and non-radial Luenberger productivity index were used to analyze the dynamic change in the total-factor carbon emission performance of industrial land use (TCPIL) [11]. This method can also be used to measure green development performance [4] and evaluate the total factor energy efficiency [10]. In addition, there is a classic environment Kuznets curve (EKC) between IPPU CO_2 emissions and the economy. The EKC hypothesis is usually verified at the regional or industry level [1]. The development of spatial econometrics and the use of the spatial panel data model incorporate spatial dynamics into the analysis of the trends and drivers of China's carbon emissions [7]. Finally, a paper evaluates the potential for reducing CO_2 emissions by co-integration analysis and scenario analysis [12].

The linear regression or linear transformation of multivariate variables is introduced as well. Using the spatial lag term by employing the regional Durbin model, the strategic interactions among regions is tested. How pollution intensity is influenced by the direct and spatial spillover effects of environmental regulation was also explored [6]. One paper uses a logit model to examine purchasing propensities [19], another takes the establishment of Poyang Lake Eco-Economic Zone in 2009 as a quasi-natural experiment because of its exogeneity and utilizes the DID method to evaluate its influence on agricultural labor productivity [17].

Linear programming is also used in this special issue. A static CGE model is constructed and the social accounting matrix (SAM) of eight sectors of China is compiled to simulate the economic impacts of power blackouts and brownouts. The CGE model is a good method to simulate the effects of policies or the shocks of exogenous events and can be understood as an input-output model [3]. One paper proposes a new multivariate EMD-based methodology for combining distant data features during the modeling and the forecasting of crude oil price. The researchers propose a time delay embedding method to analyze a higher dimension domain and introduce the multivariate EMD model to analyze and model the geometric multi-scale data features in the higher dimensional domain [16]. Unlike crude oil in China, the natural gas economy is a coal-dominated energy structure economy and suffers from serious pollution problems. A paper develops an optimizing approach to model gas production in China and conducts a sensitivity analysis of the factors affecting the path of natural gas exploitation [15]. In the real process of materials selection, a number of variables should be taken into account. Consequently, MADM methods, such as analytic hierarchy process (AHP) and grey relational analysis (GRA) are able to rank material alternatives and help designers determine the best material. A paper introduces a computational approach for materials selection in the context of green design by using GRA, integrated with AHP [18]. Another paper used information entropy analysis and GRA to analyze the changes in water utilization structure and its impact factors using water consumption data for agricultural, industrial, domestic, and ecological areas in the city of Tianjin [14].

Due to the interdisciplinary character of energy economics and management, diverse approaches could, and should, be imported and used for better understanding of multivariate complexity, as well as for the generation of more systematic implications and suggestions [20].

4. Conclusions

Environmental problems related to carbon emissions have become increasingly serious in China, pacing increased levels of consumption, because China's energy consumption structure has given priority to coal in recent years. Thus, saving energy and reducing consumption is necessary for sustainable development, because of the non-renewability and scarcity of fossil fuels, such as coal and gas. The optimization of energy production systems is a relevant issue that must be considered if China is to adhere to reduction policies for fossil fuel consumption and regulations on CO_2 emissions ([13], p. 271). Therefore, the Chinese Government has proposed targets for reducing the energy consumption, optimizing the energy structure by lowering the reliance on fossil fuels, and promoting marketization of energy prices by reforming energy pricing mechanisms [21,22]. These issues of energy economics and management have captured the attention of the participants at the 6th Annual Conference of Energy Economics and Management.

Coordinating economic growth alongside improvement in environmental quality may be daunting, but it is certainly not impossible [23]. We must pay close attention to the shifting patterns among energy economics and management norms and practices, and to governance institutions in China to keep pace with the world and its challenges [24]. Further, most papers in this special issue emphasize the importance of creating a new paradigm of energy economics and management, as well as sustainable management geared toward value creation. The papers in this Special Issue, thus, make important contributions in helping to understand key attributes of the changes now underway.

Acknowledgments: We thank the *Editors of Sustainability* for providing this special issue. We gratefully acknowledge the financial support from the National Natural Science Foundation of China (41461118; 71473105; 71603102), National Social Science Foundation of China (14ZDB144; 15ZDA054), GDUPS (2015) and Research Center on Low-carbon Economy for Guangzhou Region.

Author Contributions: All authors made equal contributions to the work in this editorial text.

Conflicts of Interest: The authors declare no conflict of interest.

References

1. Duan, Y.; Mu, H.; Li, N. Analysis of the Relationship between China's IPPU CO_2 Emissions and the Industrial Economic Growth. *Sustainability* **2016**, *8*, 426.
2. Peng, O.; Huang, R.; Yao, X. Economic Impact of Power Shortage. *Sustainability* **2016**, *8*, 687.
3. Chen, S.; Golley, J. Green's productivity growth in China's industrial economy. *Energy Econ.* **2014**, *44*, 89–98.
4. Li, K.; Song, M. Green Development Performance in China: A Metafrontier Non-Radial Approach. *Sustainability* **2016**, *8*, 219.
5. Li, H.; Zhou, L. Political turnover and economic performance: The incentive role of personnel control in China. *Public Econ.* **2005**, *89*, 1743–1762.
6. Huang, J.; Xia, J. Regional Competition, Heterogeneous Factors and Pollution Intensity in China: A Spatial Econometric Analysis. *Sustainability* **2016**, *8*, 171.
7. Liu, Y.; Xiao, H.; Zhang, N. Industrial Carbon Emissions of China's Regions: A Spatial Econometric Analysis. *Sustainability* **2016**, *8*, 210.
8. Wang, P.; Zhu, B. Estimating the Contribution of Industry Structure Adjustment to the Carbon Intensity Target: A Case of Guangdong. *Sustainability* **2016**, *8*, 355.
9. Sun, Z.; Luo, R.; Zhou, D. Optimal Path for Controlling Sectoral CO_2 Emissions among China's Regions: A Centralized DEA Approach. *Sustainability* **2016**, *8*, 28.
10. Lai, P.; Du, M.; Wang, B.; Chen, Z. Assessment and decomposition of Total Factor Energy Efficiency: An Evidence Based on Energy Shadow Price in China. *Sustainability* **2016**, *8*, 408.
11. Wang, W.; Xie, H.; Jiang, T.; Zhang, D.; Xie, X. Measuring the Total-Factor Carbon Emission Performance of Industrial Land Use in China Based on the Global Directional Distance Function and Non-Radial Luenberger Productivity Index. *Sustainability* **2016**, *8*, 336.
12. Lin, J.; Lin, B. How Much CO2 Emissions Can Be Reduced in China's Heating Industry. *Sustainability* **2016**, *8*, 642.
13. Pan, X.; Yan, Y.; Peng, X.; Liu, Q. Analysis of the threshold Effect of Financial Development on China's Carbon Intensity. *Sustainability* **2016**, *8*, 271.
14. Zhang, C.; Dong, L.; Liu, Y.; Qiao, H. Analysis on the Impact Factors of Water Utilization Structure in Tianjin, China. *Sustainability* **2016**, *8*, 241.
15. Xiao, J.; Wang, X.; Wang, R. Research on Factors Affecting the Optimal Exploitation of Natural Gas Resources in China. *Sustainability* **2016**, *8*, 435.
16. He, K.; Zha, R.; Wu, J.; Lai, K.K. Multivariate EMD-Based Modeling and Forecasting of Crude Oil Price. *Sustainability* **2016**, *8*, 387.
17. Wu, T.; Wang, Y. Did the Establishment of Poyang Lake Eco-Economic Zone Increase Agricultural Labor Productivity in Jiangxi Province, China? *Sustainability* **2016**, *8*, 8.
18. Zhao, R.; Su, H.; Chen, X.; Yu, Y. Commercially Available Materials Selection in Sustainable Design: An Integrated Multi-Attribute Decision Making Approach. *Sustainability* **2016**, *8*, 79.

19. Zhou, G.; Hu, W.; Huang, W. Are Consumers Willing to Pay More for Sustainable Products? A Study of Eco-Labeled Tuna Steak. *Sustainability* **2016**, *8*, 494.
20. Choi, Y.; Zhang, N. Introduction to the Special Issue on "The Sustainable Asia Conference 2014". *Sustainability* **2015**, *7*, 1595–1602.
21. Zhang, N.; Wang, B.; Liu, Z. Carbon emissions dynamics, efficiency gains, and technological innovation in China's industrial sectors. *Energy* **2016**, *99*, 10–19.
22. Zhang, N.; Wang, B.; Chen, Z. Carbon emission reductions and technology gaps in the world's factory, 1990–2012. *Energy Policy* **2016**, *91*, 28–37.
23. Zhang, N.; Wei, X. Dynamic total factor carbon emissions performance changes in the Chinese transportation industry. *Appl. Energy* **2015**, *146*, 409–420.
24. Zhang, N.; Xie, H. Toward Green IT: Modeling Sustainable Production Characteristics for Chinese Electronic Information Industry, 1980–2012. *Technol. Forecast. Soc. Chang.* **2015**, *96*, 62–70.

Economic Impacts of Power Shortage

Peng Ou, Ruting Huang and Xin Yao

Abstract: The electricity industry is a basic industry of the national economy. It has experienced several large-scale power shortages, hard power shortage and soft power shortage, which have brought a great threat to China's sustainable economic development. To solve this problem better, it is necessary to make a quantitative assessment of the economic impacts of power shortage. The CGE model is commonly used for simulating economic shocks and policy effects. It describes supply, demand and equilibrium in different markets by simulating the economic mechanism through a set of equations. Once changed, the exogenous variables will affect a certain part of the system and then the whole system, leading to changes in quantities and prices. The equilibrium state will also change from one to another. A static CGE model is built in this paper, and the Social Accounting Matrix (SAM) of eight sectors of China in 2007 is compiled, in order to simulate the economic impacts of hard power shortage and soft power shortage. Simulation results show that the negative effects of power shortage on economic development are very significant, and the effects vary in different sectors. Especially, under the background of hard power shortage, the industrial sector suffers most. The economic cost of power shortage is considerable, and the main reason for it is the specific administrative pricing system in China. The low electricity price in the long term will lead to insufficient construction and hard power shortage; moreover, that in the short run would result in soft power shortage. In order to solve the problem of power shortage completely, power system reform is inevitable.

Reprinted from *Sustainability*. Cite as: Ou, P.; Huang, R.; Yao, X. Economic Impacts of Power Shortage. *Sustainability* **2016**, *8*, 687.

1. Introduction

The electricity industry is a basic and significant industry of the national economy, which is closely related to economic development. On the one hand, electricity is a driving force of economic development. The shortage of power supply will seriously affect the healthy development of the economy and can cause large economic losses [1,2]. On the other hand, the level and speed of macroeconomic development play a decisive role in determining electricity demand [3]. The process of electric power construction normally, from planning to completion and commission, will take three or even five years. The fluctuation of the economic situation, as well as the delay of electric power investment will unavoidably lead to an unbalance between electricity supply and demand. Therefore, advance planning for electric power investment to match the economic demand becomes a vital issue in electric power development. In economics, shortage means that the effective demand of the product is greater than the actual supply. In an effective market, the equilibrium price will automatically smooth out the imbalance between demand and supply, and there should be no power shortage in an equilibrium status.

China has been experiencing massive electricity shortages and power surplus repeatedly. In spite of the rapid expansion of electricity infrastructure, electricity shortage remains as a very common problem in China [4]. In the 1950s, China's power supply and demand were in balance overall. Since the 1960s, China has experienced a long-term power shortage. In 1996, the shortage even exceeded 20% of the total energy output. After the Asian financial crisis in 1997, China's economy development slowed down; so did the growth rate of electricity demand, which reached the minimum in 1998. Then, the electricity supply changed from shortage to surplus. In 1999, China had surplus power of about 10%. Since 2000, power demand has bottomed out. With high-speed economic growth, power demand increased quickly, and the unbalance between power supply and demand appeared again. In 2002, 12 provinces in China suffered slight electricity shortages. What was worse, in 2003, 21 provinces experienced power shortages, and in 2004, the number of provinces reached 24. The electricity shortage nationwide reached 31 million kilowatts. In the first half of 2005, about 26 provinces faced power shortages. In 2006, the dilemma between power demand and supply eased, but in 2007, 16 provinces still underwent power shortages or power rationing. In 2011, China's power shortage showed new characteristic, namely widespread power shortage in the case of sufficient capacity, where the growth of capacity supply exceeds the power demand. The reason for the electricity shortage before is insufficient capacity, resulting from an imbalance between electricity demand and electric power construction. However, in 2011, the new characteristic was large-scale power shortage coupled with sufficient electricity capacity, known as "soft electricity shortage". The electricity shortages can be divided into two cases: hard electricity shortage and soft electricity shortage. Hard electricity shortage refers to insufficient capacity. The market demand for electricity cannot be met even with high utilization hours of power generating units. Soft electricity shortage is the result of low utilization hours of power generating units with sufficient capacity, since the existing installed capacity is underutilized [5,6].

The general reasons for hard electricity shortage and soft hard electricity shortage are different, but the profound reason for both lies in the monopoly of the power industry and its administrative pricing. The Chinese electricity industry has evolved since the early 1990s into a "dual system", with dominant state planning at the core and a decentralized generation system, owned by private enterprises and government organizations at different levels [7]. In 2002, the electrical power industry in China had experienced a profound electricity reform. China has introduced limited competition in generation, though the rest of the electricity chain has been regulated heavily [8]. In China, the electricity price is controlled by the state, and the coal price has been de-regulated by the government. The generation plants are paid by the on-grid power price, calculated as the costs plus the product of the allowed rate of return and the audited asset of power plants, by transmission/distribution companies rather than end-users [9]. In China, the electricity price is relatively low compared to other countries, and it cannot effectively reflect the generating cost [10]. The reasons for hard power shortage are mainly as follows. The first reason is the fast growth of electricity demand as the Chinese economy grew quickly in these years. Moreover, insufficient capacity is the major cause. In light of the special electricity system in China, the benefits of power generation plants are decided by administrative measures instead of the market. The market mechanism is distorted in the Chinese power industry. The long-term low electricity price and the single property rights hinder the free flow of capital, which cause insufficient construction and hard power shortage. As for soft power shortage, the conflict between market coal price and planned electricity price plays an important role. In China, coal is taken as the primary energy. The generation of coal-fired units accounts for about 70% of the total generation, and the installed capacity of coal-fired power plants accounts for approximately 75% of the total installed capacity in China, meaning that the power generating market is affected significantly by coal price and output [6]. Since the price of coal rises much more than the magnitude of electricity price adjustments, the more electricity power plants generate, the more the losses. Even with sufficient capacity, the power plants are unwilling to produce more electricity due to the high generation costs and low electricity price, which causes the widespread soft power shortages. To sum up, the underlying cause for power shortage is low electricity price, while the difference lies in that hard power shortage is caused by low price in the long term, and soft power shortage comes from that in the short run. The power shortage reflects discordance between China's electricity industry and overall economic development, which is not conducive to the healthy development of the electricity industry and the national economy.

Studies on the economic impact of power shortage are of practical significance. On the one hand, analysis of the relationship between economic output and power demand is conductive to making decisions on electric power construction and investment. From another perspective, the study of the economic cost of electricity shortage provides a reference for current power policies and future electricity system reform. However, there is little research on this, in particular the application of the CGE model to calculate the economic impact of power shortage, so this is the objective of our research.

The rest of this paper is organized as follows. Section 2 reviews the literature. Section 3 sets up the model for the analysis. Section 4 introduces the data, and

Section 5 presents empirical results. Section 6 concludes and proposes some policy suggestions.

2. Literature Review

Power shortage, an important problem in China, has attracted much attention. Lin [11] studied the impacts of power shortages, the short-run responses and long-term strategies of the government to solve this problem. He proposed that it was necessary to build an early warning system and to make a national systematization plan. Zhang et al. [5] analyzed the mechanism of power shortages in China with the annual data of thirty provinces from the years 2006 to 2010. They found that the capacity shortage and capacity under-utilization are the main causes for power shortages in China and that capacity shortage can be solved by a "free-investment" policy, and capacity under-utilization can only be solved by a "coal-price-inflation-internalizing" policy. Ming et al. [6] studied China's large-scale power shortages of 2004 and 2011 and their differences. Cheng [12] analyzed the impacts of the electricity power shortage on China's GDP based on an econometric analysis with the data from 1953 to 2010. They found that the growth of electricity generation would encourage GDP growth, but not vice versa. Ma et al. [13], Fisher [14], Thomson [15], Yuan [16], and so on, have also investigated recent nationwide shortages. However, the problem of power shortages in China is mostly cyclical, and the relationship between electric power and economic development is complex [17]. Many sectors would be affected, and the degree of influence varies. From one aspect, any change in the economic system will directly or indirectly affect the power industry; what is more, any change in the electricity industry itself will firstly influence a certain sector and then affect other sectors, even the whole economic system [18]. Few of the previous studies have measured the economic impacts of power shortage quantitatively and comprehensively. In order to analyze the relationship between electricity and economic development, it is necessary to establish a mathematic model that contains all of the sectors in the economic system and that can reflect the connection among these sectors. Based on the general equilibrium theory, the computable general equilibrium (CGE) model, which can effectively reflect linkages among different sectors in the economic system, presents the particular impact of a certain change on each sector and the whole system. Obviously, the CGE model can address those two problems, to a certain extent, and can also satisfy the need for research on the relationship between electricity and economic development.

Since Johansen [19] built the first CGE model in 1960, it has gradually become an important tool for economic policy simulation. The CGE model can be understood as an input-output model in which the price feedback mechanism is introduced. The CGE model is generally used for international trade, tax policy, environmental protection, economic reform and economic development policy [20–22]. The electricity sector is an important sector in the macro-economy, and many researchers have studied this problem based on the CGE model. The application of the CGE model to the electric power sector can be summarized from the following aspects.

For the previous studies, some researchers focus on the relationship between electricity supply and economics based on the CGE model [23–27], and many scholars, such as [28–32], assessed the impacts from the electricity demand side. Moreover, the impact of the increase of electricity price is of great significance, and many studies have focused on this problem in China and abroad [33–37]. In addition, the core of energy reform is to straighten out energy prices and form a proper and transparent pricing mechanism. Energy prices in China are controlled by government pricing and cross subsidies, which will result in the low efficiency of energy consumption in the long run. China, with India and Russia, has followed the patterns of reducing energy subsidies, which are characterized by two steps forward, one step back [38]; thus, further energy subsidy reform is needed. Existing research [39,40] mostly focuses on the impacts of energy subsidies on residential living, energy consumption and carbon dioxide emissions and less on the effects of eliminating the subsidy on industrial development and industrial structure. The partial equilibrium analysis and input-output models are commonly used. Hosoe [41], Akkemi [42], Li et al. [43] and Chi et al. [44] applied a CGE model to analyze the impacts of power market reform and policies.

To sum up, using the CGE model, these researchers study the impacts of the electric power sector from many aspects, and CGE is a good method to simulate policies or the shocks of exogenous events. However, much attention is focused on electricity demand or supply, price and market reform; few of them have ever analyzed the influences of power shortages on economics with the CGE model. In China, despite the rapid expansion of electricity infrastructure, electricity shortages remain common. The problem of power shortages in China is mostly cyclical, and undoubtedly, electricity shortages would hamper China's healthy development. The CGE models not only involve the relationship and diffraction effects among various sectors, which the partial equilibrium analysis cannot achieve, but also overcome the defect of fixed input coefficients in the input-output model, so it is effective in structural analysis to simulate power shortages and to calculate the equilibrium after disturbance and quantify the impacts of shocks on the economic system. Thus, in this paper, we apply the CGE model to analyze the economic impacts of power shortages.

3. Model Construction

In the CGE model, the general equilibrium theory is transformed from an abstract form into a realistic and computable one. Actually, it uses a set of equations to describe supply, demand and equilibrium in the economic system. In these equations, there are both exogenous variables (economic shocks) and endogenous variables (quantities and prices). The impacts of exogenous shocks on any sector will spread to the whole system, leading to the changes in quantities and prices. The equilibrium state changes from one to another. When external conditions (exogenous variables) change, new equilibrium quantities and prices can be obtained by solving the CGE model. We build our model based on [45]; thus, the static CGE model used in this paper consists of five parts, namely production, income, demand, price and international trade, which will be described in detail below.

3.1. The Production Part

The production function is the core in this part, and the main economic agents are producers. The CGE model assumes that each production sector consists of many similar companies, and enterprises from different sectors produce different products [46], so each sector is in complete competition. Generally, labor and capital are the main input factors in the production function. In reality, as energy becomes increasingly scarce and important, we put electricity, coal and petroleum into a nested constant elasticity of substitution (CES) production function in this paper.

A standard CES function is as follows:

$$Q = A(\delta_1 {x_1}^{\rho} + \delta_2 {x_2}^{\rho})^{\frac{1}{\rho}} \tag{1}$$

where x_1 and x_2 represent two different inputs; A represents factors affecting output, except input factors, such as technological progress, which is known as the "factor of the total-factor of production" or the "technological progress parameter". δ_1, and δ_2 indicate the shares of two inputs, and $\delta_1 + \delta_2 = 1$. ρ is used to define the elasticity of substitution between two inputs. However, it is not the substitution elasticity, and its relationship with elasticity of substitution σ is as follows:

$$\sigma = \frac{1}{1-\rho}, \left(\rho = 1 - \frac{1}{\sigma}\right) \tag{2}$$

When the CES function includes two or more inputs, the inputs will be nested into the function. For example, if there are three input factors, capital K, labor L and land E, K and L are firstly organized as a composite input X_{KL} in the form of the CES function; then X_{KL} and E are put into the final output function.

$$Q = A(\delta {X_{KL}}^{\rho} + (1-\delta)E^{\rho})^{\frac{1}{\rho}} \tag{3}$$

$$X_{KL} = A_{KL}(\delta_{KL} K^{\rho_{KL}} + (1-\delta_{KL}) L^{\rho_{KL}})^{\frac{1}{\rho_{KL}}} \tag{4}$$

The production function in the paper includes capital, labor, energy and other intermediate inputs, so we adopt a nested CES production function as follows and the specific nesting of the production function is shown in Figure 1.

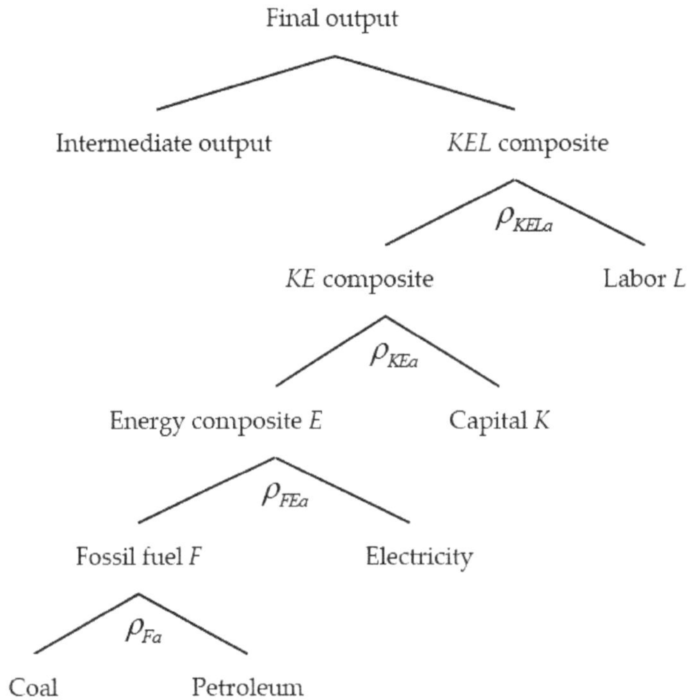

Figure 1. Nesting of the production function.

In order to explore the role of energy, particularly electricity in the economic system, we used the nested function mentioned above. The detailed analysis can be found in the relevant literature [45]. The elasticity in each nested layer will be discussed and determined in the following section in detail. Each nested function is shown as follows. The Leontief production function [47,48] is used in the first layer:

$$QA_a = \min\left(QKELA_a/ia_{1a},\ QINTA_a/ia_{2a}\right) \tag{5}$$

The proportions of input and output are:

$$QKELA_a = ia_{1a} \times QA_a,\ QINTA_a = ia_{2a} \times QA_a \tag{6}$$

When the production tax accounting for a fixed proportion of total output is added into the production function, then:

$$(1 - t_a^P) \times PA_a \times QA_a = PKELA_a \times QKELA_a + PINTA_a \times QINTA_a \tag{7}$$

The second layer is a composite function of capital, energy composite and labor, shown as follows:

$$QKELA_a = A_a^{KEL} \times \left(\delta_a^{KEL} QKEA_a^{\rho^{KEL}} + \left(1 - \delta_a^{KEL}\right) QLD_a^{\rho^{KEL}}\right)^{\frac{1}{\rho^{KEL}}} \tag{8}$$

The first-order condition for cost minimization is:

$$\frac{PKEA_a}{w} = \frac{\delta_a^{KEL}}{1 - \delta_a^{KEL}} \left(\frac{QLD_a}{QKEA_a}\right)^{1-\rho^{KEL}} \tag{9}$$

The price of the composite product is:

$$PKELA_a \times QKELA_a = PKEA_a \times QKEA_a + w \times QLD_a \tag{10}$$

The second layer is the composite function of intermediate inputs, except energy, which also has the form of the Leontief production function.

$$QINT_{cne,a} = ia_{cne,a} \times QINTA_a \tag{11}$$

Additionally, its price is:

$$PINTA_a = \sum_{cne} ia_{cne,a} \times PC_{cne} \tag{12}$$

The third layer is a composite function of capital and energy (*KE*):

$$QKEA_a = A_a^{KE} \times \left(\delta_a^{KE} QFEA_a^{\rho^{KE}} + \left(1 - \delta_a^{KE}\right) QKD_a^{\rho^{KE}}\right)^{\frac{1}{\rho^{KE}}} \tag{13}$$

Its first-order condition for cost minimization is:

$$\frac{PFEA_a}{r} = \frac{\delta_a^{KE}}{1 - \delta_a^{KE}} \left(\frac{QKD_a}{QFEA_a}\right)^{1-\rho^{KE}} \tag{14}$$

Its price is:

$$PKEA_a \times QKEA_a = PFEA_a \times QFEA_a + r \times QKD_a \tag{15}$$

The fourth layer is a composite function of fossil fuel composite and electricity (FE). The fossil fuel composite consists of coal and petroleum, shown as follows:

$$QFEA_a = A_a^{FE} \times \left(\delta_a^{FE} QFA_a^{\rho^{FE}} + \left(1 - \delta_a^{FE}\right) QEle_a^{\rho^{FE}}\right)^{\frac{1}{\rho^{FE}}} \tag{16}$$

Its first order condition for cost minimization is:

$$\frac{PFA_a}{PEle} = \frac{\delta_a^{FE}}{1 - \delta_a^{FE}} \left(\frac{QEle_a}{QFA_a}\right)^{1-\rho^{FE}} \tag{17}$$

Its price is:

$$PFEA_a \times QFEA_a = PFA_a \times QFA_a + PEle \times QEle_a \tag{18}$$

The fifth layer is the composite function of coal and petroleum (F):

$$QFA_a = A_a^{F} \times \left(\delta_a^{F} QCoal_a^{\rho^{F}} + \left(1 - \delta_a^{F}\right) QOil_a^{\rho^{F}}\right)^{\frac{1}{\rho^{F}}} \tag{19}$$

Its first order condition for cost minimization is:

$$\frac{PCoal}{POil} = \frac{\delta_a^{F}}{1 - \delta_a^{F}} \left(\frac{QOil_a}{QCoal_a}\right)^{1-\rho^{F}} \tag{20}$$

Its price is:

$$PFA_a \times QFA_a = PCoal \times QCoal_a + POil \times QOil_a \tag{21}$$

3.2. The Product Market Part

Taking input and output into consideration, products produced by different sectors are not necessarily the total supply in the domestic market. First of all, most domestic products would be sold in the domestic markets (selling), and some are exported. Secondly, some goods sold in the domestic market would be imported from overseas.

Notably, a small country assumption is adopted with regard to import and export for simplicity. That is, import and export prices are set exogenously, and domestic economic subjects are price takers.

Specific settings are shown as follows:

3.2.1. Flow of Domestic Product

Ultimate outputs (QA) of domestic production are sold in the domestic market (QDA) and exported (QE), and the substitution between them satisfies the CET conversion function:

$$QA_a = A_a^t \times (\delta_a^t QDA_a{}^{\rho_a^t} + (1-\delta_a^t)QE_a^{\rho_a^t})^{\frac{1}{\rho_a^t}}, \quad \rho_a^t > 1 \tag{22}$$

Similar to the CES production function, the parameter ρ_a^t is also determined exogenously. The price of product sold in the domestic market is PDA, and the price of exported product is PE. The first-order condition for optimization is:

$$\frac{PDA_a}{PE_a} = \frac{\delta_a^t}{1-\delta_a^t}\left(\frac{QE_a}{QDA_a}\right)^{1-\rho_a^t} \tag{23}$$

The relationship between those two prices is:

$$PA_a \times QA_a = PDA_a \times QDA_a + PE_a \times QE_a \tag{24}$$

PDA is determined by the domestic market, but PE is exogenous:

$$PE_a = (1-t_a^e) \times pwe_a \times EXR \tag{25}$$

where pwe is duty-paid free on board (FOB) calculated in a foreign currency (e.g., U.S. dollars). t^ε is the export tax rate. EXR is the exchange rate quoted as the domestic currency per unit of foreign currency. In other words, pwe is FOB including export tax, but it is not what the producer can finally obtain. It minus a certain tax is the export price that the producer can get.

3.2.2. Conversion from Domestic Product (QDA) to Domestic Commodity (QDC)

In this model, based on the social accounting matrix (SAM) table [49], it is assumed that one production department produces only one product; there is a one-to-one correspondence between products and goods. For better construction of the CGE model, we also need to convert product to commodity in various sectors. As mentioned above, a indicates sector, and c indicates commodity. For example, product produced and sold in the domestic market is represented by QDA_a, and the corresponding commodity is QDC_c. To convert QDA_a into corresponding QDC_c, the CGE model should be processed as follows.

I is an n-dimensional identity matrix; QDA is an n-dimensional column vector of the product; QDC is an n-dimensional column vector of the commodity.

PDA and *PDC* are the prices of the corresponding column vectors, and they satisfy the following relationship:

$$QDC = I \times QDA, \quad PDC = I \times PDA \tag{26}$$

For ease of programming, the above equations can be re-written as follows:

$$\begin{aligned} QDC_c &= \sum_a IDENT_{ca} \times QDA_a \\ PDC_c &= \sum_a IDENT_{ca} \times PDA_a \end{aligned} \tag{27}$$

3.2.3. Sources of Domestic Supply

Let QC_c be the amount of commodity consumed in the domestic market, and the corresponding price is PC_c. QC_c is the demand for various domestic economic subjects (consumers, companies, government), including consumption, intermediate inputs and investment.

On the other hand, the supply of QC_c comes from the domestic market (QDC_c) and import (QM_c). Substitution between them is described by the Armington condition:

$$QC_c = A_c^s \times (\delta_c^s {QDC_c}^{\rho_c^s} + (1 - \delta_c^s) QM_c^{\rho_c^s})^{\frac{1}{\rho_c^s}}, \quad \rho_a^t < 1 \tag{28}$$

The first-order condition for optimization is:

$$\frac{PDC_c}{PM_c} = \frac{\delta_c^s}{1 - \delta_c^s} \left(\frac{QM_c}{QDC_c} \right)^{1-\rho_c^s} \tag{29}$$

The relationship for price is:

$$PC_c \times QC_c = PDC_c \times QDC_c + PM_c \times QM_c \tag{30}$$

PDC_c is determined by the above-mentioned conversion relationship, while PM_c is set exogenously:

$$PM_c = (1 + t_c^m) \times pwm_c \times EXR \tag{31}$$

Similar to the previous export price, t^m is the import tax rate, and *pwm* is the CIF (cost, insurance and freight) of the commodity calculated in a foreign currency without import tax. The result of *pwm* plus the import tax is the price of imported commodity that consumers pay. The Figure 2 shows the flow of production and the supply of goods.

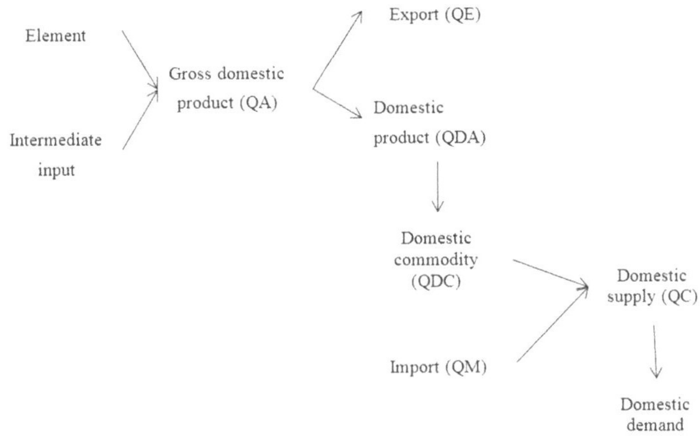

Figure 2. The flow of production and the supply of goods.

3.3. The Economic Subject Part

3.3.1. Resident Account

Residential income includes labor income, capital income, enterprise transfer payment, government transfer payment and transfer payments overseas. According to the model, all labor is from the residents, and they get all of the labor payment. Capital income is distributed among enterprises, residents and the government. For simplicity, it is supposed that the shares three subjects are fixed. The share of capital income residents is $shif^{hk}$. Transfer payments from the government and overseas are set exogenously, and transfer payments from enterprises are a fixed proportion of enterprise revenue. Residential income is as follows:

$$YH = w \times QLS + r \times shif^{hk} \times QKS + transfr^{hent} + transfr^{hgov} + transfr^{hrow} \quad (32)$$

where $shif^{hk}$, $transfr^{hgov}$ and $transfr^{hrow}$ are set to be fixed and calibrated by SAM.

Let $shif^{hent}$ be the proportion of transfer payment (fixed value, calibrated by SAM), and then, the transfer payment from enterprises to residents is shown below, given that $YENT$ is enterprise revenue:

$$transfr^{hent} = shif^{hent} \times (1 - ti^{ent}) \times YENT \quad (33)$$

Residential expenditure is primarily for individual income tax and consumption. Set ti^h to be the rate of individual income tax, and then, the disposable income of residents is $\left(1 - ti^h\right) \times YH$. If the marginal propensity to consume is mpc^h, then the income used for consumption is $mpc^h\left(1 - ti^h\right) \times YH$. ti^h and mpc^h are calibrated by the SAM table.

Suppose the utility function for residents is a Cobb-Douglas function, and then, residents' demand for commodity c is:

$$PC_c \times QH_c = shrh_c \times mpc^h \times (1 - ti^h) \times YH \tag{34}$$

Residents' savings ($HSAV$) is:

$$HSAV = (1 - mpc^h) \times (1 - ti^h) \times YH \tag{35}$$

3.3.2. Enterprise Account

Enterprise income only comes from capital gains. Set $shif^{entk}$ to be the share of capital gains, which is calibrated by the SAM table, so enterprise income is as follows:

$$YENT = r \times shif^{entk} \times QKS \tag{36}$$

Business spending includes corporate income tax (tax rate is ti^{ent}) and business transfer payments to residents ($transfr^{hent}$). Business savings are as below:

$$\begin{aligned} ENTSAV &= (1 - ti^{ent}) \times YENT - transfr^{hent} \\ &= (1 - shif^{hent}) \times (1 - ti^{ent}) \times YENT \end{aligned} \tag{37}$$

3.3.3. Government Account

Government income includes: production tax, customs duty, import VAT (import tax), capital gains, individual income tax, corporate income tax, income yield, transfer payment from overseas, and so on. Let the government income be $YGOV$, the production tax rate be t_a^p, the import tax rate be t_c^m, the individual income tax rate be ti^h, the corporate income tax rate be ti^{ent}, the share of the government's capital gains be $shif^{govk}$, the transfer payment from overseas to the government be $transfr^{govrow}$ and the bond yield be idt. These parameters are calibrated by the SAM table.

$$\begin{aligned} YGOV &= \sum_a t_a^P \times PA_a \times QA_a + \sum_c t_c^m \times pwm_c \times EXR \times QM_c \\ &+ ti^h \times YH + ti^{ent} \times YENT + transfr^{govrow} + idt \end{aligned} \tag{38}$$

Government expenditure includes consumption and transfer payments to residents and foreigners. Let the government expenditure be $EGOV$, the consumption for commodity c be QG_c, the transfer payment to residents be $transfr^{hgov}$ and the transfer payment to foreigners be $transfr^{rowgov}$, then:

$$EGOV = \sum_c PC_c \times QG_c + transfr^{hgov} + transfr^{rowgov} \tag{39}$$

where government transfer payments to residents and foreigners are exogenous variables calibrated by the SAM table. Therefore, the point is how to determine government consumption QG_c. For simplicity, some scholars set QG_c as exogenous. Generally, the government, as a regulatory agency, is independent of the economic system relatively and usually taken as exogenous. Without considering the evolution of the economic system, the set of fixed government consumptions is feasible in the static CGE model. A more complex approach is to set a utility function for the government and to derive its consumption demand. If the utility function is set, we also need the government's spending budget in order to determine the government consumption of different goods. In other words, in the government consumption part, there must be an exogenous variable (similar to residents' savings rate). In this article, the utility function is assumed as the Leontief function, where the ratio of the government demand for various goods is maintained as fixed. The government spending budget is set to be a fixed proportion of government income and determined by the SAM table.

Government savings ($GSAV$) is:

$$GSAV = YGOV - EGOV \tag{40}$$

3.3.4. Foreign Account

The foreign account mainly involves the exchange rate and trade surplus. For China, the exchange rate (EXR) is set to be exogenous and fixed, so the balance of payment (foreign savings) $FSAV$ is an endogenous variable, shown as follows:

$$FSAV = \sum_c pwm_c \times QM_c \times EXR - \sum_a pwe_a \times QE_a \times EXR \tag{41}$$

$$EXR = \overline{EXR} \tag{42}$$

3.4. Closed System Part

3.4.1. Commodity Market Clearing

As previously mentioned, the supply of domestic commodity c is QC_c, and the demand for commodity c includes intermediate inputs, resident consumption, government consumption and investment $QINV_c$.

In equilibrium:

$$QC_c = \sum_a QINT_{ca} + QH_c + QINV_c + QG_c \tag{43}$$

3.4.2. Elements Market Clearing

Aggregate demand for capital and labor equals the aggregate supply of capital and labor in equilibrium:

$$\sum_a QLD_a = QLS \tag{44}$$

$$\sum_a QKD_a = QKS \tag{45}$$

3.4.3. Investment-Savings Balance

Total savings come from resident savings (*HSAV*), corporate savings (*ENTSAV*), government savings (*GSAV*) and foreign savings (*FSAV*). Some of the savings are for investment (including desired investment and inventory investment), and the others are lent to the government (government debt income). The equations are as follows:

$$INV = \sum_c PC_c \times QINV_c \tag{46}$$

$$INV = HSAV + ENTSAV + GSAV + FSAV - idt \tag{47}$$

where debt income (*idt*) is fixed and calibrated by the SAM table (see the "government account"). Investment on commodity c ($QINV_c$) is a fixed proportion of total investment, which is determined by the SAM table.

3.4.4. Macro-Closure

Macro-closure mainly refers to the aggregate supply of capital and labor. Whether the aggregate supply of capital and labor is variable has been the focus of debate among major economic scholars, so it must be considered carefully. Generally, there are four kinds of macro-closures: neo-classicism closure, Keynes closure, Lewis closure and Koldorian closure. In China, urban and rural development are in disharmony, and there is a large amount of surplus labor in the countryside. The price for labor would be inelastic for a long time. Thus, Lewis closure is used in the model here [50], where labor supply is sufficient; wage is fixed; capital supply is fixed; and capital price is an endogenous variable.

$$\begin{aligned} w &= \overline{w} \\ QKS &= \overline{QKS} \end{aligned} \tag{48}$$

4. Data

Data in this paper are based on the SAM table of the input-output of 42 sectors in 2007. The social accounting matrix (SAM) has a close relationship with the national economy accounting. It contains detailed information of various transaction and socio-economic flows, and it is an important form to illustrate the relationship among different parts of the socio-economic system.

Referring to the sector classification of the input-output table and based on reality, we divide the 42 sectors of the input-output table into agriculture, light industry, heavy industry, construction industry, service industry, coal mining and selection industry, petroleum and gas extracting industry, electricity and heat production and supply industry, for simplicity. In order to analyze the correlation between the electricity industry and the entire economy, it is necessary to pull apart the three energy industries: electricity, coal and petroleum. The light industry produces consumer goods and makes hand tools, while the heavy industry provides material and the technical basis and other kinds of production. The classification of light and heavy industries refers to China Energy Statistical Yearbook [51]. The sources in detail are shown in Table 1.

As the SAM data come from different sources, their statistic scope and accuracy are also different. It is difficult to match them with each other, so the SAM based on this cannot guarantee that the sum of all rows is equal to that of all columns. When completing the SAM table, although we have taken some less important data as balance items if necessary, the SAM table still cannot be balanced. The SAM table still needs to be further processed. Balancing the SAM table is a mathematical operation. Due to the contradiction or defect of external information, mathematical methods are used to artificially adjust the data of the SAM table, in order to make the SAM table eventually balanced. The basic principle of balancing SAM is making the overall difference between the original SAM and adjusted SAM minimum. How to measure and control the overall difference is the focus of different methods. Currently, the methods most commonly used are the RAS [52] and cross-entropy (CE) method, and in this paper, the CE method is applied. The basic idea of cross-entropy is embedding the added information into SAM, thus making the difference between adjusted SAM and original SAM minimum. The difference could be measured by "cross-entropy distance", which is proposed by Kullback–Leibler ([53,54]).

Parameters in this CGE model are divided into two kinds: one needs to be calibrated by SAM, such as the coefficients of shares of the element inputs in the production function and the initial values of all variables; the other ones are set by certain econometric methods or experience, such as the elasticity of the production function and the conversion elasticity of import and export.

Table 1. Social accounting matrix (SAM) data sources.

Row Account	Colum Account	Corresponding Transaction	Data Source
1 activity	2 commodity	self-cropping and self-selling	Difference between total output and export; a balance account
	9 abroad	export	Input-output table in 2007
2 commodity	1 activity	Intermediate inputs	Input-output table in 2007
	2 commodity	transaction cost	Transaction cost is not considered in this table
	5 resident	resident consumption	Input-output table in 2007
	7 government	government consumption	Input-output table in 2007
	8 saving-investment	investment	Input-output table in 2007; gross capital formation, stock added and others are combined
3 element	1 activity	added value of elements	Input-output table in 2007; capital income equals fixed asset depreciation and operational surplus
	8 abroad	element reward from overseas	Capital flow statement in 2007 (physical quantity)
4 resident	3 element	resident's element income	Labor income is a balance item; data of capital reward comes from capital flow statement in 2007 (physical quantity)
	5 enterprise	transfer payment from enterprises to residents	A balance item of the resident account
	6 government	transfer payment from government to residents	China Finance Yearbook 2008
	8 abroad	transfer payment from overseas to residents	China Statistical Yearbook 2008
5 enterprise	3 element	element reward of enterprise	Input-output table in 2007
6 government	1 activity	production tax and value added tax	Input-output table in 2007
	2 commodity	sales tax and duties	Input-output table in 2007
	3 element	element tax	Element tax is not considered in this table
	4 resident	resident income tax	China Finance Yearbook 2008
	5 enterprise	enterprise tax	China Finance Yearbook 2008
	7 S-I	government debt	China Finance Yearbook 2008
7 saving-investment	4 resident	resident savings	A balance item of resident account
	5 enterprise	enterprise savings	A balance item of enterprises account
	6 government	government savings	A balance item of the government account
	8 abroad	net foreign savings	A balance item of the foreigner account
8 abroad	2 commodity	import	Input-output table in 2007
	6 government	transfer payment from government to foreigners	China Finance Yearbook 2008

5. Simulation Results

China's electricity shortage can be divided into hard electricity shortage and soft electricity shortage. Hard electricity shortage refers to insufficient capacity caused by insufficient investment and construction of electric power. Soft electricity shortage is the result of the unwillingness of power enterprises to generate electricity even with full capacity, and the cause of it is the conflict between coal market and planned electricity. Although both hard electricity shortage and soft electricity shortage are power shortages, their causes, time span and economic impacts are different. The CGE model is used to simulate the economic impacts of these two kinds of power shortages.

5.1. Model Setting for "Hard Electricity Shortage"

In reality, "shortage" refers to the fact that the demand for a certain commodity is not met, which can be addressed only by queuing and quotation. From the view of economics, the so-called shortage means that at a certain price, effective demand of a certain commodity is greater than the actual supply. In an efficient market system, long-term shortage would not exist. Even if short-term shortage occurs, the market can quickly adjust the supply and demand to make them balanced through the price mechanism. Electricity shortage for months or even years in China is very common, which has something to do with the special situation of China's electric power industry. On the one hand, electricity construction itself takes a long time, so it is difficult to keep up with the pace of economic growth. On the other hand, the monopoly of China's electricity market and administrative pricing make the price mechanism fail.

Under the condition of hard electricity shortage, the capacity of the electric power industry has already reached the saturation point, where generation hours are very long and power output almost reaches installed capacity. We believe that the power industry's output is at a rigid level, and the electricity supply curve is a vertical line. The supply and demand of electricity is shown in Figure 3.

As shown in the figure, the electricity prices is set to be P1 administratively; the corresponding equilibrium point should be at B; and the corresponding supply is Q2 in equilibrium. When hard electricity shortage occurs, the construction of electricity is not enough, and the actual supply curve shifts left. At that time, power output has already reached saturation, and the power supply curve is a vertical line. In shortage, under the market condition, the electricity market should achieve equilibrium at Point E with the equilibrium price of P0. However, the electricity price is fixed at P1 by the government, lower than the equilibrium price P0, so a power shortage exists. The actual power supply is Q1, so the amount of electricity shortage is Q2 − Q1.

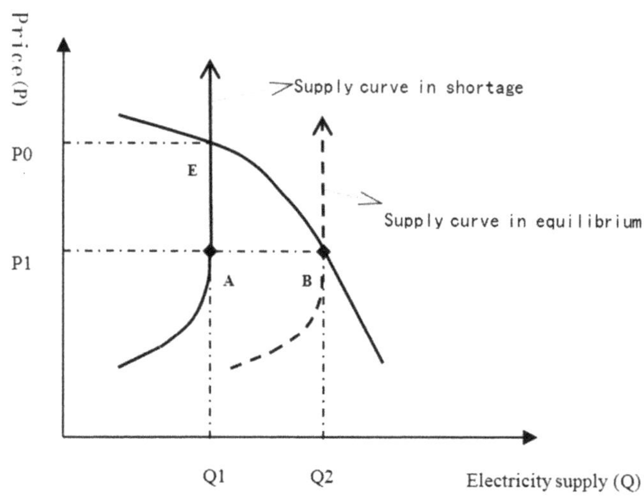

Figure 3. Hard electricity shortage.

It is assumed that power supply is in equilibrium (Point B above) in the reference state. When simulating the effects of hard electricity shortage, the power supply curve is moved left by a certain proportion (indicated by (Q2 − Q1)/Q1). Besides, the inputs and outputs of the power industry are fixed. What is more, it is supposed that there is a gap between effective demand and actual demand. The simulation results are shown below.

The impacts of electricity shortage vary with sectors, and the importance of different sectors in the national economy differs. The impacts of hard electricity shortage on different sectors may not be the same. Limited to the availability of data, it is assumed that electricity shortage accounts for a certain proportion of effective power demand for simplicity. Based on the proportion of electricity consumption of different sector in the base year, we set the amount of electricity shortage of different sectors. We set up four cases for the proportion of power shortage: 3%, 7%, 11% and 15% (that is, the electricity supply curve shifts from Q2 to Q1). After simulation in the CGE model, power shortages of different production sectors (except the electricity production sector) are shown in the Table 2.

As can be seen from the above table, when the total power shortage is set to be 3%, 7%, 11% and 15%, respectively, the proportions of electricity shortage in various industries are different, but the differences are not very large on the whole.

Table 2. Electricity shortage of various sectors under the condition of "hard electricity shortage"; unit: %.

Proportion of Power Shortage	3	7	11	15
Primary industry	4.45	9.96	15.04	19.75
Heavy industry	4.53	10.39	16.03	21.44
Light industry	4.50	10.21	15.59	20.67
Construction industry	4.44	9.92	14.94	19.54
Service industry	4.47	10.05	15.27	20.17
Coal mining and selection industry	4.53	10.36	15.92	21.18
Petroleum and gas extracting industry	4.42	9.81	14.67	19.06

5.1.1. Macro Impacts

As a basic industry of the national economy, the electricity industry plays an irreplaceable role to support economic development. The impacts of power shortage on GDP should be put in the first place in order to know the power shortage better. Under hard electricity shortage, the impacts of different levels of power shortage on GDP are shown in the Table 3.

Table 3. Impacts of hard electricity shortage on GDP and social welfare; unit: %.

Proportion of Power Shortage	3	7	11	15
GDP	−0.98	−2.51	−4.31	−6.39
Welfare of consumers	−0.62	−1.40	−2.14	−2.92
Price	0.63	1.51	2.41	3.33

As shown in the previous table, the negative impacts of power shortage on GDP are very significant. GDP loss accelerates as the proportion of power shortage increases. When the proportion is 15%, GDP decreases by 6.39%. The effects of electricity shortage on GDP are the result of the output reduction of the electricity sector itself; moreover, the output of other sectors also falls to different extents with limited electricity. Lin and Liu [55] point out that, in the stage of the industrialization and urbanization process, there is a relatively stable 1:1 relationship between the growth rates of Chinese GDP and energy consumption in the long term, which does not conflict with the results in this paper. On the one hand, we focus much on the stock relationship between GDP and power demands, not the flow relationship. On the other hand, the CGE model offers equilibrium results, where industries would move from high power-consuming sectors to low power-consuming ones in the case of a power shortage, easing rigid demand for electricity to some extent.

Changes in consumer welfare are also presented in the above table. Consumer welfare is measured by Hicksian equivalent variation, and its expression is:

$$EV = \sum_i PA_i^b XAC_i^s - \sum_i PA_i^b XAC_i^b \quad (49)$$

where EV indicates the Hicksian equivalent variation of consumer welfare; PA^b is the price before changes. XAC^b and XAC^s are the amount of consumption before and after changes. Consumer welfare decreases with an accelerating rate with more power shortages, but much smaller compared to GDP. Consumer spending habits would be changed by hard power shortage.

In addition, a power shortage causes prices to rise. The output of various sectors would be reduced due to the lack of electricity, and the supply reduces more than demand does.

5.1.2. Impacts on the Output of Various Sectors

The impacts of hard electricity shortage on the output of various sectors are shown in Table 4. As can be seen, the impacts of hard electricity power on various sectors are different. The heavy industry and the light industry the most affected, whose output reduction increases as power shortage increases. When the power shortage proportion is 15%, the output of heavy industry and light industry decreases by 11.39% and 10.12%, respectively. The primary industry, the construction industry and the service industry are less affected, but the coal mining and selection industry are also heavily affected. In contrast to other sectors, the output of the petroleum and gas extracting industry increases slightly.

Table 4. Impacts of "hard electricity power" on the output of various sectors; unit: %.

Loss Relative to Equilibrium Output	3	7	11	15
Primary industry	−0.22	−0.59	−1.04	−1.58
Heavy industry	−1.51	−4.10	−7.39	−11.39
Light industry	−1.32	−3.62	−6.54	−10.12
Construction industry	−0.14	−0.42	−0.81	−1.31
Service industry	−0.13	−0.38	−0.72	−1.16
Coal mining and selection industry	−1.46	−3.37	−5.23	−7.01
Petroleum and gas extracting industry	0.48	1.15	1.86	2.62
Electricity and heat production and supply industry	−3.00	−7.00	−11.00	−15.00

The results mentioned above illustrate several economic effects of hard electricity shortage.

(1) As a basic input of the national economy, power shortage makes the output of almost every sector decline.

(2) The reduction in output varies greatly among different sectors. The heavy industry and the light industry are the most affected, while other sectors are less affected, indicating that under the background of a long-term power shortage, industries transfer to low power-consuming sectors.

(3) As energy inputs, a contrary trend is observed for coal output and oil output. This is because coal is taken as the primary energy in China, so electricity shortage leads to a reduction in coal output rigidly. Coal can substitute for electricity to some extent, but at this point, the complementation effect is clearly greater than the substitution effect. As for the power supply, only a small part of the electricity supply is from petroleum and gas, so the oil department of electricity mainly reflects the substitution effect. For the power shortage in 2011 in China, when many small and medium-sized enterprises used their own diesel engines for power generation, the oil output did not fall, but rose.

5.2. Model Setting for Soft Electricity Shortage

Different from hard electricity shortage, the electric power industry has sufficient capacity in the case of soft electricity shortage. Since the administratively-regulated price is less than the equilibrium price, the effective electricity demand is greater than the actual supply, and the difference between them is the amount of power shortage. As shown in the following figure, in the case of soft electricity shortage, the supply curve is an oblique upward curve. The equilibrium point is E, and the equilibrium price is P0. P1 is less than P0, which results in insufficient power output and power shortage.

It is assumed that, in the reference state, power supply is in equilibrium (Point E below). When simulating soft electricity shortage, it is supposed that there is a gap (indicated by (Q2 − Q1)/Q1) between effective demand and actual demand. Besides, the power output is changeable. Simulation results are shown in Figure 4.

Different from hard electricity shortage, soft electricity shortage is the result of a low price in the short run. The initial equilibrium point in the above figure is not Point B, but E. We set Point E as a reference state of the CGE model. Similarly, it is assumed that electricity shortage accounts for a certain proportion of effective power demand. Based on the proportion of the electricity consumption of different sectors in the base year, we set the amount of electricity shortage of different sectors. We set up four cases for the proportion of power shortage: 3%, 7%, 11% and 15%. After simulation in the CGE model, power shortages of different production sectors (except the electricity production sector) are shown in the Table 5.

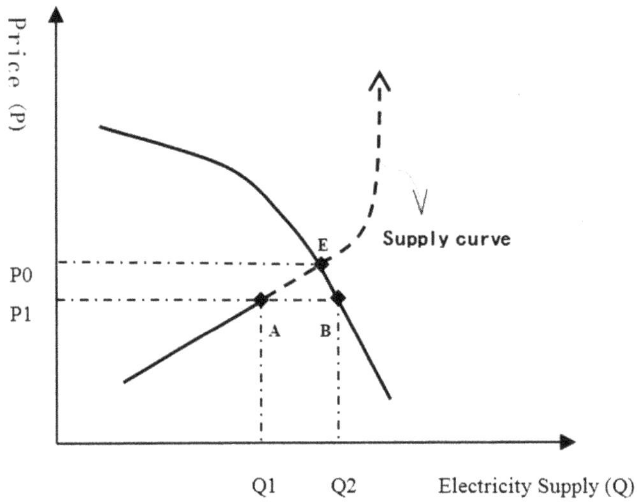

Figure 4. Soft electricity shortage.

Table 5. Electricity shortage of different sectors under the condition of "soft electricity shortage"; (unit: %).

Proportion of Power Shortage	3	7	11	15
Primary industry	3.61	8.70	14.12	19.92
Heavy industry	3.60	8.63	13.91	19.42
Light industry	3.66	8.95	14.78	21.21
Construction industry	3.61	8.65	13.99	19.64
Service industry	3.69	9.15	15.33	22.39
Coal mining and selection industry	3.75	9.49	16.16	23.83
Petroleum and gas extracting industry	3.70	9.15	15.23	21.96

5.2.1. Macro Impacts

Under the case of soft electricity shortage, the impacts of different proportions of power shortage on GDP are shown in the Table 6.

Table 6. Impacts of soft electricity shortage on GDP and social welfare (unit: %).

Proportion of Power Shortage	3	7	11	15
GDP	−1.60	−3.84	−6.21	−8.78
Welfare of consumers	−0.62	−1.66	−2.99	−4.67
Price	0.85	1.96	3.04	4.06

As can be seen, the effects of soft electricity shortage on GDP, consumer welfare and prices are similar, but differ in degree. Compared to hard electricity shortage, the negative impacts of soft electricity shortage on GDP and consumer welfare are more significant. There are two reasons:

(1) Although we use the same reference point to simulate two kinds of power shortage, the specific operation modes are not the same. Moreover, the hard electricity shortage results from inadequate investment, which tends to be the result of economic development for a longer time. The soft electricity shortage is the result of industrial structure imbalance in a shorter period. The economic explanations behind them are different, and there is no full comparability, so a horizontal comparison between their macro-economic impacts only has a reference significance.

(2) During the soft electricity shortage, the time involved is much shorter. Substitution effects among inputs are not so obvious, neither is the substitution among different goods. Industry structure and consumption tendency do not change too much. As the relationship between GDP and electricity consumption is more rigid, the negative effects of power shortage on economic development are more obvious.

5.2.2. Effects on the Output of Various Sectors

The effects of soft electricity shortage on the output of various sectors are shown in Table 7. The output of all sectors drops less than that under the case of hard electricity shortage. The coal production industry is affected most, because demand for coal is rigidly decreased by the reduction of the power output. Meanwhile, oil output does not increase, but begins to fall, because the substitution of oil for electricity is not obvious. The outputs of heavy industry and light industry decrease greatly. When the proportion of short electricity shortage is 15%, the outputs of heavy industry and light industry decrease by 12.32% and 11.3%, respectively. The construction industry is least affected.

Table 7. Effects of soft electricity shortage on the output of various sectors; unit: %.

Loss Relative to Equilibrium Output	**3**	**7**	**11**	**15**
Primary industry	−1.37	−3.23	−5.16	−7.20
Heavy industry	−2.32	−5.49	−8.81	−12.32
Light industry	−2.02	−4.86	−7.93	−11.30
Construction industry	−0.82	−1.95	−3.13	−4.37
Service industry	−1.33	−3.20	−5.18	−7.34
Coal mining and selection industry	−3.12	−7.10	−10.89	−14.50
Petroleum and gas extracting industry	−2.18	−4.99	−7.72	−10.36
Electricity and heat production and supply industry	−2.88	−6.66	−10.37	−14.01

6. Conclusions and Suggestions

In this paper, a CGE model is used to simulate the impacts of hard electricity shortage and soft electricity shortage on China's economy, and we have these research findings. Firstly, the macro-economy, such as GDP and the welfare of consumers, is negatively affected by electricity shortage. With more electricity shortage, macro-economic loss accelerates, due to the rigid demand for energy, as China is currently in the stage of industrialization and urbanization. Moreover, the negative impacts of soft electricity shortage are more obvious than that of hard electricity shortage. Regardless of differences in model setting, the rigid relationship between economic development and electricity demand is becoming more obvious in the case of soft electricity shortage. Finally, as for the outputs, the light industry, heavy industry and coal production industry are influenced most. This is easy to understand. The first two industries rely more heavily on electricity, and coal is a major resource for electricity generation. It is worth noting that the output loss of various sectors in the case of hard electricity shortage differs greatly, while this is not the case of soft electricity shortage. Theoretically speaking, long-term power shortage under the background of hard electricity shortage forces industries to transfer to low electricity-consuming industries. However, in reality, it seems that power supply is guided by the economic situation and industry pattern, but not vice versa. The effects of power shortage on the industry pattern are still questionable, which need be further studied and improved.

Through the simulation, we understand the economic costs of electricity shortage, and there is no doubt that the economic cost of avoiding electricity shortage is far less than that of power shortage. To solve the problem of Chinese power shortage better, we propose the following suggestions.

Firstly, to solve the power shortage, it is very critical for effective electric power demand forecasting to plan power investment rationally. In these years, China has emphasized much the electricity infrastructure construction; thus, the threat of hard power shortage becomes less. There is a new problem, and that is over-capacity. In China, the plans of electricity power facilities construction and grid infrastructure should be audited and determined by the National Development and Reform Commission (NDRC). One of the reasons for the hard power shortage around 2004 is the mistakes of electricity prediction and planning due to the specific electric system. Accordingly, the NDRC or the government should pay more attention to the relationship between electricity consumption and GDP growth, coordinate capacity and inter-regional transmission network investments and, thus, plan the construction of electricity rationally.

Secondly, as mentioned above, we know that the main cause for soft power shortage is the conflict between market coal and planned electricity, and thus, to figure out the relationship between the prices of them is essential and inevitable.

In 2004, China's government has begun to implement the policy of the linkage between coal price and entered-network price, namely if the increase of coal prices reaches or surpasses a certain degree, then the electricity price will be increased. During these years, the policy has been blocked repeatedly, and the dilemma of coal and electricity has not been settled appropriately. However, further advancing this policy will ease the conflict between market coal and planned electricity and ensure the profits of the generation plants in the condition of sufficient marketization. If implemented effectively, the policy of the coal-electricity price linkage is a good way to solve the soft power shortage.

Finally, the two suggestions above can just be temporary solutions. In the long-term, to solve power shortage completely, electric price reform is essential and inevitable. Nowadays, the Chinese economy has entered "the new normal", where economic growth slows down and the coal price falls. The two-track system of coal has been integrated, and the contradiction between power supply and demand is mitigated. It is a great time to advance and deepen the electric price reform. It is common sense that China's electric price is in need of reform, but in which direction? The important things of the reform are to clarify the transmission and distribution prices, then to make the power grid enterprises be service providers instead of middlemen; and moreover, to introduce competition into the generation and retail side. The on-grid prices are decided by true price bidding instead of the so-called price bidding at present. In this way, electricity is an important approach to adjust market demand instead of the macro-adjustment tool of the government. The connection between generation plants and end-user could be built, and the generation plants could be paid and compensated for their efforts by prices that work as transmission channels. The flow of capital into the electricity sector is free. The investments in the electric sector and the benefits of generation plants could be decided by the market instead of the government; what is more, demand side management will become possible. Power shortage and other electric problems could be better mitigated and solved finally.

Acknowledgments: The paper is supported by the National Natural Science Foundation of China (No. 71203186), the Program for Major Projects in Philosophy and Social Science Research of the Ministry of Education of China (No. 14JZD031) and the major projects of the National Social Science Fund (No. 14ZDB144).

Author Contributions: Peng Ou constructed the CGE models, performed all calculations and wrote most of the paper. Ruting Huang collected data and wrote the literature review. Xin Yao proposed research ideas, analyzed the research results and offered policy suggestions.

Conflicts of Interest: The authors declare no conflict of interest.

References and Notes

1. Woo, C.K.; Pupp, R.L. Costs of service disruptions to electricity consumers. *Energy* **1992**, *17*, 109–126.
2. Sun, T.; Wang, X.; Ma, X. Relationship between the economic cost and the reliability of the electric power supply system in city: A case in Shanghai of China. *Appl. Energy* **2009**, *86*, 2262–2267.
3. Shiu, A.; Lam, P.L. Electricity consumption and economic growth in China. *Energy Policy* **2004**, *32*, 47–54.
4. Tsai, C.M. The reform paradox and regulatory dilemma in China's electricity industry. *Asian Surv.* **2011**, *51*, 520–539.
5. Zhang, L.; Ruan, J.; Ding, J. The institutional power shortage in China: Capacity shortage or capacity under-utilisation? *Appl. Energy* **2014**, *136*, 480–494.
6. Ming, Z.; Song, X.; Ling, Y.; Wang, Y.; Wei, Y.; Li, Y. China's large-scale power shortages of 2004 and 2011 after the electricity market reforms of 2002: Explanations and differences. *Energy Policy* **2013**, *61*, 610–618.
7. Cherni, J.A.; Kentish, J. Renewable energy poliy and electricity market reform in China. *Energy Policy* **2007**, *35*, 3616–3629.
8. Ngan, H.W. Electricity regulation and electricity market reforms in China. *Energy Policy* **2010**, *38*, 2142–2148.
9. Li, H.Z.; Tian, X.L.; Zou, T. Impact analysis of Coal-electricity pricing linkage scheme in China based on stochastic frontier cost function. *Appl. Erengy* **2015**, *151*, 296–305.
10. Song, M.L.; Cui, L.B. Economic evaluation of Chinese electricity price marketization based on dynamic computational general equilibrium model. *Comput. Ind. Eng.* **2016**.
11. Lin, B.Q. Power Shortage, Short-run Response, and Long-run Consideration. *Econ. Res. J.* **2004**, *3*, 28–36.
12. Cheng, Y.S.; Wong, W.K.; Woo, C.K. How much have electricity shortages hampered China's GDP growth? *Energy Policy* **2013**, *55*, 369–373.
13. Ma, H.; Oxley, L.; Gibson, J. China's energy situation in the new millennium. *Renew. Sustain. Energy Rev.* **2009**, *13*, 1781–1799.
14. Fisher-Vanden, K.; Mansur, E.T.; Wang, Q.J. Electricity shortages and firm productivity: Evidence from China's industrial firms. *J. Dev. Econ.* **2015**, *114*, 172–188.
15. Thomson, E. Power shortages in China: Why? *China Int. J.* **2005**, *3*, 155–171.
16. Yuan, J.; Lei, Q.; Xiong, M.; Guo, J.; Zhao, C. Scenario-Based Analysis on Water Resources Implication of Coal Power in Western China. *Sustainability* **2014**, *6*, 7155–7180.
17. Apergis, N.; Payne, J.E. A dynamic panel study of economic development and the electricity consumption-growth nexus. *Energy Econ.* **2011**, *33*, 770–781.
18. Yang, Y.; Chen, Y.E.; Liu, Z. Energy constraints and China's economic development. *J. Econ. Policy Reform* **2007**, *10*, 343–354.
19. Johansen, L. A multi-sectoral study of economic growth. *Am. Econ. Rev.* **1961**, *51*, 436–438.

20. Jiang, L.; Wu, F.; Liu, Y.; Deng, X.Z. Modeling the impacts of urbanization and industrial transformation on water resources in china: An integrated hydro-economic CGE analysis. *Sustainability* **2014**, *6*, 7586–7600.
21. Boccanfuso, D.; Savard, L.; Estache, A. The distributional impact of developed countries' climate change policies on senegal: A macro-micro CGE application. *Sustainability* **2013**, *5*, 2727–2750.
22. Li, N.; Wang, X.; Shi, M.; Yang, H. Economic Impacts of Total Water Use Control in the Heihe River Basin in Northwestern China—An Integrated CGE-BEM Modeling Approach. *Sustainability* **2015**, *7*, 3460–3478.
23. Scaramucci, J.A.; Perin, C.; Pulino, P.; Bordoni, O.F.; Da Cunha, M.P.; Cortez, L.A. Energy from sugarcane bagasse under electricity rationing in Brazil: A computable general equilibrium model. *Energy Policy* **2006**, *34*, 986–992.
24. Welsch, H.; Ochsen, C. Dismantling of nuclear power in Germany: Sectoral and macroeconomic effects. *Energy Policy* **2001**, *29*, 279–289.
25. Cansino, J.M.; Cardenete, M.A.; González-Limón, J.M.; Román, R. The economic influence of photovoltaic technology on electricity generation: A CGE (computable general equilibrium) approach for the Andalusian case. *Energy* **2014**, *73*, 70–79.
26. Cansino, J.M.; Cardenete, M.A.; Gonzalez, J.M.; Pablo-Romero, M.D.P. Economic impacts of solar thermal electricity technology deployment on Andalusian productive activities: A CGE approach. *Ann. Reg. Sci.* **2013**, *50*, 25–47.
27. Cai, Y.; Arora, V. Disaggregating electricity generation technologies in CGE models: A revised technology bundle approach with an application to the US Clean Power Plan. *Appl. Energy* **2015**, *154*, 543–555.
28. Tarancón, M.A.; del Río, P.; Albinana, F.C. Assessing the influence of manufacturing sectors on electricity demand. A cross-country input-output approach. *Energy Policy* **2010**, *38*, 1900–1908.
29. Rodrigues, R.; Linares, P. Electricity load level detail in computational general equilibrium—Part I—Data and calibration. *Energy Econ.* **2014**, *46*, 258–266.
30. Rodrigues, R.; Linares, P. Electricity load level detail in computational general equilibrium—Part II—Welfare impacts of a demand response program. *Energy Econ.* **2015**, *47*, 52–67.
31. Holmoy, E. The Anatomy of Electricity Demand: A CGE Decomposition for Norway. Statisics Norway, Research Department. 2005. Available onlinde: https://brage.bibsys.no/xmlui/handle/11250/180277 (accessed on 16 July 2016).
32. Wiskich, A. Implementing a load duration curve of electricity demand in a general equilibrium model. *Energy Econ.* **2014**, *45*, 373–380.
33. Nguyen, K.Q. Impacts of a rise in electricity tariff on prices of other products in Vietnam. *Energy Policy* **2008**, *36*, 3145–3149.
34. Liu, M.H.; Margaritis, D.; Zhang, Y. Market-driven coal prices and state-administered electricity prices in China. *Energy Econ.* **2013**, *40*, 167–175.

35. He, Y.X.; Zhang, S.L.; Yang, L.Y.; Wang, Y.J.; Wang, J. Economic analysis of coal price-electricity price adjustment in China based on the CGE model. *Energy Policy* **2010**, *38*, 6629–6637.
36. He, Y.X.; Yang, L.F.; He, H.Y. Electricity demand price elasticity in China based on computable general equilibrium model analysis. *Energy* **2011**, *36*, 1115–1123.
37. Hwang, W.S.; Lee, J.D. A CGE analysis for quantitative evaluation of electricity market changes. *Energy Policy* **2015**, *83*, 69–81.
38. Dansie, G.; Lanteigne, M.; Overland, I. Reducing energy subsidies in China, India and Russia: Dilemmas for decision makers. *Sustainability* **2010**, *2*, 475–493.
39. Lin, B.Q.; Jiang, Z.J. Estimates of energy subsidies in China and impact of energy subsidy reform. *Energy Econ.* **2011**, *33*, 273–283.
40. Wang, Y.; Almazrooei, S.A.; Kapsalyamova, Z.; Diabat, A.; Tsai, I.T. Utility subsidy reform in Abu Dhabi: A review and a Computable General Equilibrium analysis. *Renew. Sustain. Energy Rev.* **2016**, *55*, 1352–1362.
41. Hosoe, N. The deregulation of Japan's electricity industry. *Jpn. World Econ.* **2006**, *18*, 230–246.
42. Akkemik, K.A.; Oğuz, F. Regulation, efficiency and equilibrium: A general equilibrium analysis of liberalization in the Turkish electricity market. *Energy* **2011**, *36*, 3282–3292.
43. Li, J.F.; Wang, X.; Zhang, Y.X.; Kou, Q. The economic impact of carbon pricing with regulated electricity prices in China—An application of a computable general equilibrium approach. *Energy Policy* **2014**, *75*, 46–56.
44. Chi, Y.; Guo, Z.; Zheng, Y.; Zhang, X. Scenarios analysis of the energies' consumption and carbon emissions in China based on a dynamic CGE Model. *Sustainability* **2014**, *6*, 487–512.
45. Lofgren, H.; Harris, R.L.; Robinson, S. A Standard Computable General Equilibrium (CGE) Model in GAMS. International Food Policy Research Institute. 2002. Available online: http://www.un.org/en/development/desa/policy/mdg_workshops/training_material/lofgren_lee_and_robinson_2002.pdf (accessed on 14 July 2016).
46. Some CGE models allow that one sector produces a variety of different products, and that different sectors produce the same product, but it has little effect on results. For the sake of simplicity, the CGE model in this article assumes one sector produces one product.
47. The Leontief production function is a special form of CES production function when ρ equals $-\infty$.
48. Roland-Holst, D. General Equilibrium Techniques for Policy Modeling. 2009. Available online: http://are.berkeley.edu/~dwrh/FAO_ECTAD_FMD_Cambodia/Documents/Modeling_Manual_RHVDM.pdf (accessed on 14 July 2016).
49. A Social Accounting Matrix (SAM) represents flows of all economic transactions that take place within an economy (regional or national).
50. Rattso, J. Different macro-closures of the original Johansen model and their impact on policy evaluation. *J. Policy Model.* **1982**, *4*, 85–97.

51. Chinese Energy Statistical Yearbook edited by Energy Statistics is an annual statistical publication, which covers very comprehensive data in energy construction, production, consumption, equilibrium of supply and demand in an all-round way.
52. RAS method is a kind of methods which are used to amend the technical coefficient matrix of the orgnial input-output table and thus to compile the input-output table in the planning period with some data controlled.
53. Pyatt, G. A SAM approach to modeling. *J. Policy Model.* **1988**, *10*, 327–352.
54. Robinson, S.; Cattaneo, A.; El-Said, M. Updating and Estimating a Social Accounting Matrix Using Cross Entropy Methods. *Econ. Syst. Res.* **2001**, *13*, 47–64.
55. Lin, B.Q.; Liu, C. Why is electricity consumption inconsistent with economic growth in China? *Energy Policy* **2016**, *88*, 310–316.

How Much CO_2 Emissions Can Be Reduced in China's Heating Industry

Jing Lin and Boqiang Lin

Abstract: China's heating industry is a coal-fired industry with serious environmental issues. CO_2 emissions from the heating industry accounted for an average 6.1% of China's carbon emissions during 1985–2010. The potential for reducing emissions in China's heating industry is evaluated by co-integration analysis and scenario analysis. The results demonstrate that there is a long-run equilibrium relationship among CO_2 emissions and the influencing factors, including energy intensity, industrial scale, labor productivity, and energy productivity. Monte Carlo technique is adopted for risk analysis. It is found that the CO_2 emissions reduction potential of the heating industry will be 26.7 million tons of coal equivalent (Mtce) in 2020 and 64.8 Mtce in 2025 under the moderate scenario, compared with 50.6 Mtce in 2020 and 122.1 Mtce in 2025 under the advanced scenario. Policy suggestions are provided accordingly.

Reprinted from *Sustainability*. Cite as: Lin, J.; Lin, B. How Much CO_2 Emissions Can Be Reduced in China's Heating Industry. *Sustainability* **2016**, *8*, 642.

1. Introduction

The vigorous growth of the Chinese economy has increasingly connected China to the global economy. Meanwhile, China's energy consumption has grown rapidly, making her the world's leading emitter of greenhouse gases. The Chinese government has faced unprecedented pressure to reduce greenhouse gas emissions such as CO_2 emissions. China's CO_2 emissions per capita exceed the world's average. China's CO_2 emissions per capita stands at seven tons of CO_2 equivalent, compared with the global average of 6.4 tons [1]. Therefore, the Chinese government is prioritizing the development of sustainable green economy and energy conservation and ecological industry development.

Why is China's heating industry the focus of this research? Firstly, China's heating industry has been preferentially developed by the government as a part of infrastructure construction. As a necessity for the promotion of life quality during winter, more residents of south China demand central heating. Secondly, we estimated that energy consumption of coal-related energy sources, on average, accounted for 89.4% of the total energy consumption in China's heating industry during 1985–2010. In fact, the coal consumption of the heating industry kept an average annual growth rate of 8.86% over 1985–2010. In 1985, coal consumption in China's heating industry amounted to 87.44 Mtce, and reached 411.72 Mtce in 2010. In addition, CO_2 emissions of China's heating industry grew at an average annual rate of 6.39% over 1985–2010. In 1985, CO_2 emissions of the heating industry accounted for 4.64% of China's CO_2 emissions, while China's CO_2 emissions accounted for 9.29% of global CO_2 emissions. In 2010, CO_2 emissions in the heating industry accounted for 5.01% of China's total CO_2 emissions, while CO_2 emissions in China was responsible for 24.85% of global CO_2 emissions.

Figure 1 shows that CO_2 emissions in China's heating industry grew rapidly, but its share in China's CO_2 emissions fluctuated between 5% to 8.5% during 1985–2010.

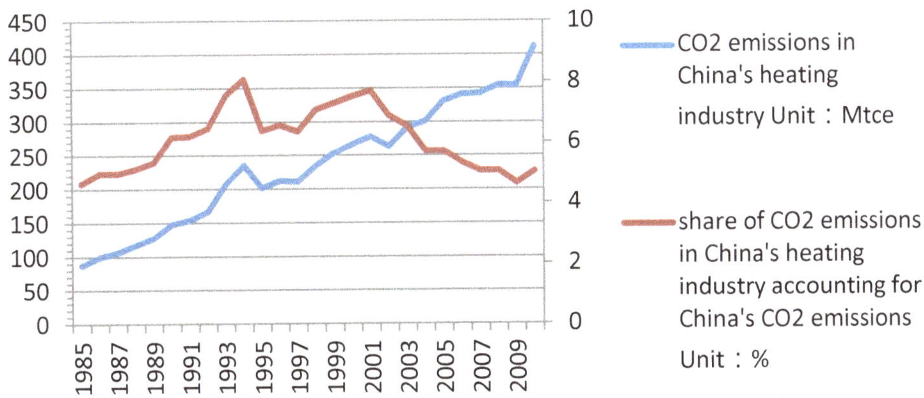

Figure 1. China's heating industry's CO_2 emissions and its share in China's CO_2 emissions (1985–2010).

Investigating the potential of CO_2 emissions reduction in China's heating industry is necessary. For the purpose of prediction, we select CO_2 emissions (CE) of the heating industry as the major objective of this research, and then select four main influencing factors: energy intensity (EI), industrial scale (IS), labor productivity (LP), energy productivity (P). We specify four measurement indexes, corresponding to the four factors as independent variables. The co-integration analysis approach is utilized to ascertain the long equilibrium relationship among CO_2 emissions and the four independent variables. For the purpose of evaluating carbon emissions and future CO_2 emissions reduction potential, we design differentiated conditions in three scenarios. To certify the results of the prediction are reliable, we apply the Monte Carlo simulation to carry out risk analysis.

The remainder of this paper is structured as follows. A brief literature review on CO_2 emissions mitigation is provided in Section 2. Section 3 presents the research methodology applied in this paper. In Section 4, we describe the independent variables and the relevant data. The results are discussed in Section 5. Section 6 provides policy suggestions and the conclusion of the paper.

2. Literature Review

From the literature review, we find that there have been many studies on CO_2 emissions reduction, whereas there are few studies on the potential of CO_2 emissions mitigation in China's heating industry. Chen et al. [2] analyzed energy conservation and CO_2 emissions reduction potential of different models for urban heating in China. Niu et al. [3] analyzed energy consumption of rural household heating under comfortable temperature in the Loess Plateau of Gansu Province in China. Their results proposed improvement of rural household's thermal comfort. Johannes et al. [4] assessed and compared environmental impacts of existing new technologies for space heating in Germany. Mehdi et al. [5] analyzed the socio-eco-environmental impact on simple and combined heat and power generation. Chang et al. [6] found that building-integrated solar thermal systems is the dominant factor for disseminating solar water heating system in metropolitan areas. The study also find that fiscal policies are necessary to incent industrial heating saving energy. Ioan et al. [7] estimated energy conservation potential of heating stations and provided a set of reasonable suggestions to minimize the environmental impact of heating stations. Barkenbus [8] compared different income groups in America and found that the lower-income group set the thermostat higher in heating season than other income groups. It concluded that it is difficult to make Americans set their indoor temperature lower to realize energy conservation, so the use of programmable thermostats is not of much use for energy conservation. It is suggested that a social marketing strategy should be adopted to realize energy conservation. Graham [9] examined the relationship between energy efficiency of space heating and greenhouse mitigation and peak demand. It was found that in the long term, energy savings from efficient measures will be contracted with the efficiency rebound.

Wang et al. [10] decomposed CO_2 emissions of the residential and industrial energy consumption of China over 1996–2012. The results showed that the determinants influencing the CO_2 emissions is economic activity; and in order to reduce CO_2 emissions, energy intensity should be reduced, and the energy mix and industrial structure should be adjusted. Xue et al. [11] employed the Granger causality test to explore the relationship among CO_2 emissions, energy consumption, and economic development in nine European countries over 1970–2008. Risk analysis was applied to evaluate the impacts of CO_2 emission reduction on local economic development. Nan et al. [12] studied the feasibility of adjusting both energy mix and industrial structure to realize the CO_2 emissions reduction targets. Through comprehensive simulation, feasible measurements for better sustainable development by 2020 were proposed. Lin et al. [13] estimated CO_2 emissions mitigation in China's textile industry by evaluating the CO_2 intensity under different designed scenarios. Lin et al. [14] assessed the potential of emissions reduction in construction industry through empirical study on the determinant factors influencing carbon emissions in buildings.

Scenario analysis was employed to investigate the influences of information communications technology on CO_2 emissions at the macroeconomic level [15]. Victor [16] discussed the impacts of scale and intensity on greenhouse gas emissions under several designed macroeconomic scenarios in Canada using low growth simulation model. Greiner et al. [17] conducted simulations using scenarios representing combinations of climatic, market, institutional and technological assumptions to assist with industry planning across the whole of Northern Australia pastoral industry. Hao et al. [18] adopted scenario analysis to analyze energy consumption and carbon emissions from passenger transport vehicles in China. Samuel et al. [19] studied the uncertainty analysis of CO_2 storage capacity for different methods under three storage scenarios in China. Ajay et al. [20] assessed the development of China's future energy system by scenario analysis for China's CO_2 emissions in 2050.

3. Methodology

3.1. Co-Integration

The co-integration method is a very important econometric research method, which is usually applied to probe into the long equilibrium relation among variables. Engle and Granger [21] showed that if independent variable series which are integrated of the same order *d*, denoted as I (*d*), in addition, their residuals of the linear regression are integrated of the order *d-b*, denoted as I (*d-b*), in that way, these independent variable series can be considered to be co-integrated of the order *d, b*, denoted as CI (*d, b*). This method enhances the forecasting stability of the model. When using time series to find long-term co-integrated relationships, there is no need to transform the series.

Before co-integration analysis, it is required to test time series stationarity. The linear combination with the variables could be stationary, which means that the long equilibrium relation between variables exists. It is essential to ensure that the series are stationary before conducting the co-integration analysis. Therefore, the unit root test is required.

Unit-root test methods are commonly applied to conduct stationary test, including the ADF test (Said, 1984) [22] or the PP test (Phillips et al., 1988) [23].

The OLS regression method is the basis of the ADF test:

$$\Delta z_t = \beta_0 + \alpha_0 t + \alpha_1 z_{t-1} + \sum_{i=1}^{m} \beta_i \Delta z_{t-i} + \varepsilon_t \quad (1)$$

where z_t is the variable in the sample period t; Δz_{t-1} is $z_{t-1} - z_{t-2}$; ε_t denotes the independent and identically distributed disturbance term with mean 0 and variance 1; t is the linear time trend; and m is the optimal bandwidth.

The null hypothesis that the unit root in the test of z_t (i.e., $z_t \sim I(1)$) equals the null hypothesis that $\alpha_1 > 0$ in Equation (1). If $\alpha_1 < 0$ is significant, then the rejection of the null hypothesis of the unit root could be accepted.

The PP test model and the ADF test model are similar, but PP test model is less sensitive to the autocorrelation and heteroscedasticity of residuals.

Thus, both ADF test and PP test are adopted for unit root tests.

If each series is integrated of the same order, then it is suitable to investigate the long equilibrium relationship among variables. Engle-Granger two-step approach [21], Johansen-Juselius approach [24] and Johansen approach [25] are the most frequently adopted approaches. Generally, the Johansen maximum likelihood test of co-integration is applied to where there is possibility of more than two co-integrating vectors. Hence, the Johansen-Juselius method is adopted.

3.2. Risk Analysis

In this paper, future values of the variables are estimated according to the historical values of the variables in the co-integration model; therefore, there is only one static prediction. Monte Carlo simulation can avoid the shortcomings of the static prediction process. Thus, we adopt the Monte Carlo simulation to conduct the risk analysis over the probabilities of prediction results.

Monte Carlo simulation was first developed by Metropolis and Ulam [26], and numerical results were obtained by random sampling. The risk analysis addressed by Monte Carlo simulation includes two parts: the possible values of the annual growth rates of the variables in the prediction period and their corresponding probability.

Four specific steps are carried out in the Monte Carlo simulation:

(i) Ascertain the probability distribution of independent variables' annual growth rates according to available data series.
(ii) Substitute the annual growth rates of independent variables with a set of random values.
(iii) Calculate independent variables' possible values and their corresponding probability distribution in the prediction period.
(iv) Acquire independent variables' probability distribution, then, describe their probability distribution histograms and cumulative probability curve.

In other words, to ensure the forecast is reliable, risk analysis via Monte Carlo simulation is adopted to testify the co-integration model and depict future variation trends in the annual growth rates of the independent variables from a probability theory perspective.

This paper uses MATLAB to carry out the Monte Carlo simulation and repeats the simulation process 100 thousand times.

4. Data

To carry out the research objective, we select CO_2 emissions (CE), energy intensity (EI), industrial scale (IS), labor productivity (LP), and energy productivity (P) as the basic research variables. Since not all variables can be directly obtained, some of them need to be derived by calculation of relative variables. We apply time series of each variable during the period of 1985–2010 for the empirical analysis. Nominal data, such as GDP, are all converted to constant price (1985 = 100).

4.1. CO_2 Emissions (CE)

Statistical data of CO_2 emissions in China's heating industry are not directly available, so we estimated CO_2 emissions through multiplying energy consumption by emissions coefficients. Emissions coefficients of the CO_2 emissions of the varied types of energy sources are estimated according to the IPCC [1] coefficients. There are a lot of factors that can influence the CO_2 emissions in China's heating industry. We emphatically analyze energy intensity, industrial scale, labor productivity, and energy productivity. The four factors considered in this paper are labeled as EI, IS, LP, and P, respectively, for brevity.

4.2. Energy Intensity (EI)

Energy intensity (energy consumption/economic output) is one of the most important factor influencing CO_2 emissions. Lian [27] found that the increase in energy intensity can increase carbon intensity. We suppose that energy intensity is positive to the CO_2 emissions of China's heating industry. We use energy consumption per unit of industrial value added to represent EI. The data of industrial value added in China's heating industry come from CEIC China Database. The data of energy consumption of China's heating industry (unit: Mtce) come from the *China Energy Statistical Yearbook* [28].

4.3. Industrial Scale (IS)

Industrial scale can directly influence carbon emissions in China. Dong et al. [29] concluded that the growth of energy-related consumption caused by industrial value added is the major reason for the rapid growth of greenhouse gas emissions, and the effect has offset the carbon emission reduction from change in energy structure. With decreasing primary energy demand for industrial sectors, the demand for secondary energy sources such as heat will increase. It is particularly critical to emphasize the importance of energy saving and CO_2 emissions mitigation in the heating industry.

Industrial scale is the direct cause of obvious increment in China's carbon emissions. China's heating industry has developed rapidly as a key part of China's infrastructure construction. With its rapid development, the industry's energy demand has increased. Increase in energy demand and economic activity can result in increase in carbon dioxide emissions. Industrial expansion could increase CO_2 emissions, and industrial scale is expected to influence CO_2 emissions positively. In this paper, the share of industrial value added of China's heating industry to China's GDP is selected to represent the variable industrial scale IS. The data come from *CEIC Database* and *China Statistical Yearbook* [30].

4.4. *Labor Productivity (LP)*

Labor productivity can reflect the level of mechanization which can influence CO_2 emissions. According to ONS [31], decline in labor productivity accompanied by a reduction in unemployment from over 8% to 5.6% led to decline in greenhouse gas emissions by almost 12% in Britain during 2008–2013. According to USIEA [32] and Trading Economics [33], there is also a simultaneous slowing of labor productivity and a reduction in unemployment and carbon emissions in the United States and Germany. Mazzanti et al. [34] used CO_2 per unit of industrial value added to represent environmental efficiency. It was found that labor productivity and environmental efficiency were positive to most air emissions. Studying the Australian baking industry, Ferguson [35] illustrated that productivity growth was not good for sustainability, and low labor productivity was good for employment and the environment. In summary, labor productivity improvement can result in unemployment and increase in carbon emissions to a certain extent. During the process of mechanization, labor productivity improvement could replace workers with machines. In other words, improvement in labor productivity can bring further mechanization, which will lead to higher energy consumption. Therefore, we suppose that labor productivity is positive to the CO_2 emissions of China's heating industry. The industrial value added per capita is selected to represent LP. Considering that heating power supply serves as a necessity in people's life, we choose population size within the heating supply areas as the index of labor input. Similarly, the data come from *CEIC China Database* [36], and *China Statistical Yearbook* [30].

4.5. *Energy Prodctivity (P)*

Energy productivity is an important factor influencing energy efficiency which is directly related to CO_2 emissions. Patterson [37] found that energy productivity can be considered as a useful indicator for the energy efficiency of an industry. John [38] explored the relationship between energy productivity improvement and energy consumption rebound effect, and found that the energy caused by the rebound effects might not counteract all the realized energy conservation progresses in carbon emissions in a short time, whereas, over a long time, increasing energy demand might offset completely realized energy saved in carbon emissions. Wang et al. [39] found that technology had more stronger influence on energy productivity growth than energy efficiency in China. Fan et al. [40] proved that the impact of technological advance in reducing energy intensity is limited due to the rebound effect. Zha et al. [41] found that with energy efficiency improvement, energy efficiency rebound effect may occur. That means the expected benefits of energy conservation from energy efficiency improvement may partially be offset by the new increase in energy demand. Zha et al. [42] found that the improvement of energy efficiency in electricity can achieve better carbon emissions reduction. However, because of the rebound effect, the reduction of carbon emissions from energy efficiency improvement could be partly offset. The carbon emissions would not decrease proportionally.

Increase in energy productivity improves energy efficiency. Energy productivity improvement will encourage industrial production. If energy price remains the same and energy demand increases, even though energy efficiency improves, energy consumption will still increase. The extra energy consumption promoted by economic activity will offset or even exceed the energy saved from the improvement of energy efficiency; that is, energy consumption will rebound. According to the above analysis, we assume that as the energy productivity of China's heating industry improve, the carbon dioxide emissions would increase. Because the relevant data are unavailable in China, we have to select an index for energy productivity of China's heating industry. Since central heating supply is the main heating supply method in China, energy efficiency of central heating supply could be chosen as index for energy productivity. Considering service properties of the central heating supply, we choose per capita area of central heating supply as the index. That is, we divide the area of central heating supply by the population density of the area of central heating supply. In other words, we choose per capita area of central heating supply as index for energy productivity of China's heating industry. The data are obtained from *China Statistical Yearbook* [30].

4.6. Other Factors

In addition, other factors such as energy price may also have an impact on the carbon emissions. Nevertheless, energy price is highly regulated and kept relatively low by the government in China. In this case, consumers are not sensitive to the price, so we do not take per capita income into consideration. We therefore ignored these two factors in our paper.

Energy substitution is another important factor influencing CO_2 emission. However, historical data on clean energy consumption of the heating industry are currently unavailable. Therefore, we are unable to take this factor into consideration in this paper.

5. Results and Discussion

5.1. Results of Co-Integration Analysis

5.1.1. Results of the Unit-Root Tests

Both the ADF and the PP test are adopted to examine the stationary of all the variables. The tests show that the test for the five variables are in levels and first differences (in Table 1).

Table 1. Unit Root Test.

SERIES	ADF		PP	
	NO TREND	WITH TREND	NO TREND	WITH TREND
LNCE	−1.3319	−3.4903 **	−2.6160	−3.2462 *
LNEI	1.8044	−0.9492	1.2075	−1.8736
LNIS	−0.2268	−6.5477 ***	−0.7484	−6.5478 ***
LNLP	2.9474	−0.3332	−0.5974	−2.7223
LNP	−0.8136	−3.2064	−0.8996	3.2976
⊿LNCE	−6.2821 ***	−6.5610 ***	−7.5876 ***	−12.6140 ***
⊿LNEI	−8.1557 ***	−9.4483 ***	−7.7050 ***	−9.6404 ***
⊿LNIS	−11.2215 ***	−10.9608 ***	−20.4410 ***	−21.3237 ***
⊿LNLP	−6.3254 ***	−5.2194 ***	−7.4951 ***	−8.0719 ***
⊿LNP	−6.03820 ***	−5.8717 ***	−7.2316 ***	−7.1338 ***

Note: ***, **, and * indicate significance level of 1%, 5% and 10%, respectively.

As shown in Table 1, both the ADF test and the PP test verify that the five variables are first-difference stationary. As a result, we further test whether the long equilibrium relationship exists.

First, the Johansen co-integration rank test is carried out, which indicates the number of the linearly independent co-integration vectors. The consequence of rank test is listed in Table 2.

Table 2. Unrestricted co-integration rank test of trace.

Hypothesized No. of CE(s)	Eigenvalue	Trace Statistic	0.05 Critical Value	*p*-Value
None *	0.9344	146.0862	60.0614	0.0000
At most 1 *	0.7337	67.0751	40.1749	0.0000
At most 2 *	0.4133	28.7066	24.2760	0.0129
At most 3 *	0.3553	13.2444	12.3209	0.0349
At most 4	0.0176	0.5156	4.1299	0.5355

Note: Trace test indicates four co-integrating equation(s) at the 0.05 level. * denotes rejection of the hypothesis at the 0.05 level.

5.1.2. Johansen–Juselius Co-Integration Rank Test

In Table 2, trace tests verify that four linearly independent co-integration vectors exist at the 5% level. In Table 3, maximum eigenvalue tests suggest that three linearly independent co-integration vectors exist at the 5% level.

Table 3. Unrestricted co-integration rank test of maximum eigenvalue.

Hypothesized No. of CE(s)	Eigenvalue	Max-Eigen Statistic	0.05 Critical Value	p-Value
None *	0.9344	79.0112	30.4396	0.0000
At most 1 *	0.7337	38.3684	24.1592	0.0003
At most 2	0.4133	15.4622	17.7973	0.1085
At most 3 *	0.3553	12.7288	11.2248	0.0270
At most 4	0.0176	0.5156	4.1299	0.5355

Note: Max-eigenvalue test indicates three co-integrating equation(s) at the 0.05 level.
* denotes rejection of the hypothesis at the 0.05 level.

5.1.3. Selection of Lag Intervals for VAR Model

To proceed with the co-integration analysis, the vector auto regression model (VAR) consisting of LnCE, LnEI, LnIS, LnLP and LnP is built. After that, it is required to determine lag intervals of the VAR model. According to the Sequential modified LR test statistic (LR), Final prediction error criteria (FPE), Akaike information criterion (AIC), Schwarz information criterion (SC), and Hannan–Quinn information criterion (HQ), the optimal lag order is chosen. The results indicate that a lag of 1 should be selected (in Table 4).

Table 4. Selection order criteria.

Lag	LogL	LR	FPE	AIC	SC	HQ
0	26.9572	NA	1.10×10^{-7}	−1.8298	−1.5843	−1.7647
1	135.7087	163.1272 *	1.08×10^{-10} *	−8.8091 *	−7.3365 *	−8.4184 *
2	153.8009	19.5999	2.64×10^{-10}	−8.2334	−5.5337	−7.5172

Note: Endogenous variables: LnCE, LnEI, LnIS, LnLP and LnP; exogenous variables: C; sample: 1985–2010. * indicates lag order selected by the criterion.

Based on the analysis above, the co-integration equation can be obtained. In order to better explain CE, we select and normalized the co-integrating vector, which is consistent with our expectations and meets the economic reality. The normalized co-integration equation is obtained as follows:

$$LnCE = 0.169710LnEI + 0.273970LnIS + 0.447645LnLP + \\ 0.628290LnP + 2.888034(0.0500)\ (0.0786)\ (0.0148)\ (0.0000)\ (0.0505) \quad (2)$$

The standard errors are listed in parentheses.

From the standardized Equation (2), several conclusions can be obtained as below:

Firstly, a long equilibrium relationship exists among CO_2 emissions and the influencing factors, which include energy intensity, industrial scale, labor productivity, and energy productivity during the period of 1985–2010.

Secondly, the positive coefficients of the variables demonstrate that an increase in energy intensity, industrial scale, labor productivity, and energy productivity increase the CO_2 emissions in the long term, which is consistent with our expectation and the social economic reality in China. Furthermore, the elasticity coefficients indicate that a 1% rise in energy intensity, industrial scale, labor productivity, and energy productivity can result in a 0.170%, 0.274%, 0.448%, and 0.628% increase in CO_2 emissions, respectively.

Thirdly, both energy productivity and labor productivity improvements have the strongest impact on CO_2 emissions. Industrial scale is closely associated with CO_2 emissions. The elasticity coefficient of energy intensity is lower than the elasticity coefficient of the other three variables. In China, rapid economic development facilitates urbanization and industrialization. With GDP growth, industrial scale expands. Rapid development of industrialization results in energy consumption expansion. Mechanization is the inevitable result of rapid industrialization. Mechanization can result in more machines and less labor input, and can improve labor productivity. Improvement in labor productivity could lead to machines taking the place of workers. While energy productivity improves and energy price remains the same, more energy can be consumed than before. Improvements in energy productivity and labor productivity accelerate mechanization and industrialization, leading to more energy consumption. Meanwhile, energy price has been relatively low in China as it is regulated by the government. As a result, the underestimated energy price is unable to contribute to energy conservation. Heating fees is also controlled to low levels by the government, which limits the impact of heat price on carbon dioxide emission reduction.

In summary, we regard this model as reasonable, because it is consistent with our expectations and it adequately explains China's current economic reality. Subsequently, we conduct the stability test for the co-integration model.

5.2. Stability Test

The results of the stability test presented (in Table 5) show that all the eigenvalues of adjoined matrix lying inside the unit circle. Thus, the VAR model is stable and reliable.

Table 5. Stability Condition.

Root	Modulus
0.9725 − 0.0161i	0.9727
0.9725 + 0.0162i	0.9727
0.0880 − 0.7883i	0.7932
0.0880 + 0.7883i	0.7932
−0.7046	0.7046
0.5600 − 0.2145i	0.5996
0.5600 + 0.2145i	0.5996
0.01081 − 0.3412i	0.3414
0.01081 + 0.3412i	0.3414
−0.2361	0.2361

5.3. Model Fitting Accuracy

We firstly verify the accuracy of the co-integration equation. As Figure 2 displays, the estimated values of the model equation fit the actual values well. Next, we conduct verification on the forecast function of the model.

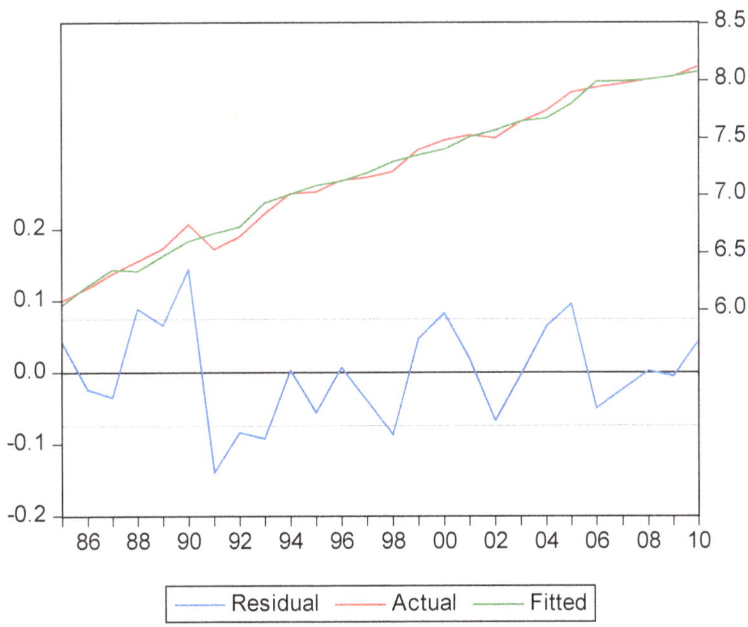

Figure 2. Actual, fitted and residual for the co-integration equation.

To ensure the forecast is reliable, energy intensity, industrial scale, labor productivity and energy productivity data during the period 1985–2010 are substituted into Equation (2) and the estimates of CO_2 emissions of the heating industry during the past 26 years are obtained (Figure 3). It is found that the curve of the fitted values fits the curve of actual values. Hence, we can proceed to forecast CO_2 emissions in China's heating industry.

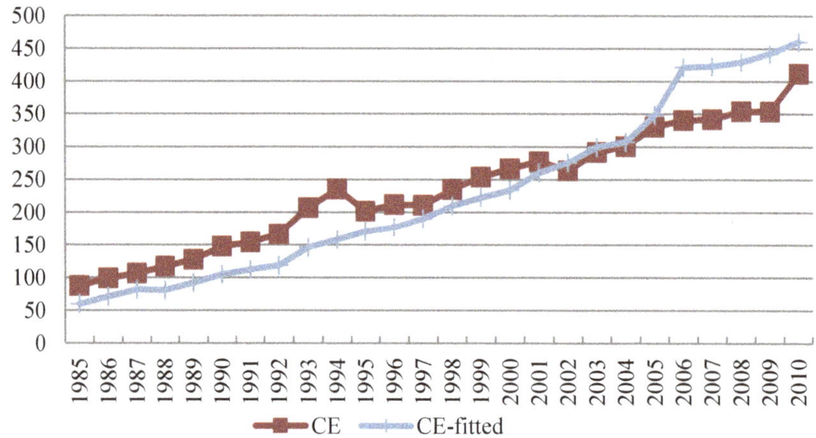

Figure 3. Actual and fitted values of CO_2 emissions of China's heating industry (Unit: Mtce).

Future CO_2 emissions in China's heating industry are able to be predicted by the co-integration equation. Before prediction, we have to confirm the tendency of independent variables.

Figure 4 shows the changing trends of carbon dioxide emissions (CE) and energy intensity (EI), industrial scale (IS), labor productivity (LP) and energy productivity (P). Based on the trend, we suppose the average annual growth rates of these variables during 1985–2010 as the baseline scenario (hereinafter referred to as BAU). Specifically, under the BAU condition, all the variables are assumed to maintain their historical average growth rate. This makes it possible to predict the future CO_2 emission in China's heating industry. That is, under the BAU condition, average annual growth rates of EI, IS, LP and P are −8.51%, 4.3%, −18.3% and 26.5%, respectively.

Our forecast period is from 2011 to 2025 (12th Five-Year Plan, 13thFive-Year Plan, and 14th Five-Year Plan), because this period is critical for China's transition to sustainable development, and the prediction of CO_2 emissions during this period can provide references for policies on reducing energy consumption and carbon emissions in China's heating industry.

Upon the historical growth rates of all the independent variables mentioned above and the co-integration equation (Equation (2)), the CO_2 emissions in China's heating industry in 2020 and 2025 are predicted as 786.85 Mtce and 1028.98 Mtce, respectively (in Figure 5).

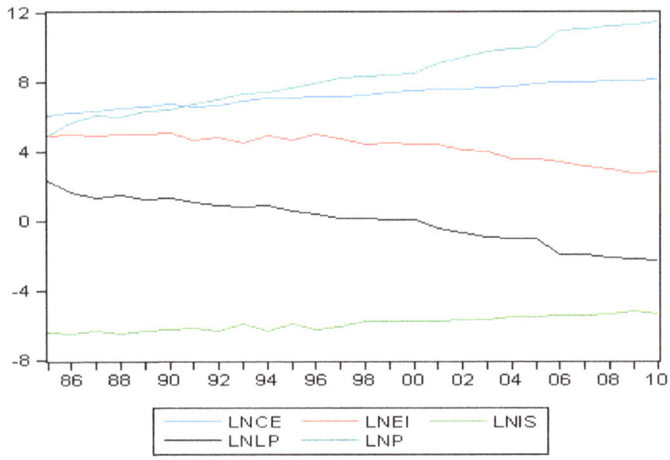

Figure 4. Tendency charts of independent variables (1985–2010).

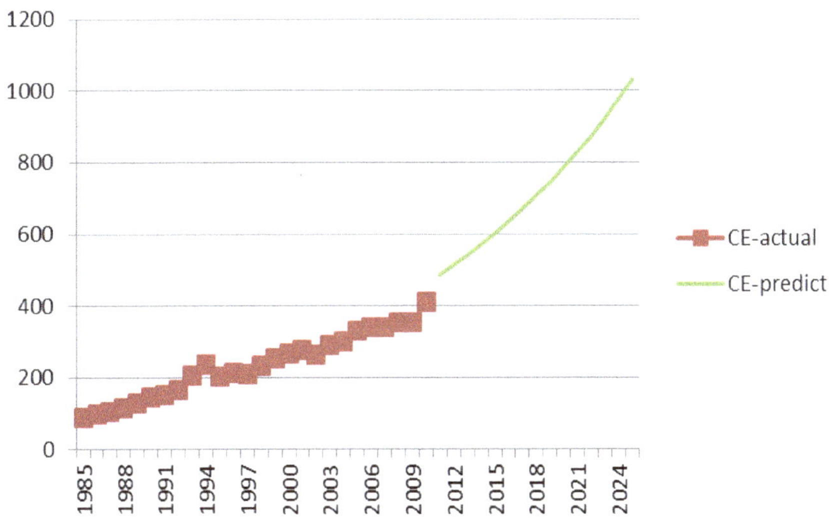

Figure 5. CO_2 emissions of China's heating industry during 2015–2025 (Unit: Mtce).

5.4. Mitigation Potential of CO_2 Emissions in Different Scenarios

5.4.1. Scenario Design

In addition to BAU, we set two more carbon dioxide emissions reduction scenarios: the moderate scenario and the advanced scenario. The three scenarios are set as follows: (i) The BAU scenario is set on the basis of the historical growth rates of the variables during 1985–2010. The BAU scenario reflects a possible trend of the growth of carbon dioxide emission if the industry develops in its current way.

The BAU scenario is the base for setting the other two scenarios. (ii) The moderate scenario is an intermediate situation that the government implement a set of mild policies to guide the industry to reduce CO_2 emissions, and it is between the BAU and the advanced scenario. (iii) The advanced scenario is a strict circumstance, in which the government implements a set of strict policies to force the industry to reduce the largest amount of CO_2 emissions within the economic reality. As previously mentioned, the other two scenarios are based on the BAU scenario. Considering that a set of carbon emission reduction policies may lower the development of the industry, we set the growth rates of the variables a bit lower than the historical growth rate. According to the 13th Five-Year Plan, GDP growth is expected to be 6.5%–7%, which floats within the range of 0.5%. Considering the heating industry is closely related to people's daily life, it is necessary to maintain a smooth and stable development of the industry. Therefore, we set the change of the growth rates of the variables within 0.5% under the designed scenarios. The advantage of the scenario analysis is that we can have a relatively comprehensive analysis on the possible results of CO_2 emissions, so that different policies on energy utilization and carbon dioxide emissions can be compared. The purpose of setting these three scenarios are not to provide precise estimates of the specific economic conditions but to make references for the major factors that contribute to the future CO_2 emissions reduction. All of the hypothetical variables are shown in Table 6.

Table 6. Hypothetical variables under different scenarios.

Variables	BAU	Moderate Scenario	Advanced Scenario
EI	−8.51%	−9%	−9.5%
IS	4.3%	3.8%	3.3%
LP	−18.3%	−18.5%	−19%
P	26.5%	26%	25.5%

5.4.2. Risk Analysis

The above estimation is made on the historical trends of all the independent variables during the sample period. However, the average annual growth rates of independent variables in the future is not yet known. Reasonable predictions should consider multiple probabilities and their corresponding consequences. Therefore, we adopt risk analysis, and focus on the most possible carbon dioxide emissions of the heating industry and its corresponding probability in 2025.

We use the Monte Carlo simulation to generate random numbers in accordance with the distributions of the annual growth rates of independent variables. The estimates of the dependent variable (CO_2 emissions of China's heating industry) in 2025 can be generated upon these random numbers. Conducting the simulations 100 thousand times, 100 thousand groups of random numbers are generated, and accordingly 100 thousand possible values of carbon dioxide emissions in 2025 and their corresponding probability distributions are calculated. From the Monte Carlo simulated analysis, the distribution histogram and cumulative probability curves of CO_2 emissions of China's heating industry in 2025 are presented in Figure 6.

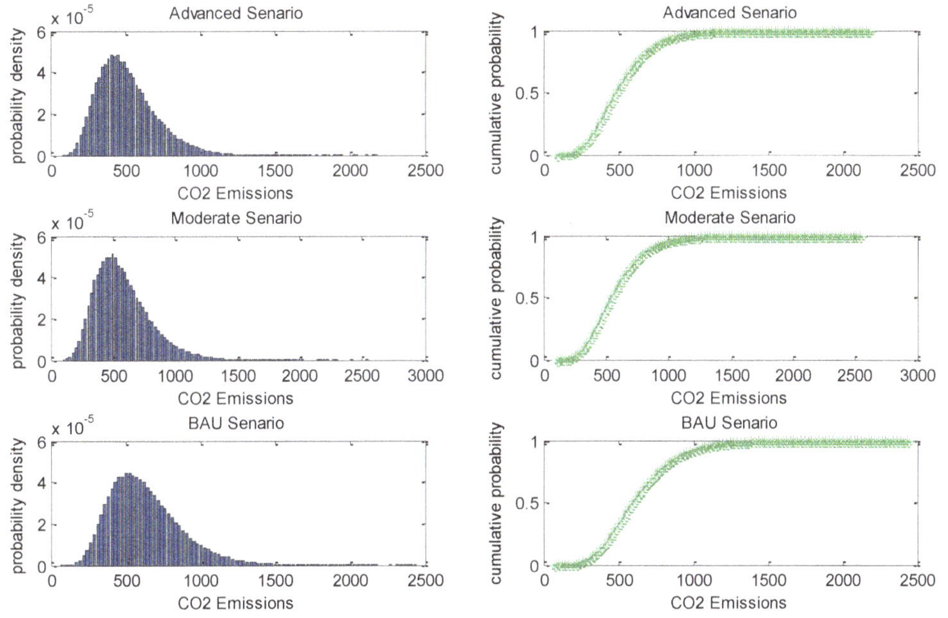

Figure 6. Distribution histogram and Cumulative probability curves of CO_2 emissions in China's heating industry in 2025.

Figure 6 shows that under the BAU scenario, [155, 1120] Mtce is the probability distributions for carbon emissions in 2025 with the significance level of 5%, and the probability of carbon emissions in 2025 reaching 1120 Mtce is approximately 100%. Under the moderate scenario, [141, 1025] Mtce is the probability distributions for carbon emissions in 2025 with the significance level of 5%, and the probability of carbon emissions in 2025 reaching 1025 Mtce is approximately 100%. Under the advanced scenario, [122, 919] Mtce is the probability distributions for carbon emissions in 2025 with the significance level of 5%, and the probability of carbon emissions in 2025 reaching 919 Mtce is approximately 100%. The predicted values of the carbon emissions are 1028.98 Mtce, 978.354 Mtce and 906.881 Mtce in 2025 under the BAU scenario, moderate scenario and advanced scenario, respectively. Compared with the probability distributions, we find that the predicted values of carbon emissions in 2025 under the three scenarios lie in the maximum probability distribution.

The Monte Carlo simulated analysis results satisfy our initial prediction.

5.4.3. Reduction Potential of CO_2 Emissions

Using the co-integration equation (Equation (2)), the growth trends of the independent variables under the differentiated scenarios mentioned above can be obtained, and we can thus predict future values of CO_2 emissions in China's heating industry under the differentiated scenarios. Figure 7 and Table 7 present the prediction results.

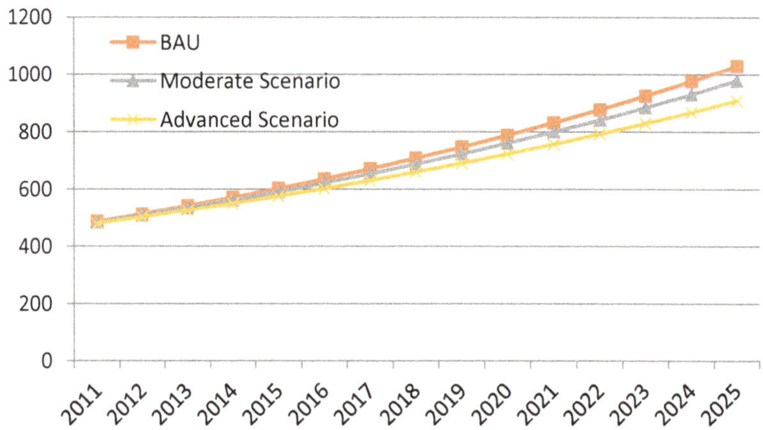

Figure 7. Prediction on CO_2 emissions of China's heating industry under three differentiated scenarios (Unit: Mtce).

Table 7. Forecast on carbon dioxide emissions of China's heating industry in 2020 and 2025 (Unit: Mtce).

Scenario	2020	2025
BAU Scenario	786.849	1028.980
Moderate Scenario	760.155	978.354
Advanced Scenario	722.037	906.881

Table 7 shows the prediction results. CO_2 emissions in China's heating industry in 2020 will be 760.155 Mtce under the moderate scenario, and 722.037 Mtce under the advanced scenario, decreasing by 3.39% and 8.24%, respectively, compared with the BAU scenario. CO_2 emissions in China's heating industry in 2025 will reach 978.354 Mtce under the moderate scenario and 906.881 Mtce under the advanced scenario, decreasing by 4.92% and 11.87%, respectively, compared with the BAU scenario.

After the estimates of potential of CO_2 emissions reduction in China's heating industry, estimating the amount of CO_2 emissions reduction and its impact on China's CO_2 emissions is possible. In the moderate emission scenario, CO_2 emissions in China's heating industry transit from the BAU to the moderate situation. In the same way, in the advanced emissions reduction scenario, CO_2 emissions in China's heating industry transits from the BAU to the advanced stage.

Based on the estimates of CO_2 emissions in China's heating industry under the designed scenarios listed in Table 7, the amount of CO_2 emissions reduction and its impact on China's CO_2 emissions under different scenarios are obtained (presented in Table 8).

From Table 8, we can see that there is considerable potential of CO_2 emissions reduction in China's heating industry in 2020 and 2025. From the BAU to the moderate situation, 26.694 Mtce CO_2 emissions can be reduced in 2020, accounting for 0.181% of China's carbon dioxide emissions and 3.392% of the industry's carbon dioxide emissions. In addition, 64.812 Mtce CO_2 emissions can be reduced in 2025, accounting for 0.327% of China's total CO_2 emissions and 8.237% of the industry's CO_2 emissions. From the BAU to the advanced scenario, 50.626 Mtce CO_2 emissions can be reduced in 2020, accounting for 0.342% of China's CO_2 emissions and 4.920% of the industry's carbon dioxide emissions. In addition, 122.099 Mtce CO_2 emissions can be reduced in 2025, accounting for 0.615% of China's CO_2 emissions and 11.866% of the industry's CO_2 emissions. The results indicate that there is a considerable CO_2 emissions mitigation potential in China's heating industry. Therefore, the realization of CO_2 reduction in China's heating industry is achievable, which contributes to China's carbon dioxide emissions reduction strategy.

Table 8. CO_2 emissions reduction in China's heating industry and its impact on China's CO_2 emissions under different carbon emission reduction scenarios.

Year	Moderate Scenario			Advanced Scenario		
	CO_2 Emissions Reduction (Mtce)	Impact on CO_2 Emissions in the Heating Industry	Impact on China's CO_2 Emissions	CO_2 Emissions Reduction (Mtce)	Impact on CO_2 Emissions in the Heating Industry	Impact on China's CO_2 Emissions
2020	26.694	3.392%	0.181%	50.626	4.920%	0.342%
2025	64.812	8.237%	0.327%	122.099	11.866%	0.615%

6. Conclusions

This paper focuses on the estimation of potential of carbon emissions reduction in China's heating industry. We adopt the co-integration method to investigate the relationship between CO_2 emissions and factors including energy intensity, industrial scale, labor productivity and energy productivity. The results indicate that a 1% growth in energy intensity, industrial scale, labor productivity, and energy productivity lead to a 0.170%, 0.274%, 0.448%, and 0.628% increase in the industry's CO_2 emissions, respectively. China's heating industry has great potential for carbon emission reduction.

Policy implications of this paper can be provided for China's heating industry as follows.

Firstly, there is need to lower the energy intensity of China's heating industry. Efforts should be made to enhance technical transformation and eliminate obsolete equipment with serious pollution and intensive energy consumption to achieve improvement of energy efficiency in the heating industry. The low-carbon concept should also be promoted in the heating industry. Meanwhile, energy conservation and carbon dioxide emissions reduction policies should be promoted.

Secondly, as an energy intensive consumption industry, China should not irrationally expand its heating industry but should ensure appropriate development by optimizing the resource allocation, upgrading the industry and developing heat recovery technologies. Highly-efficient heating systems and a variety of clean energy-source heating systems such as absorption heat pumps, geothermal heat pump systems, air source heat pump systems, and drain-water heat pump systems should be encouraged.

Thirdly, there is need to adjust the energy mix of China's heating industry and change the situation of heavy reliance on a single source of energy. China's heating industry should make carbon dioxide emissions reduction practically feasible and gradually reduce the consumption of coal-related energy sources by substituting coal with other low pollution energy resources or clean energy, such as natural gas, solar power, biomass energy, ground-source, water-source, air-source and nuclear power.

Fourthly, the improvement of labor productivity in China's heating industry mainly comes from the rapid development of industrialization and mechanization. It is necessary to put great efforts into technology innovation. This should include the importation of advanced technologies, development of new products, and enhancement of technical skills, especially the training of skilled high-tech labors.

Fifthly, in order to mitigate energy consumption rebound, it is required to realize energy conservation while improving energy efficiency of China's heating industry. The government should implement a set of policies combing tax preference with pricing mechanism to guide the heating industry to switch to less carbon-intensive energy, such as natural gas or coal bed gas.

Lastly, although energy price is controlled to low levels by the government, its impact on CO_2 emissions cannot be ignored. In the future, market-oriented energy pricing reform is necessary. Consequently, considerable reductions in CO_2 emissions in China's heating industry could be realized, which would significantly contribute to China's carbon dioxide emissions reduction.

Acknowledgments: The paper is supported by the Grant for Collaborative Innovation Center for Energy Economics and Energy Policy (No: 1260-Z0210011), Xiamen University Flourish Plan Special Funding (No: 1260-Y07200), and Newcastle University Joint Strategic Partnership Fund.

Author Contributions: Boqiang Lin and Jing Lin conceived and designed the research; Jing Lin analyzed the data and prepared the draft paper; and Boqiang Lin and Jing Lin revised the paper together. All authors read and approved the final manuscript.

Conflicts of Interest: The authors declare no conflict of interest.

References

1. IPCC Contribution of Working Groups I, II and III. *IPCC Fifth Assessment Report (AR5) 2014: Climate Change 2014: Synthesis Report*; IPCC: Geneva, Switzerland, 2014.
2. Chen, X.; Wang, L.; Tong, L.; Sun, S.; Yue, X.; Yin, S.; Zheng, L. Mode selection of China's urban heating and its potential for reducing energy consumption and CO_2 emission. *Energy Policy* **2014**, *67*, 756–764.
3. Niu, S.W.; Li, Y.X.; Ding, Y.X.; Qin, J. Energy demand for rural household heating to suitable levels in the Loess Hilly region, Gansu province, China. *Gen. Inf.* **2010**, *35*, 2070–2078.
4. Henkel, J.; Kunde, R.; Gaderer, M.; Erdmann, G. Assessment of global emissions, local emissions and immissions of different heating systems. *Sustainability* **2009**, *1*, 494–515.
5. Aliehyaei, M.; Atabi, F.; Khorshidvand, M.; Rosen, M. Exergy, economic and environmental analysis for simple and combined heat and power IC engines. *Sustainability* **2015**, *7*, 4411–4424.
6. Chang, K.-C.; Lin, W.-M.; Chung, K.-M. Sustainable development for solar heating systems in Taiwan. *Sustainability* **2015**, *7*, 1970–1984.
7. Sarbu, I.; Valea, E. Energy savings potential for pumping water in district heating stations. *Sustainability* **2015**, *7*, 5705–5719.
8. Barkenbus, J. Indoor thermal comfort: The behavioral component. *Sustainability* **2013**, *5*, 1680–1699.
9. Palmer, G. Does energy efficiency reduce emissions and peak demand? A case study of 50 years of space heating in Melbourne. *Sustainability* **2012**, *4*, 1525–1560.
10. Wang, G.; Chen, X.; Zhang, Z.; Niu, C. Influencing factors of energy-related CO_2 emissions in China: A decomposition analysis. *Sustainability* **2015**, *7*, 14408–14426.
11. Xue, B.; Geng, Y.; Müller, K.; Lu, C.; Ren, W. Understanding the causality between carbon dioxide emission, fossil energy consumption and economic growth in developed countries: An empirical study. *Sustainability* **2014**, *6*, 1037–1045.
12. Xiang, N.; Xu, F.; Sha, J. Simulation analysis of China's energy and industrial structure adjustment potential to achieve a low-carbon economy by 2020. *Sustainability* **2013**, *5*, 5081–5099.
13. Lin, B.; Moubarak, M. Mitigation potential of carbon dioxide emissions in the Chinese textile industry. *Appl. Energy* **2014**, *113*, 781–787.
14. Lin, B.; Liu, H. CO_2 emissions of China's commercial and residential buildings: Evidence and reduction policy. *Build. Environ.* **2015**, *92*, 418–431.
15. Erdmann, L.; Hilty, L.M. Scenario Analysis. *J. Ind. Ecol.* **2010**, *14*, 826–843.
16. Victor, P.A. Growth, degrowth and climate change: A scenario analysis. *Ecol. Econ.* **2012**, *84*, 206–212.

17. Greiner, R.; Puig, J.; Huchery, C.; Collier, N.; Garnett, S.T. Scenario modelling to support industry strategic planning and decision making. *Environ. Model. Softw.* **2014**, *55*, 120–131.
18. Hao, H.; Liu, Z.; Zhao, F.; Li, W.; Hang, W.; Lund, H.; Kaiser, M.J. Scenario analysis of energy consumption and greenhouse gas emissions from China's passenger vehicles. *Energy* **2015**, *91*, 151–159.
19. Höller, S.; Viebahn, P. Facing the uncertainty of CO_2 storage capacity in China by developing different storage scenarios. *Energy Policy* **2016**, *89*, 64–73.
20. Gambhir, A.; Schulz, N.; Napp, T.; Tong, D.; Munuera, L.; Faist, M.; Riahi, K. A hybrid modelling approach to develop scenarios for China's carbon dioxide emissions to 2050. *Energy Policy* **2013**, *59*, 614–632.
21. Engle, R.F.; Granger, C.W.J. Co-integration and error correction: Representation, estimation, and testing. *Econometrica* **1987**, *55*, 251–276.
22. Said, S.E. Testing for unit roots in autoregressive-moving average models of unknown order. *Biometrika* **1984**, *71*, 599–607.
23. Phillips, P.C.B.; Perron, P. Testing for a unit root in time series regression. *Biometrika* **1988**, *75*, 335–346.
24. Johansen, S.; Juselius, K. Maximum likelihood estimation and inference on cointegration—With applications to the demand for money. *Oxf. Bull. Econ. Stat.* **1990**, *52*, 169–210.
25. Johansen, S. Likelihood-based inference in cointegrated vector autoregressive models. *Gen. Inf.* **1995**, *14*, 517–524.
26. Metropolis, N.; Ulam, S. The monte carlo method. *J. Am. Stat. Assoc.* **1949**, *44*, 335–341. PubMed]
27. Liang, G.H. Empirical study of factors in China carbon intensity. *J. Ind. Technol. Econ.* **2015**, *3*, 79–84. (In Chinese)
28. National Bureau of Statistics of China; National Energy Administration of China. *China Energy Statistical Yearbook*; China Statistics Press: Beijing, China, 1987. (In Chinese)
29. Dong, J.; Zhang, X. Decomposition of carbon emissions and low carbon strategies for industrial sector energy consumption in China. *Resour. Sci.* **2010**, *32*, 1856–1862.
30. National Bureau of Statistics of China. *China Statistical Yearbook*; China Statistics Press: Beijing, China, 2014. (In Chinese)
31. Office for National Statistics (ONS). *Labor Market Statistics*, 2014 ed.; Office for National Statistics: London, UK, 2014.
32. US Energy Information Administration (USEIA). *International Energy Statistics*; US Energy Information Administration: Washington, DC, USA, 2015.
33. Economics Tradings. Productivity and Employment Data: United States and Germany. 2015. Available online: http://www.tradingeconomics.com/ (accessed on 7 July 2016).
34. Mazzanti, M.; Zoboli, R. Environmental efficiency and labour productivity: Trade-off or joint dynamics? A theoretical investigation and empirical evidence from Italy using NAMEA. *Ecol. Econ.* **2009**, *68*, 1182–1194.

35. Ferguson, P. Productivity growth as a barrier to a sustainability transition. *Environ. Innov. Soc.Trans.* **2015**, in press.
36. National Bureau of Statistics of China. *China's Economic and Social Development Statistical Database*; National Bureau of Statistics of China: Beijing, China. (In Chinese)
37. Patterson, M.G. What is energy efficiency?: Concepts, indicators and methodological issues. *Energy Policy* **1996**, *24*, 377–390.
38. Dimitropoulos, J. Energy productivity improvements and the rebound effect: An overview of the state of knowledge. *Energy Policy* **2007**, *35*, 6354–6363.
39. Wang, H.; Zhou, P.; Zhou, D.Q. Scenario-based energy efficiency and productivity in China: A non-radial directional distance function analysis. *Energy Econ.* **2013**, *40*, 795–803.
40. Fan, R.; Luo, M.; Zhang, P. A study on evolution of energy intensity in China with heterogeneity and rebound effect. *Energy* **2016**, *99*, 159–169.
41. Zha, D.L.; Zhou, D.Q. The research on China's energy efficiency rebound effect based on CGE model. *J. Quant. Tech. Econ.* **2010**, *12*, 39–53. (In Chinese)
42. Zha, D.L.; Zhou, D.Q. Root Cause: No Reduce to Energy Consumption with the Increase of Energy Efficiency. *Manag. Rev.* **2012**, *24*, 45–51. (In Chinese)

Are Consumers Willing to Pay More for Sustainable Products? A Study of Eco-Labeled Tuna Steak

Guzhen Zhou, Wuyang Hu and Wenchao Huang

Abstract: A high demand for seafood leads to overfishing, harms the long-term health of seafood stocks, and threatens environmental sustainability in oceans. Sustainability certification is one of the major sustainability movements and is known as eco-labeling. For instance, in the tuna industry, leading tuna brands have committed to protecting sea turtles by allowing the tracing of the source of their tuna "from catch to can." This paper relies on an Internet survey on consumers from Kentucky conducted in July 2010. The survey investigates household-level tuna steak (sashimi grade) consumption and examines consumer preferences for eco-labeling ("Certified Turtle Safe" (CTS) in this study) while mimicking individuals' seafood procurement processes. A random parameter logit model is utilized, and willingness-to-pay measures are calculated based on model estimation results. It was found that respondents on average preferred turtle-safe-labeled tuna steak and were likely to pay more for it; however, they were less likely to purchase wild-caught species, and insignificant results were found for pre-frozen. Moreover, significant heterogeneities were found across individuals regarding tuna steak purchases. The findings indicate evidence of public support for environmental friendliness, particularly with regard to eco-labeling.

Reprinted from *Sustainability*. Cite as: Zhou, G.; Hu, W.; Huang, W. Are Consumers Willing to Pay More for Sustainable Products? A Study of Eco-Labeled Tuna Steak. *Sustainability* **2016**, *8*, 494.

1. Introduction

Global catches of tuna and tuna-like commodity species increased from less than 0.6 million tons in the 1950s to 6.6 million tons in 2010 (2010 is the last year for which tuna fisheries provided data on nominal tuna catches by fishing gear, species, stock, country, and year) [1], including all seven principal species of tuna: albacore, bigeye, bluefin (3 species), skipjack, and yellowfin. The society is seeing more demand for fish than ever before, especially due to the widely known health benefits of fish. Both the industry as well as academic communities have gradually noticed the necessity of ensuring the long-term health of all tuna stocks. As a response, protective measures, such as eco-labels (or environmental sustainability labels) and traceability, have been taken to mitigate the overfishing problem.

Since 2009, global industry standards have been implemented that allow U.S. tuna companies to trace the source of all their tuna "from catch to can." Some leading U.S. tuna brands even work in partnership with governments worldwide, the scientific community, and leading globalized conservations or organizations, including the World Wildlife Fund (WWF), to maintain sustainable tuna production. Economic studies have also investigated consumption and consumer preference regarding tuna; however, relatively little is known about eco-labels' impact on consumer demand for tuna despite a number of studies having been performed for fishery products [2–9]. Consequently, this study aims to examine consumer preference for tuna steak products, from which the results will provide empirical evidence for market design and policy-making.

As for our research interests and purpose, a household survey on the purchasing habits and preference for tuna of consumers from Kentucky in the U.S. is utilized for this paper. Kentucky is a typical midland state in the U.S., the residents of which are general consumers who might not face immediate environmental sustainability issues compared to those who live close to an ocean. The survey included a choice-based conjoint experiment to investigate consumers' choices between wild-caught (conventional) and farm-raised tuna fish, the results of which help to predict future markets and assist in policy-making.

Most importantly, the impact of environmentally friendly eco-labels is another key element in this study. Participants were asked both qualitative and quantitative questions for the "Certified Turtle Safe" (CTS) label [10] on tuna products. Prior to the choice experiment, respondents were instructed with the following information: "Certified Turtle Safe by definition is fish harvested by fisheries under stringent controls to avoid sea turtle by-catch."

2. Research Background

2.1. High Tuna Demands and Calls for Fishery Management

Tuna is one of the most popular seafoods because of its health benefits. For example, seafood is considered a good source of omega-3 fatty acids, eicosapentaenoic acid (EPA), and docosahexaenoic acid (DHA), and these nutrients have been proven to be beneficial compounds for heart health and early neurological development. Seafood is also low in fat and cholesterol, and rich in protein, vitamins, and minerals. According to the 2010 Dietary Guidelines for Americans [11], eating approximately 8 ounces per week of a variety of seafood, which provides approximately 250 mg intake per day of EPA and DHA, can help to reduce cardiac death among individuals with and without pre-existing cardiovascular disease.

Numerous published works have also shown the valuable effects of fish as a unique and rich food source of healthy nutrients [12–16]. Tuna species are on the top of the list for seafood choices. Their meat contains almost no fat and has all of the essential amino acids needed by the human body along with the B vitamins niacin, B1, and B6. Moreover, tuna is an oily fish that has high-quality protein and is a good source of omega-3 fatty acids. Hence, tuna's nutritious benefits uphold the high demand for it. The U.S. per capita consumption of canned tuna was 2.7 lbs in 2011 and was 2.6 lbs in 2010, and the U.S. has the second highest seafood consumption in the world [17].

However, a high demand for tuna leads to its overfishing. Tuna are fished in over 70 countries as the world's most valuable commercial species and then marketed in fresh, frozen, or canned form. Japan and the U.S. are the two largest tuna-consuming countries [14]. Among the seven principal tuna species in the world, 33.5% were estimated to be overexploited, 37.5% were fully exploited, and 29% were not fully exploited, as reported by the Food and Agriculture Organization (FAO) in 2012 [1]. In the long term, tuna stocks and, thus, tuna catches may deteriorate unless there are significant improvements in tuna fishery management or in the cultivation of a substantial tuna fishing and supply system. Due to overfishing, some tuna species are even at risk of extinction [18,19]. Recognizing the environmental emergency surrounding tuna, it requires conservation and the sustainable use of tuna species worldwide. One challenge is that tuna-fishing fleets and their catches have been growing, often unsustainably [20]. Consequently, restrictive measures are necessary to control potential overfishing.

2.2. *Sustainability Issues and Eco-Labels*

Tuna populations and resource sustainability issues directly influence the nature of tuna production and the fishing industry. Sustainability certification is one of the major sustainability movements. One type of sustainability certification is eco-friendly labels (also known as environmentally friendly, nature friendly, or green), which are also known as eco-labels [21]. These labels are intended to introduce environmental awareness for consumers who care about the environment and to help identify consumer products that are ecologically superior. Some labels quantify pollution or energy consumption by way of index scores or units of measurement; others simply assert compliance with a set of practices or minimum requirements for sustainability or reduction of harm to the environment [22]. Eco-labels were initiated in the 1990s and have been introduced in the fisheries sector, and they are increasingly being applied to tuna fisheries.

Theoretically, consumers are believed to derive eco-labels' utility from using products produced with specific processes, such as environmentally friendly practices, which are reflected with eco-labels [23]. Some researchers [22,24] found that the market success of eco-friendly food products requires not only environmental but other verifiable attributes, such as better taste or higher safety. Other researchers also noted that the adoption of eco-labels may earn a premium price for food products [25–27]. In this study, we will examine the impact of an eco-label, "Certified Turtle Safe," on consumer preference. This label is not yet seen on the market, but it may help alleviate one of the controversies associated with the tuna-fishing industry: sea turtles may be harmed or even killed during the fishing.

2.3. *Wild-Caught or Farm-Raised?*

An alternative way of supplementing the supply of wild-caught fish (also known as commercial catches) is aquaculture or farm-raised fish [28,29]. Some consumers prefer wild-caught fish for higher quality and better taste than the farm-raised option [30], while other consumers are concerned with the negative impacts to the environment posed by aquaculture [28]. However, the benefits of aquaculture include lower cost and year-round availability compared to wild-caught fish [30,31]. As discussed above, a high demand for tuna has threatened tuna stock and the environment. Raising fish on a farm could be a solution to this problem. The first tuna farm was approved in Hawaii [21] and promised to create an environmentally friendly open-ocean farm for bigeye tuna. Consequently, this study would be the first to examine consumers' acceptance and willingness to pay for this new technique and to evaluate the market potential, especially before significant money and resources are implemented.

2.4. *Other Tuna Attributes: Storage Mode and Price*

According to Lancaster's "new theory of consumer demand" [32], consumers are not seeking to acquire goods *per se*, but their characteristics. Product demand is affected by attributes that may include flavor/taste, freshness, size, or fish form. Referring to freshness or storage mode, this survey attempts to compare consumers' preference between fresh, which is never frozen, and previously frozen tuna [33–35]. Secondly, two types of product processes are considered: farm raised and wild-caught [34,35]. Therefore, this study will examine consumer preference for the attributes of farm-raised tuna and contribute to an understanding of the future development of tuna farms. Finally, the prices of tuna are also included. A detailed description of the attributes used and the choice of experiment design are provided later in Section 4: Choice Experiment and Model.

3. Survey and Data

3.1. *Survey Sample Statistics*

For the purposes of this study, an online survey was utilized. It was conducted via the Internet in July 2010 through zoomerang.com, which is operated by a professional survey company called MarketTool, Inc. The survey participants were residents from a typical land-locked state in the U.S.—the state of Kentucky. A total of 421 completed questionnaires were returned and usable for analysis in this study. The questionnaire was designed to examine household level tuna steak (sashimi grade) consumption and purchase preference, especially consumers' preference for eco-labeled food products and raising origins. The survey consisted of three parts: general seafood purchasing habit questions, aiming to establish a basic understanding of consumer demand and attract respondents' attention [23]; a choice experiment; and questions on demographics information. The descriptive statistics of the sample are provided in Table 1.

Table 1. Sample descriptive statistics.

Variable	Group	Dist.	Sample Mean	Kentucky Average *	Description
URBAN	-	-	0.55	-	Dummy Variable; = 1 if live in an urban area
FEMALE	-	-	0.71	51.60%	Dummy Variable; = 1 if female
AGE	-	-	52.2	48.5	Continuous Variable; in years
EDU	Less than High School High School Only Some College, no degree Associate's Degree Bachelor's Degree Master's Degree Professional Degree Doctorate	2.38% 25.42% 30.40% 11.88% 14.73% 10.93% 3.09% 1.19%	14	KY College Attainment Rate among adults aged 25–64 is 30.5% in 2009	Continuous Variable; in years
EMPLOY			0.50	55.30%	Dummy Variable; = 1 if employed
INCOME	0–$14,999 $15,000–$24,999 $25,000–$49,999 $50,000–$74,999 $75,000–$99,999 $100,000–$14,9999 >$150,000	7.13% 13.78% 40.14% 19.95% 11.40% 5.46% 2.14%	5.13	17.90% 13.20% 26.90% 17.80% 10.90% 8.90% 4.50%	Continuous Variable; in $10 k, of household income
WHITE	-	-	0.92	88.9%	Dummy Variable; = 1 if race is Caucasian
HHSIZE	-	-	2.60	2.48	Continuous Variable
COAST	-	-	0.09	-	Dummy Variable; = 1 if grew up within 50 miles of the seacoast

* Source: 2010–2014 KY State Average [36].

In general, the demographical statistics compare closely to the Kentucky state average. A total of 71% of the respondents were female, which was slightly more than the state average (56.6%) because it was acceptable that the survey participants were more likely to be female respondents who might be more interested in grocery shopping. The mean age for this sample was 52 years old, older than the 48.5-year-old state average. Almost half of the respondents were employed either full-time or part-time, which was close to the state average level (55.3%). In the U.S. Census, Kentucky's college attendance rate among adults aged 25–64 was 30.5% in 2009. In this sample, 30.40% of "some college experience" was observed. The median annual household income for Kentucky was $43,342 from 2010 to 2014, which was below the average in the U.S. Census. However, the income distribution was more closely representative. For example, 26.9% of the population had a household income ranging from $25,000 to $49,999, which was the largest group in the Kentucky according to the 2010 to 2014. In comparison, a larger portion (40.14%) was found to belong to this range in the sample, which was also the biggest group. A greater number of Caucasian respondents (92%) took the survey compared to the state average (88.9%). A household size of 2.6 members observed in the sample was consistent with the state average of 2.48. It was noted that the sample average was slightly different from the KY census statistics; however, this was within reason. Although the survey was conducted via the Internet, individuals who were female, older, and wealthier—older individuals were more likely to be married, and their household incomes were thus on average higher than single individuals—were more likely to take the survey, which concerned grocery shopping, than males and younger individuals.

Consumers' household location was also examined and labeled as URBAN. This is a dummy variable indicating that respondents lived in either urban or suburban areas. In this survey, 55.11% of the respondents were urban residents, and 44.89% were rural residents. Childhood experiences concerning seafood were also considered with the item, "whether (they) grew up within 50 miles of the seacoast," and a dummy COAST was created if the answer was yes. Approximately 9% of the respondents lived near the coast during their early childhood years, which meant that the sample comprised mostly land-locked consumers in whom the study was initially interested.

3.2. Perceptions and Attitudes Statistics

Beyond demographics, the questionnaire also inquired about respondents' perceptions and attitudes regarding seafood consumption. For instance, respondents were asked whether they were able to "differentiate between wild-caught and aquaculture/farm-raised fish, aside from labeling, either pre- or post-consumption." According to the results, only 3% of the respondents self-reported that they could always recognize wild-caught and farm-raised fish, and another 26% were able to differentiate either "most of the time" or "sometimes," compared to the 33% unsure. Figure 1 shows the details. In the subsequent sections, a dummy variable was created to examine how such a perception of fish could affect consumers' patron preference and willingness to pay. The dummy variable was created as variable "DIFFER" with a value of 1, indicating "being always able to differentiate wild-caught and farm-raised fish," "most of the time," or at least "sometimes."

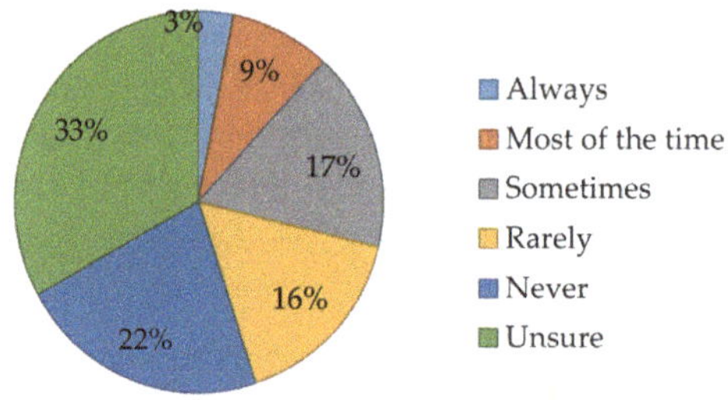

DIFFER=1 if choose "Always", "Most of the time", "Sometimes";
0 if choose "Rarely", "Never" and "unsure".

Figure 1. "Are you able to differentiate between wild-caught and farm-raised fish?"

Secondly, respondents were asked whether they "have noticed supermarket labels specifying whether seafood is farm-raised or wild-caught." Feedback from this question would provide references and suggestions for future labeling designs for both tuna marketers and policy makers. According to the results, 41% of the respondents had noticed labels specifying that seafood was farm-raised or wild-caught. However, a larger portion (42% "had not" and 17% were "unsure") reported that they had not or were unsure about such information when purchasing seafood. The dummy variable "Labelnotice" was then generated (Figure 2).

In addition to perception issues, the survey also asked for information on consumers' attitudes toward seafood labels. Figure 3 gives a summary of their answers as to "whether labels on a product will affect your decision to purchase seafood," and the variable "Labelinfluence" was generated for later econometric analysis. One third (34.68%) of the respondents reported a neutral attitude on this issue, and a fourth (23.04%) indicated that they cared a great deal about labels when buying seafood. However, 13.78% of the respondents stated that labels do not affect their purchasing decisions at all.

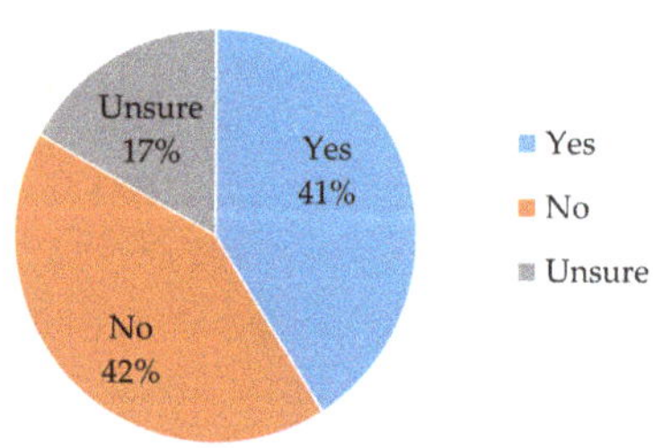

Label notice= 1 if Yes
0 if No or Unsure

Figure 2. "Have you noticed supermarket labels specifying whether seafood is farm-raised or wild-caught?"

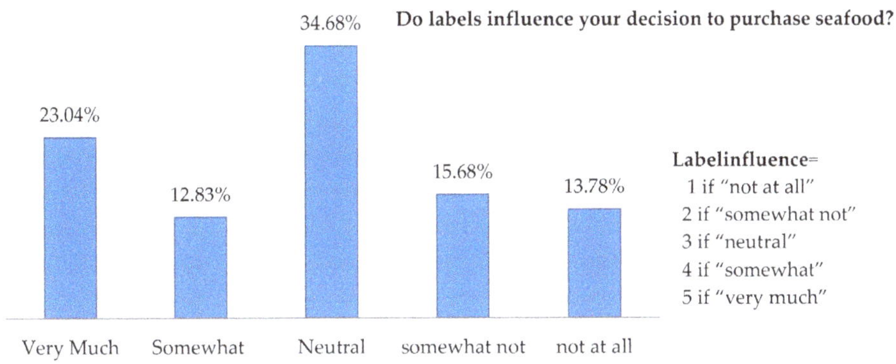

Figure 3. "Do labels influence your decision to purchase seafood?"

Lastly, environmental friendliness factors (dummy variables "Env Friendly_WC" and "Env Friendly_FR") were also created. Respondents revealed their attitudes when answering the Likert scale question ranging from 1 (not at all) to 5 (very important). Over 40% of the respondents said that environmental friendliness was a "very important" attribute. Meanwhile, 23.52% considered it an "important" attribute, and another 23.75% believed it was moderately important. Figure 4 shows the details.

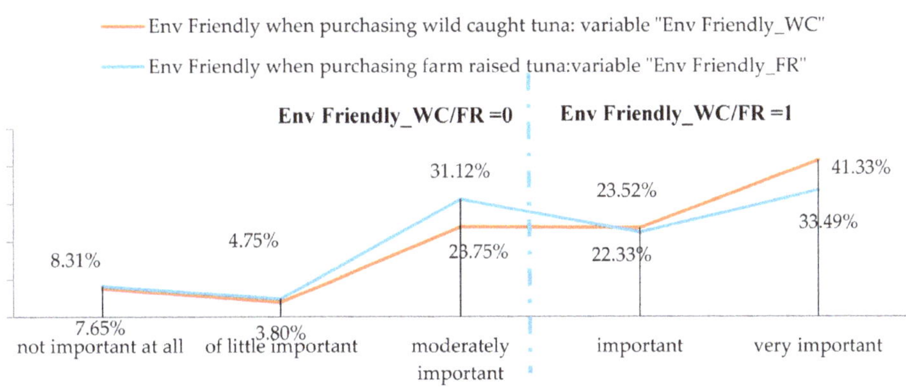

Figure 4. Environmental friendliness when purchasing tuna.

4. Choice Experiment and Models

4.1. Choice Experiment

A choice experiment was employed in this survey that is widely used to elicit consumers' preferences for food products [37–42]. It presented consumers/respondents with a set of alternatives that differed in tuna steak attributes and asked consumers to choose an alternative they prefer. The responses were used to elicit Willingness-to-Pay (WTP) for different tuna steak attributes, as discussed in Section 2.2 through 2.4. Table 2 provides descriptions of the attributes and levels.

Table 2. Tuna (steak form and sashimi grade) attributes and levels.

Attribute	Alternatives	Variable Name	Descriptions
Origin	Wild-Caught Farm-Raised	Wild-Caught	Refers to the origin where the tuna was raised.
Storage Mode	Previously Frozen Fresh and Never Frozen	Pre-Frozen	Refers to the method used for storage.
Eco-labeled	Certified Turtle Safe * None	Turtle Safe	Refers to whether the tuna is labeled as turtle safe or not.
Price($/lb)	8.99 14.49 19.99 25.49	PRICE	Refers to tuna steak price (in sashimi grade) in the retail grocery store where the respondent typically shops.

* Certified turtle safe by definition is fish harvested by fisheries under stringent controls to avoid sea turtle by-catch.

As presented, tuna fish may be wild-caught or farm-raised. Additionally, storage mode can be either "previously frozen" or "fresh and never frozen". The eco-label "Certified Turtle Safe" may be used for a product. Finally, price was also included as an attribute, which is critical for future WTP estimates. Four price levels were used that were obtained after researching comparable products at regular chain retail markets in Lexington, KY: $8.99/lb; $14.49/lb; $19.99/lb; and $25.49/lb. These price levels were chosen to ensure that they covered the lowest and highest possible prices. Corresponding variables for these attributes are Wild-Caught (hereinafter, WC), Pre-Frozen (Pfr), Turtle Safe (TS), and PRICE, as given in Table 1.

Using an orthogonal design and adding the "buy neither" option, four choice situations were generated. In each choice situation, respondents were asked to choose one from options A, B, and C in each scenario. Then, a dummy variable BUYNO was used to indicate the third alternative in each choice set. The choice experiment was designed using an orthogonal design (with SPSS software) in which eight choice sets and four situations were created (Origin (2) $\times$ Storage Mode (2) $\times$ Eco-labeling (2) $\times$ Prices (4) = 32). A pilot study was also conducted in August 2009 in which 17 respondents from Charlie's Fresh Seafood Market in Kentucky were targeted. An example of the choice scenario is displayed in Figure 5. The respondents were told before entering the choice scenarios, for example, that the two options (please refer to option A and B in Figure 5) were identical for all other characteristics rather than those described and instructed not to compare across scenarios. These were important assumptions for our later model specifications.

Figure 5. The Choice Question Example. Note: * Turtle Safe: Fish harvested by fisheries under stringent controls to avoid sea turtle by-catch.

4.2. Theoretical Framework

Supposing individual i faces a choice consisting of different attributes and chooses j (where, j = 1, 2, 3, ... , J) among all the alternatives in the t-th choice situation, the attributes can be represented by x_{ijt} (x = BUYNO, WC, Pfr, TS, PRICE). It is assumed that the consumer will choose alternative j if and only if the associated utility is greater than or at least equal to any other alternatives, *ceteris paribus*. Mathematically, utility can be represented in a random utility framework (McFadden, 1974), namely,

$$U_{ijt} = \mathbf{X}_{ijt}\boldsymbol{\beta} + \varepsilon_{ijt}, \tag{1}$$

where U_{ijt} refers to the indirect utility obtained by individual i, which is a linear function of the observable vector of attributes $\mathbf{X}_{ijt}$ and their coefficients, vector $\boldsymbol{\beta}$, which are to be estimated; and ε_{ijt} represents the random error that captures all other unobservable factors that influence the choice process. McFadden showed that, if the error terms follow an *i.i.d.* maximum extreme value Type I distribution, the utility maximization process leads to the choice probability of alternative j chosen in choice set t:

$$P_{ijt} = \frac{\exp\left(\mathbf{X}_{ijt}\boldsymbol{\beta}\right)}{\sum_{k=1}^{J}\exp\left(\mathbf{X}_{ikt}\boldsymbol{\beta}\right)}. \tag{2}$$

This is the form of the conditional logit model; however, it suffers from two major limitations: (1) It cannot represent random taste variation, and (2) it does not avoid the restrictive substitution pattern suggested by the property of Independence of Irrelevant Alternatives (IIA) [43]. A mixed logit model addresses these limitations. Recent improvements in the computational package have promoted empirical applications of mixed logit, and this study used STATA 12.

Following Train (2003) [43], the parameters in vector $\boldsymbol{\beta}$ are assumed as random variables and may vary across individuals in the sample rather than fixed coefficients in conditional logit. Supposing the distribution of $\boldsymbol{\beta}$ is specified as $\boldsymbol{\beta} \sim \mathbf{H}(\boldsymbol{\theta}, \Delta)$, H can be the individual probability distribution function, and parameters $\boldsymbol{\theta}$ and Δ are the mean and variance. Benefits from the mixed distribution is that unobserved variation can be represented in the form of any appropriate distribution by specifying the form of function H. Among the commonly used distributions are normal, lognormal, and uniform distributions. Given the random parameter context, the choice probability is updated as:

$$P_{ijt} = \int \frac{\exp\left(\boldsymbol{X}_{ijt}\boldsymbol{\beta}\right)}{\sum_{k=1}^{J} \exp\left(\boldsymbol{X}_{ikt}\boldsymbol{\beta}\right)} h\left(\boldsymbol{\beta}\right) d(\boldsymbol{\beta}), \tag{3}$$

where $h\left(\boldsymbol{\beta}\right)$ is the (jointly) density function of H for parameters $\boldsymbol{\beta}$. The integral can be approximated by simulation. Consequently, instead of $\boldsymbol{\beta}$, parameters $\boldsymbol{\theta}$ and Δ are to be estimated.

In addition to product attribute variables in a basic mixed logit model, other factors may also affect the decision process. A natural extension of the model would be to consider respondents' individual demographic characteristics [38] and their perceptions and attitudes. Therefore, our specification of the mixed logit model can be augmented with interactions:

$$U_{ijt} = V_{ijt}(\boldsymbol{X}_{ijt}, \boldsymbol{D}, \boldsymbol{P};\ \boldsymbol{\beta}, \gamma) + \ \varepsilon_{ijt}\ , \tag{4}$$

$$V_{ijt} = \boldsymbol{X}_{ijt}\boldsymbol{\beta} + \gamma^{D}\left(\boldsymbol{X}_{ijt} * \boldsymbol{D}_{i}\right) + \gamma^{P}\left(\boldsymbol{X}_{ijt} * \boldsymbol{P}_{i}\right), \tag{5}$$

$$\boldsymbol{X}_{jt} = \left[BUYNO,\ WC,\ Pfr,\ TS,\ PRICE\right]_{jt}, \tag{6}$$

$$\boldsymbol{D}_{i} = Demograpics, \text{and} \tag{7}$$

$$\boldsymbol{P}_{i} = perceptions\ /\ attitudes\ variables. \tag{8}$$

The choice probability function is:

$$P_{ijt} = \int \frac{\exp\left(\boldsymbol{X}_{ijt}\boldsymbol{\beta}+\gamma^{D}\left(\boldsymbol{X}_{ijt}*\boldsymbol{D}_{i}\right)+\gamma^{P}\left(\boldsymbol{X}_{ijt}*\boldsymbol{P}_{i}\right)\right)}{\sum_{k=1}^{J} \exp\left(\boldsymbol{X}_{ijt}\boldsymbol{\beta}+\gamma^{D}\left(\boldsymbol{X}_{ijt}*\boldsymbol{D}_{i}\right)+\gamma^{P}\left(\boldsymbol{X}_{ijt}*\boldsymbol{P}_{i}\right)\right)} h\left(\boldsymbol{\beta}\right) d\left(\boldsymbol{\beta}\right) \qquad \boldsymbol{\beta} \sim \mathbf{H}(\boldsymbol{\theta}, \Delta)\ . \tag{9}$$

The marginal value or WTP for an attribute is given by the ratio of the attribute coefficient to the price coefficient, which is set to be fixed as above, such that:

$$Marginal\ Value\ /\ WTP = -\frac{\boldsymbol{\beta}_{X} + \gamma_{X}^{D} * \boldsymbol{D} + \gamma_{X}^{P} * \boldsymbol{P}}{\boldsymbol{\beta}_{PRICE} + \gamma_{PRICE}^{\boldsymbol{D}} * \boldsymbol{D} + \gamma_{PRICE}^{\boldsymbol{P}} * \boldsymbol{P}}\ . \tag{10}$$

In this study, we assume the coefficients for BUYNO and PRICE are fixed parameters for ease of calculation. Other coefficients for tuna attributes are specified as random parameters, including WC, Pfr, and TS. Therefore, the denominator in the calculation of WTP contains fixed coefficients only. The numerator is identified as random, which includes random coefficients and interaction terms between attributes and demographic or perception variables. The standard errors of WTP measures incorporate both mean and standard deviation results, providing a better description of WTP distribution. An alternative approach is to report the distribution of the WTP as the distribution of the attributes or interactions coefficient scaled by the fixed price coefficient rather than a single representative WTP when holding demographics and other factors at sample average levels [44].

5. Estimation Results and Discussions

Tables 3 and 4 show the results from mixed logit models that were with and without interactions. Both models specified all random coefficients as normally distributed, namely variables WC, Pfr, and TS. We used 200 Halton draws per iteration in the simulated maximum likelihood estimation. Model fitness criteria identified improvement by adding interactions between attributes and other decision-making factors. Overall, the model presented in Table 4 with demographics and perception/attitude variable interactions obtains a higher log likelihood value (from −1398.6354 to −1307.0789) and lower AIC score but a higher BIC score. The McFadden R-squares were also reported in both tables.

Table 3. Mixed logit model results.

	Coef.		Std. Err.	*p* Value
MEAN				
PRICE	−0.2098	***	0.0178	0.00
BUYNO	−2.2710	***	0.2135	0.00
Wild-Caught	−1.0077	***	0.2233	0.00
Pre-Frozen	−0.0630		0.2030	0.76
Turtle Safe	1.1899	***	0.2421	0.00
Std. Dev.				
Wild-Caught	2.2589	***	0.2877	0.00
Pre-Frozen	3.1139	***	0.3031	0.00
Turtle Safe	2.6780	***	0.2804	0.00
Log Likelihood	−1398.6354			
McFadden R^2	0.3146			
AIC	2813.27			
BIC	2865.49			
Obs	1684 [a]			

*** represents the 1% significant level. [a] # of Obs = 421 respondents × 4 situations per respondent.

Table 4. Mixed logit model results with interactions.

	Coefficient		Std. Err.	*p* value
MEAN				
PRICE	−0.3466	***	0.0291	0.00
BUYNO	−2.5464	***	0.2295	0.00
Wild-Caught	−1.9861		1.8709	0.29
Pre-Frozen	0.0426		1.8677	0.98
Turtle Safe	3.9285	**	1.7898	0.03
	Demographic Interactions			
WC × URBAN	−0.1238		0.3851	0.75
WC × FEMALE	−0.1482		0.4245	0.73
WC × AGE	−0.0281		0.0192	0.14
WC × EDU	0.0410		0.0808	0.61
WC × EMPLOY	−0.2818		0.4080	0.49
WC × INCOME	0.0653		0.0577	0.26
WC × WHITE	0.6206		0.7274	0.39
WC × HHSIZE	−0.1015		0.1781	0.57
WC × COAST	1.7324	***	0.5790	0.00
Pfr × URBAN	−0.6464	*	0.3964	0.10
Pfr × FEMALE	−1.4963	***	0.4456	0.00
Pfr × AGE	0.0146		0.0190	0.44
Pfr × EDU	0.0335		0.0823	0.68
Pfr × EMPLOY	0.0801		0.4191	0.85
Pfr × INCOME	0.0156		0.0622	0.80
Pfr × WHITE	−0.0487		0.7175	0.95
Pfr × HHSIZE	0.1721		0.1803	0.34
Pfr × COAST	−0.5906		0.6718	0.38
TS × URBAN	−0.0176		0.3695	0.96
TS × FEMALE	−0.7748	*	0.4162	0.06
TS × AGE	−0.0375	**	0.0182	0.04
TS × EDU	−0.0157		0.0770	0.84
TS × EMPLOY	−0.2972		0.3919	0.45
TS × INCOME	0.0583		0.0577	0.31
TS × WHITE	0.3036		0.6796	0.66
TS × HHSIZE	−0.1632		0.1727	0.35
TS × COAST	1.1347	*	0.5941	0.06
(con't)				
Differ Ability				
WC × Differ	0.2611		0.4335	0.55
Pfr × Differ	0.4036		0.4487	0.37
TS × Differ	−0.3116		0.4544	0.49
PRICE × Differ	0.0662	***	0.0166	0.00

Table 4. *Cont.*

	Coefficient		Std. Err.	*p* value
	Whether Notice Labels When Purchasing			
WC × Labelnotice	0.3610		0.4402	0.41
Pfr × Labelnotice	0.3342		0.4519	0.46
TS × Labelnotice	0.2162		0.4572	0.64
PRICE × Labelnotice	0.0282	*	0.0157	0.07
	Whether Label Influences Purchase Decision			
WC × Labelinfluence	0.4038	**	0.1693	0.02
Pfr × Labelinfluence	−0.4099	**	0.1741	0.02
TS × Labelinfluence	−0.2554		0.1806	0.16
PRICE × Labelinfluence	0.0249	***	0.0065	0.00
	Environmental Priority			
WC × Env Friendly_WC	−0.3226		0.4493	0.47
Pfr × Env Friendly_WC	0.2187		0.5531	0.69
Pfr × Env Friendly_FR	0.8429		0.5306	0.11
TS × Env Fri × ndly_WC	1.5643	***	0.5795	0.01
TS × Env Friendly_FR	−0.4191		0.5487	0.45
PRICE × Env Friendly_WC	0.0192		0.0201	0.34
PRICE × Env Friendly_FR	0.0010		0.0191	0.96
Std. Dev.				
Wild Caught	2.1166	***	0.2742	0.00
Pre Frozen	3.0787	***	0.2996	0.00
Turtle Safe	2.5399	***	0.2761	0.00
Log Likelihood	1307.0789			
McFadden R^2	0.3439			
AIC	2722.158			
BIC	3074.645			

*, **, *** represents 10%, 5% and 1% significant levels, respectively.

Estimation results are reported in two parts in Table 3: mean and standard deviation (SD hereinafter). Variable BUYNO stands for the alternative specific constant for the no-choice option. A statistically significant negative coefficient for BUYNO (−2.2710, significant at 1% confident level) suggests that consumers would encounter a loss in utility if they did not choose any products offered in a choice set. Consumers would choose nothing only when the first two alternatives were undesirable. The PRICE variable also had a significantly negative coefficient, indicating that consumers were likely to choose products with a lower price while all other factors were held constant.

All tuna attribute variables revealed significant results except the variable Prf (Pre-frozen seafood). For instance, respondents were less likely to purchase wild-caught tuna than farm-raised species, which could be explained by the fact that Kentuckians living in a land-locked area preferred farm-raised seafood to wild-caught seafood. The coefficient for CTS tuna (1.1899) was significantly positive at the 1% level. Consumers supported this environmentally friendly labeled product. Heterogeneity existed for all three tuna attributes according to the results shown in the group of SD because all estimates were significantly different from zero at the 1% level.

WTP derived from a basic mixed logit model was summarized in Table 5. If an individual chose the BUYNO option, s/he would lose $10.82 per pound. Distributions of WTP for tuna attributes are also displayed in the table. A significant price premium of $5.67 per pound was found for tuna products if they were labeled as "Certified Turtle Safe." However, negative WTP were observed for both the wild-caught and pre-frozen attributes, which were −$4.80 and −$0.30 per pound, respectively. The results in both Tables 3 and 4 show a heterogeneous preference across individuals. Further analysis is necessary to delve into consumers' taste heterogeneities, and interactions were created between tuna attributes and consumer demographics as well as their perception and attitudes.

Table 5 demonstrates mixed logit model results when tuna attribute variables interacted with respondents' demographic information as well as their perceptions or attitudes towards seafood. The estimate included not only the main effect of the attributes but also the effects of interaction. Moreover, standard deviation estimates were statistically significant at the 1% level, which is consistent with the previous results in Table 3.

Table 5. Willingness-to-Pay (WTP) estimates.

	WTP		Std. Dev.	*p* Value	(95% Conf. Interval)	
BUYNO	−$10.82	***	$0.46	0.00	−$11.73	−$9.92
Wild-Caught	−$4.80	***	$1.10	0.00	−$6.95	−$2.66
Pre-Frozen	−$0.30		$0.97	0.76	−$2.20	$1.60
Turtle Safe	$5.67	***	$0.99	0.00	$3.73	$7.61

*** represents the 1% significant level.

5.1. *Interactions with Demographics*

Demographic characteristics played a somewhat important role in differentiating consumers based on their taste preference for seafood, especially tuna. For example, gender, age, location of residence, and whether consumers grew up near the coast were all significant with respect to consumers' seafood consumption. Younger consumers preferred tuna products labeled as "Certified Turtle Safe" more than older consumers. Moreover, compared to male consumers, female consumers were less likely to purchase or pay more for tuna that was either labeled "Certified Turtle Safe" or pre-frozen. However, females were not significantly different from males concerning the choice of wild-caught tuna. Recalling the descriptive results, the majority of the respondents were female because the main grocery shoppers tended to be female. Therefore, our findings that female individuals were less likely to purchase pre-frozen tuna imply that opening or investing more shelves for fresh seafood in stores would positively influence profit.

Consumers were asked whether they grew up within 50 miles of the seacoast, and less than 10% replied "yes." Nevertheless, the childhood memories still have a significant effect on current purchasing attitudes. Those who had a coastal childhood were more likely to choose CTS-labeled tuna as well as wild-caught tuna, the coefficients of which are significant at 5% and 1%, respectively. No significant result is observed for pre-frozen tuna, but present location was shown to matter. Individuals living in urban or suburban areas are less likely to choose pre-frozen tuna. These findings provide useful information for future market segmenting.

5.2. *Interactions with Perception/Attitude*

In addition to demographics, this study also examines the interactions between tuna attributes (WS, Pfr, TS) and consumers' perceptions/attitudes. The previous section introduced four groups of consumers' perception and attitude information. Consequently, four dummies were created as follows: Differ, Labelnotice, Env Friendly_WC/Env Friendly_FR, and a Likert scale variable Labelinfluence. Consumers were then segmented according to their perception or attitude.

In the results, respondents were more sensitive to the retail price of tuna when they were able to differentiate between wild-caught and farm-raised seafood. Consumers were also more sensitive to price when they noticed labels indicating that seafood was wild-caught/farm-raised than those who never noticed labels. Additionally, respondents who self-reported that labeling would influence their final decision to purchase seafood were also associated with a strong sensitivity to price. As shown in the model analysis results, the more that labeling influenced decision to purchase, the greater the probability was that the consumer would also purchase wild-caught tuna, but the probability that the consumer would purchase pre-frozen tuna was lower. This study also investigates consumers' preferences for environmentally friendly and eco-labeled products. The coefficient for the interaction term Turtle Safe × Env Friendly_WC is strongly significant at 1% in the results, which implies that individuals who claim that environmental friendliness is an important attribute for wild-caught seafood are more likely to choose tuna labeled as "Certified Turtle Safe."

5.3. WTP Distributions

After determining consumers' purchasing preferences from the mixed logit model in Tables 6–8 it was natural to investigate willingness-to-pay estimates for associated tuna attributes. Based on the results from mixed logit model, WTP can be inferred from nonlinear combinations of coefficients of non-price variables over the price variable, calculated with the STATA command (nlm). The results are divided into three parts: WTP distributions for interaction with perception/attitude; interaction with demographics; and tuna attributes only, corresponding to Table 5 and presented in Tables 6–8.

Table 6 displays the WTP for variables that are interacting tuna attributes with consumer perceptions or attitudes, which aim to differentiate consumers by their attitude toward labels as well as by their environmental awareness. If an individual rates the influence of labels for seafood purchase at 5, where a Labelinfluence of 5 is very important, then s/he may pay $12.82 more per pound for wild-caught tuna compared to consumers with a neutral attitude, as shown in Table 6. However, s/he would pay $13.02 less for pre-frozen compared to the fresh fish. Moreover, if an individual cares about the environmental friendliness feature of seafood, s/he would likely pay $9.93 more for CTS tuna.

Table 7 summarizes the WTP distributions for attributes when interacting with demographics. The sample average for age is 52.2 years; for education, 14 years, obtaining an associate's degree. Distributions are displayed for 25, 35, and 55 years old. Education can also be used to segment consumers; in this study, the results are grouped as 13 years (some college, no degree), 16 years (master's degree), and 22 years (doctoral degree). Taking attribute CTS (Certified Turtle Safe) as an example, female consumers would pay $4.92 per pound less for eco-labeled tuna. In addition, consumers aged 55 years would pay $13.10 less for eco-labeled than non-labeled tuna, and a younger consumer aged 25 years will also pay less but only $5.96 less. Surprisingly, the higher the individual's education, the less s/he would pay for eco-labeled tuna. Respondents who lived in a coastal area during their childhood would pay more for CTS tuna, approximately $7.21 per pound.

Table 6. WTP Estimates: interactions with perception/attitude.

	WTP		Std. Err.	P Value	(95% Conf. Interval)	
Differ Ability						
WC × Differ	$1.66		$2.72	0.54	−$3.66	$6.98
Pfr × Differ	$2.56		$2.85	0.37	−$3.03	$8.16
TS × Differ	−$1.98		$2.94	0.50	−$7.74	$3.78
Whether Notice Labels When Purchasing						
WC × Labelnotice	$2.29		$2.77	0.41	−$3.15	$7.73
Pfr × Labelnotice	$2.12		$2.87	0.46	−$3.51	$7.75
TS × Labelnotice	$1.37		$2.89	0.64	−$4.30	$7.04
Whether Label Influences Purchase Decision						
Labelinfluence = 5						
WC × Labelinfluence	$12.82	**	$5.52	0.02	$1.99	$23.65
Pfr × Labelinfluence	−$13.02	**	$5.63	0.02	−$24.06	−$1.98
TS × Labelinfluence	−$8.11		$5.70	0.16	−$19.28	$3.06
Labelinfluence = 3						
WC × Labelinfluence	$7.69	**	$3.31	0.02	$1.20	$14.19
Pfr × Labelinfluence	−$7.81	**	$3.38	0.02	−$14.43	−$1.19
TS × Labelinfluence	−$4.87		$3.42	0.16	−$11.57	$1.83
Environmental Priority						
WC × Env Friendly_WC	−$2.05		$2.89	0.48	−$7.71	$3.61
Pfr × Env Friendly_WC	$1.39		$3.52	0.69	−$5.50	$8.28
Pfr × Env Friendly_FR	$5.35		$3.41	0.12	−$1.33	$12.03
TS × Env Friendly_WC	$9.93	***	$3.74	0.01	$2.60	$17.27
TS × Env Friendly_FR	−$2.66		$3.51	0.45	−$9.55	$4.22

*, **, *** represents the 10%, 5%, and 1% significant levels, respectively.

Table 7. WTP Estimates: interaction with demographics.

	WTP	Std. Err.	*P* Value	(95% Conf. Interval)	
Average AGE = 52.2 years, EDU = 14 years Associate Degree					
WC × URBAN	−$0.79	$2.44	0.75	−$5.58	$4.00
WC × FEMALE	−$0.94	$2.70	0.73	−$6.23	$4.34
WC × AGE: *25 years old*	−$4.45	$3.07	0.15	−$10.47	$1.56
35 years old	−$6.24	$4.30	0.15	−$14.66	$2.19
55 years old	−$9.80	$6.75	0.15	−$23.04	$3.44
WC × EDU: *Some College*	$3.38	$6.67	0.61	−$9.70	$16.46
Bachelor's Degree	$4.16	$8.21	0.61	−$11.93	$20.26
Doctoral Degree	$5.72	$11.29	0.61	−$16.41	$27.85
WC × EMPLOY	−$1.79	$2.60	0.49	−$6.88	$3.30
WC × INCOME	$2.13	$1.89	0.26	−$1.57	$5.82
WC × WHITE	$3.94	$4.61	0.39	−$5.10	$12.98
WC × HHSIZE	−$1.93	$3.40	0.57	−$8.60	$4.73
WC × COAST	$11.00***	$3.76	0.00	$3.64	$18.37
Pfr × URBAN	−$4.10	$2.56	0.11	−$9.13	$0.92
Pfr × FEMALE	−$9.50**	$3.09	0.00	−$15.56	−$3.45
Pfr × AGE: *25 years old*	$2.32	$3.02	0.44	−$3.59	$8.24
35 years old	$3.25	$4.22	0.44	−$5.03	$11.53
55 years old	$5.11	$6.64	0.44	−$7.90	$18.12
Pfr × EDU: *Some College*	$2.77	$6.80	0.68	−$10.56	$16.09
Bachelor's Degree	$3.40	$8.37	0.68	−$13.00	$19.81
Doctoral Degree	$4.68	$11.51	0.68	−$17.87	$27.24
Pfr × EMPLOY	$0.51	$2.66	0.85	−$4.71	$5.73
Pfr × INCOME	$0.99	$3.95	0.80	−$6.76	$8.74
Pfr × WHITE	−$0.93	$13.67	0.95	−$27.72	$25.86
Pfr × HHSIZE	$3.28	$3.46	0.34	−$3.50	$10.05
Pfr × COAST	−$3.75	$4.28	0.38	−$12.13	$4.63
TS × URBAN	−$0.11	$2.35	0.96	−$4.71	$4.49
TS × FEMALE	−$4.92*	$2.71	0.07	−$10.23	$0.39
TS × AGE: *25 years old*	−$5.96**	$2.95	0.04	−$11.74	−$0.17
35 years old	−$8.34**	$4.13	0.04	−$16.43	−$0.24
55 years old	−$13.10**	$6.49	0.04	−$25.82	−$0.38
TS × EDU: *Some College*	−$1.29	$6.36	0.84	−$13.75	$11.16
Bachelor's Degree	−$1.59	$7.82	0.84	−$16.92	$13.74
Doctoral Degree	−$2.19	$10.75	0.84	−$23.27	$18.89
TS × EMPLOY	−$1.89	$2.49	0.45	−$6.78	$3.00
TS × INCOME	$3.70	$3.69	0.32	−$3.54	$10.94
TS × WHITE	$5.78	$12.96	0.66	−$19.62	$31.19
TS × HHSIZE	−$3.11	$3.31	0.35	−$9.59	$3.37
TS × COAST	$7.21 *	$3.83	0.06	−$0.31	$14.72

*, **, *** represents the 10%, 5%, and 1% significant levels, respectively.

Lastly, willingness-to-pay distributions for tuna attributes are provided in Table 8. The results are higher than those in Table 5. For instance, buying nothing, as referred to by the BUYNO option, is associated with a loss of utility compared to choosing any of these two hypothetical products. In the previous basic mixed logit model, the dollar amount for this loss is $10.82 per pound. However, the new estimate when considering heterogeneity and when integrated with interactions results in a higher amount, $16.17, if the BUYNO option is chosen. Willingness-to-pay for tuna attributes is also demonstrated in the rest of the table rows. If consumers have a neutral attitude toward labels, where the variable Labelinfluence equals 3, a positive WTP is observed for CTS-labeled tuna, while it is negative for pre-frozen fish. If there is a higher score for Labelinfluence, the WTP is also higher for wild-caught seafood and has a distribution range from -$0.70 to $16.84. However, it becomes insignificant for pre-frozen products, while the WTP is diminishingly positive for CTS-labeled tuna, as shown in the results.

Table 8. WTP Estimates: main effect.

	WTP		Std. Err.	P Value	(95% Conf. Interval)	
BUYNO	−$16.17	***	$1.67	0.00	−$19.45	−$12.90
Wild-Caught						
Labelinfluence = 1	−$2.19		$5.31	0.68	−$12.60	$8.23
Labelinfluence = 3	$2.94		$4.39	0.50	−$5.65	$11.54
Labelinfluence = 5	$8.07	*	$4.48	0.07	−$0.70	$16.84
Pre-Frozen						
Labelinfluence = 1	$3.55		$5.74	0.54	−$7.70	$14.80
Labelinfluence = 3	−$1.66		$4.92	0.74	−$11.31	$7.99
Labelinfluence = 5	−$6.86		$5.07	0.18	−$16.80	$3.08
Turtle Safe						
Labelinfluence = 1	$17.17	***	$5.31	0.00	$6.77	$27.57
Labelinfluence = 2	$15.55	***	$4.75	0.00	$6.23	$24.87
Labelinfluence = 3	$13.93	***	$4.43	0.00	$5.24	$22.61
Labelinfluence = 4	$12.31	***	$4.39	0.01	$3.70	$20.91
Labelinfluence = 5	$10.68	**	$4.64	0.02	$1.60	$19.77

*, **, *** represents the 10%, 5%, and 1% significant levels, respectively.

All coefficients for willingness-to-pay for CTS tuna are positive and statistically significant, implying that respondents strongly favor tuna labeled as "Certified Turtle Safe." More interestingly, if an individual reports a higher score for Labelinfluence, then s/he would likely pay $10.68 more per pound for eco-labeled tuna. However, the amounts for WTP decrease as the scores increase. For example, the WTP of a neutral individual for CTS-labeled tuna is $13.93 per pound. However, if an individual believes that the label does not affect his/her seafood purchasing and the score for this is therefore lower, s/he will likely pay $17.17 more for eco-labeled tuna. Furthermore, the distribution for this group of consumers is presented at a 95% confidence interval, from $6.77 to $27.57 per pound.

6. Conclusions

This paper relies on an Internet-based survey, representing household-level tuna steak (sashimi grade) consumption and purchase preference in a midland state in the U.S. Each respondent was asked about his/her seafood purchasing habits over the past two months as well as his/her demographic information. This study also investigated consumer preference for the attributes of tuna and the impact of eco-labels, particularly CTS labels, on consumer demand. Additionally, this study examined individuals' perceptions and attitudes toward farm-raised and wild-caught tuna species.

A choice-based conjoint experiment was employed in the survey as a series of hypothetical choices between pairs of products and a third choice to purchase neither product. These alternatives were provided in the context, which differed in major product attributes (wild-caught, farm-raised, eco-label, price, and previously frozen or not). To fulfill the objective of this study to assess consumer preferences concerning eco-friendly labeling, a CTS label was introduced as one of the product attributes. A mixed logit model was utilized to examine purchasing propensities, and the estimation results were used to elicit willingness-to-pay. The interactions between tuna attributes and individual-specific information may serve to identify consumer segments.

It was found that participants on average preferred CTS-labeled tuna steak and were likely to pay more for it but were less likely to purchase wild-caught species, and insignificant results were found for pre-frozen. This may lead to a better understanding of the relationship between consumers' behavioral tendencies and purchasing preference for tuna steak. For instance, consumers' favoring of eco-labeled (turtle safe in this study) products promises a niche market for the industry and provides optimistic support for ocean sustainability movements via the promotion of more eco-labeling.

These results supplied evidence of public support for environmental friendliness, particularly for eco-labels ("Certified Turtle Safe" in this study), and explored and unearthed market potential for tuna eco-labeling. Additionally, an econometric analysis presented quantitative and monetary estimates for specific tuna attributes, which can be beneficial not only for marketing strategies but also for policy evaluation.

This study based on consumers from a land-locked state adds to the general discussion of preferences and the consumption culture surrounding seafood. One of the main contributions of this study is to provide perspectives on consumer demand for commercial tuna fish. Comparison between wild-caught and farm-raised species also contributes to evaluating consumer perceptions of product sources pertaining to environmental concerns. The results of the study can inform tuna producers and marketers about future product marketing strategies and promotions. In addition, the premium on eco-friendly labeling suggests consumers' growing desire for ecological well-being and sustainability.

This study also has several limitations. Firstly, the study was based on a stated preference method that, like any study using a similar method, may suffer from hypothetical bias. It asked hypothetical questions about hypothetical products, mostly unmarketed goods, as we examined in this study. There is a series of non-hypothetical methods that can be used for food marketing analysis. Even though these methods are generally more difficult to apply in a case where products that do no exist on the market are examined, applying such methods may prove useful. Secondly, the mixed logit model compiles a large number of attributes and interaction terms. This may reduce the efficiency of the model to some extent, especially when the research was designed to investigate a relatively wide range of issues of interest. In the future, a piece-wise estimation process may be used to offer more targeted estimates of the various effects.

Acknowledgments: We would like to thank Minling Pan (Minling Pan is an Economist in Pacific Islands and Fisheries Science Center, National Marine Fisheries Service, National Oceanic and Atmospheric Administration (NOAA)) and Devie Poerwanto Chang (Devie Poerwanto Chang is a former gradaute student at University of Kentucky and now works as a Market Research Analyst and Marketing Budgets & Metrics Coordinator for Alltech in Nicholasville, KY), both of whom indicated no coauthorship for this paper, but contributed to the survey design. Devie Poerwanto Chang also contributed to data collection for this study. More information about these contributors and their foundational work can be found in the report via http://www.pifsc.noaa.gov/economics/economic_relationship_between_aquaculture_and_wild_capture_fisheries.php.

Author Contributions: Guzhen Zhou is the lead author of the paper, co-conceptualized the research idea with the second author, and was also responsible for model estimation and results interpretation. Wuyang Hu developed and conducted the entire survey, and contributed to the development and writing of the paper. Wenchao Huang provided consultation for research background about tuna and tuna fishing techniques, and provided guidance throughout the result discussions.

Conflicts of Interest: The authors declare no conflict of interest.

References

1. Food and Agriculture Organization of the United Nations (FAO). *The State of World Fisheries and Aquaculture*; Food and Agriculture Organization of the United Nations: Rome, Italy, 2012.
2. Thøgersen, J.; Haugaard, P.; Olesen, A. Understanding consumer responses to ecolabels. *Eur. J. Mark.* **2008**, *44*, 1787–1810.

3. Larceneux, F.; Benot-Moreau, F.; Renaudin, V. Why might organic labels fail to infuence consumer choices? Marginal labelling and brand equity. *J. Consum. Policy* **2012**, *35*, 85–104.
4. Carlucci, D.; Nocella, G.; Devitiis, B.D.; Viscecchia, R.; Bimbo, F.; Nardone, G. Consumer purchasing behaviour towards fish and seafood products: Patterns and insigts from a sample of international studies. *Appetite* **2015**, *84*, 212–227.
5. Nguyen, T.T.; Haider, W.; Solgaard, H.S.; Ravn-Jonsen, L.; Roth, E. Consumer willingness to pay for quality attributes of fresh seafood: A labeled latent class model. *Food Qual. Preference* **2015**, *41*, 225–236.
6. Jaffry, R.; Pickering, H.; Ghulam, Y.; Whitmarsh, D.; Wattage, P. Consumer choices for quality and sustainability labelled seafood products in the uk. *Food Policy* **2004**, *29*, 215–228.
7. Fonner, R.; Sylvia, G. Willingness to pay for multiple seafood labels in a niche market. *Mar. Resour. Econ.* **2015**, *30*, 51–70.
8. Teisl, M.; Roe, B.; Hicks, R.L. Can eco-labels tune a market. *J. Environ. Econ. Manag.* **2002**, *43*, 339–359.
9. Uchida, H.; Onozaka, Y.; Morita, T.; Managi, S. Demand for ecolabeled seafood in the Japanese market. A conjoint analysis of the impact of information and interaction with other labels. *Food Policy* **2014**, *44*, 68–76.
10. Lohman, K.; Wang, J.; Boles, L.; McAlister, J.; Lohmann, C.; Higgins, B. Development of Turtle-Safe Light Sticks for Use in Longline Fisheries. In *Sea Turtle and Pelagic Fish Sensory Biology: Developing Techniques to Reduce Sea Turtle Bycatch in Longline Fisheries*; National Oceanic and Atmospheric Administration, U.S. Department of Commerce: Honolulu, HI, USA, 2006.
11. United States Department of Agriculture (USDA). *2010 Dietary Guindelines for Americans*; U.S. Government Printing Office: Washington, DC, USA, 2011.
12. European Food Safety Authority (EFSA). Opinion of the scientific panel on contaminants in the food chain on a request from the european parliament related to the safety assessment of wild and farmed fish. *Eur. Food Saf. Auth. J.* **2005**, *236*, 1–118.
13. Food and Drug Administration (FDA). *Summary of Published Research on the Beneficial effects of Fish Consumption and omega-3 Fatty Acids for Certain Neurodevelopmental and Cardiovascular Endpoints*; FDA: Silver Spring, MD, USA, 2009.
14. Gibbs, E. Rhode Island Sea Grant. Fact Sheet: Tuna. P1412. Available online: http://seagrant.gso.uri.edu:80/factsheets/tuna.html (accessed on 16 may 2016).
15. He, K.; Daviglus, M.L. A few more thoughts about fish and fish oil. *J. Am. Diet. Assoc.* **2005**, *105*, 350–351.
16. He, K.; Song, Y.; Daviglus, M.L.; Horn, L.V.; Dyer, A.R.; Greenland, P. Accumulated evidence on fish consumption and coronary heart disease mortality: A meta-analysis of cohort studies. *Circulation* **2004**, *109*, 2705–2711.
17. Meijer, E.; Rouwendal, J. Measuring welfare effects in models with random coefficients. *J. Appl. Econ.* **2006**, *21*, 227–244.

18. Black, R. Last Rites for a Marine Marvel. Available online: http://davehakes.com/weblog/wp-content/uploads/2007/11/10-17-07_bbc_last_rites_for_a_marine_marvel.pdf (accessed on 16 may 2016).
19. Ito, M. Does Japan's affair with tuna mean loving it to extinction? *Japan Times* **2010**, 3.
20. Allen, R. *International Mangement of Tuna Fisheries: Arrangement, Challenges and a Way Forward*; FAO: Rome, Italy, 2010.
21. Hamilton, A.; Lewis, A.; McCoy, M.A.; Havice, E.; Campling, L. *Major Tuna Industry Status Report: Market and Industry Dynamics in the Global Tuna Supply Chain*; Pacific Islands Forum Fisheries Agency: Strengthening National Capacity and Regional Solidarity for Sustainable Tuna Fisheries: Honiara, Solomon Islands, 2011.
22. Bougherara, D.; Combris, P. Eco-labelled food products: What are consumers paying for? *Eur. Rev. Agric. Econ.* **2009**, *36*, 321–341.
23. Köster, E.P. Diversity in the determinants of food choice: A psychological perspective. *Food Qual. Preference* **2009**, *20*, 70–82.
24. Grolleau, G.; Caswell, J.A. Interaction between food attributes in markets: The case of environmental labeling. *J. Agric. Resour.* **2006**, *31*, 471–484.
25. Delmas, M.A.; Grant, L.E. *Eco-Labeling Strategies: The Eco-Premium Puzzle in the Wine Industry*; Institute for Social, Behavioral and Economic Research in UC Santa Barbara: Santa Barbara, CA, USA, 2008.
26. Ferraro, P.J.; Uchida, T.; Conrad, J.M. Price premium for eco-friendly commodies: Are 'green' markets the best way to protect endagered ecosystems? *Environ. Resour. Econ.* **2005**, *32*, 419–438.
27. Loureiro; Mccluskey, J.J. Will consumers pay a premium for eco-labeled apples? *J. Consumer Aff.* **2002**, *36*, 203–219.
28. Tidwell, J.H.; Allan, G.L. Fish as food: Aquaculture's contribution. *EMBO Rep.* **2001**, *2*, 958–963.
29. Nilssen, F. *Consumers and Aquaculture, New Products—New Worries*; Springer: Berlin, Germany, 2008; pp. 235–244.
30. O'Dierno, L.; Govindasamy, R.; Puduri, V.; Myers, J.J.; Islam, S. *Consumer Perceptions and Preferences for Organic Aquatic Products: Results from the Telephone Survey*; New Jersey Agricultural Experiment Station, Rutgers University Department of Agricultural, Food and Resource Economics: New Brunswick, NJ, USA, 2006.
31. Asche, F.; Bjorndal, T.; Young, J.A. Market interaction for aquaculture products. *Aquac. Econ. Manag.* **2001**, *5*, 303–318.
32. Lancaster, K.J. A new approach to consumer theory. *J. Political Econ.* **1966**, *74*, 132–157.
33. Davidson, K.; Pan, M.; Perwanto, D.; Hu, W. A survey of demand preferences for aquaculture across geographically distinct markets. In Proceedings of the National Aquaculture Extension Conference 2011, Memphis, TN, USA, 5–7 June 2011.
34. Kalberg, K. Evaluation of the geographic impact on consumer preferences for aquaculture and wild captured seafood; an interaction analysis approach to the conjoint choice model. In Proceedings of the World Aquaculture Society Aquaculture 2013, Economics and Marketing (WAS 2013), Nashville, TN, USA, 24 February 2013.

35. Davidson, K.; Pan, M.; Hu, W.; Poerwanto, D. Consumers' willingness to pay for aquaculture fish products *vs.* wild caught seafood—A case study in hawaii. *Aquac. Econ. Manag.* **2012**, *16*, 136–154.
36. United States. Census Bureau. QuickFacts. http://quickfacts.census.gov/qfd/states/21000.html (accessed on 16 May 2016).
37. Gracia, A.; Magistris, T.D. Preferences for lamb meat: A choice experiment for spanish consumers. *Meat Sci.* **2013**, *95*, 396–402.
38. Asioli, N.S.T.; Øvrum, A.; Almli, V.L. Comparison of rating-based and choice-based conjoint analysis models. A case study based on preferences for iced coffee in norway. *Food Qual. Preference* **2016**, *48*, 174–184.
39. Lusk, J.L.; Schroeder, T.C. Are choice experiments incentive compatible? A test with quality differentiated beef steaks. *Am. J. Agric. Econ.* **2004**, *86*, 467–482.
40. Loo, E.J.V.; Caputo, V.; Nayga, R.M.; Meullenet, J.F.; Ricke, S.C. Consumers' willingness to pay for organic chicken breast: Evidence from choice experiment. *Food Qual. Preference* **2011**, *22*, 603–613.
41. Louviere, J.J.; Hensher, D.A.; Swait, J.F.; Adamowicz, W. *Stated Choice Methods: Analysis and Applications*; Cambridge University Press: Cambridge, UK, 2000.
42. Hole, A.R. Estimating mixed logit models by using maximum simulated likelihood. *Stata J.* **2007**, *7*, 388–401.
43. Train, K. *Discrete Choice Model with Simulations*; Cambridge University Press: Cambridge, UK, 2003.
44. Hole, A.R.; Riise, J. Mixed logit estimation of willingness to pay distributions: A comparison of models in preference and wtp space using data from a health-related choice experiment. *Empir. Econ.* **2012**, *42*, 445–469.

Research on Factors Affecting the Optimal Exploitation of Natural Gas Resources in China

Jianzhong Xiao, Xiaolin Wang and Ran Wang

Abstract: This paper develops an optimizing model for the long-term exploitation of limited natural gas reserves in China. In addition to describing the life cycle characteristics of natural gas production and introducing the inter-temporal allocation theory, this paper builds the optimal exploitation model of natural gas resources within a gas field in the Ordos Basin as an example to analyze its exploitation scale and how influence factors, such as recovery rate, discount rate and the gas well exhausting cycle, affect the optimal exploration path of this gas field. We determine that an increase in the discount rate stimulates investors to invest more aggressively in natural gas exploitation in the early period due to the lower discounted value, thereby increasing the pace of the exploitation of natural gas and the exhaustion of gas fields. A higher recoverable factor implies more recoverable reserves and greater potential of increasing the output of gas fields. The exhaustion rate of gas wells affects the capability of converting capacity to output. When exhaustion occurs quickly in gas wells, the output will likely increase in the output rising period, and the output will likely decrease at a faster rate in the output reduction period. Price reform affects the economic recoverable reserves of gas fields.

Reprinted from *Sustainability*. Cite as: Xiao, J.; Wang, X.; Wang, R. Research on Factors Affecting the Optimal Exploitation of Natural Gas Resources in China. *Sustainability* **2016**, *8*, 435.

1. Introduction

In this study, we develop an optimizing model for the long-term exploitation of limited natural gas reserves in China, which is a coal-dominated energy structure economy and suffers from serious pollution problems. Nevertheless, natural gas is currently promoted by Chinese authorities due to its relatively low carbon dioxide emissions. Therefore, the Chinese government confirmed its strong determination to increase domestic supply and to narrow the increasing gap between demand and production [1,2].

A large proportion of natural gas supply activities that include exploration, extraction and transportation of gas in China are currently overseen by the National Oil Company (NOC). In the current liberalized environment, investment from the private sector is encouraged. Upstream activities (exploration and extraction) are in dire need of investment, and because these investments are scarce resources, they need to be spread among various potentially productive basins.

According to Dai *et al.* [3], conventional gas geological resources in China have increased from (5.4–7) $\times 10^{12}$ m^3 in 1981 to 63 $\times$ 10^{12} m^3 in 2010; these data have been assessed by various scholars and divisions over the past thirty years and, thus, are validated. China recorded (39.0–39.2) $\times$ 10^{12} m^3 recoverable conventional gas resources; this is an immense amount for any type of major gas producer and will provide a solid foundation for the sustainable development of China's natural gas industry. However, large gas fields in China are mainly distributed in the mid-west and include the Ordos Basin, the Sichuan Basin and the Tarim Basin. The "gold" buried at the depths of reserves is deeper than that abroad with low pressure and a low discovery rate, *i.e.*, the reserves buried at depths of 3000–4500 m account for 46.11% of the total reserves of China. Hence, their exploration will be more difficult, generate higher costs and have a greater environmental influence. As a result, the well-head price of gas in China is equivalent to the border prices of Europe's imported gas from Russia. The influence of factors, such as geology, technology, policy and regulation, makes it difficult to effectively convert China's gas reserves into economic output, thereby preventing natural gas enterprises from generating more profits. The effective exploitation of existing natural gas under varying conditions has become a primary concern for major oil-gas enterprises. Thus, it is vital to analyze corresponding incentives, including economic, technological and regulation factors, which shape the optimizing paths in China.

Given the lack of peer-reviewed studies regarding the Chinese natural gas industry, this study aims to fill this gap by developing an optimizing approach to model gas production in China. Furthermore, this study contributes to investment planning and policy-making processes related to the gas production in China. This paper is organized as follows: A literature review is provided in Section 2. Section 3 presents the model's structure and its general description. Section 4 explains the optimal exploitation path of natural gas resources, Section 5 describes the primary results of the simulation analysis, and Section 6 provides the conclusion.

2. Literature Review

The issues of optimal nonrenewable resource extraction were first proposed by Hotelling [4], whose basic model predicted that the shadow price of resource stock, which is a classic measure of the scarcity of the resource, should increase at the rate of interest. Since the time of Hotelling's study, economists have expanded on his theoretical framework to allow for more realistic features. The impact of exploration activities and an extension of the resource reserves on the Hotelling framework were first suggested by Meadows *et al.* [5], and Solow's publication on Hotelling's model [6] boosted interest in the theory of nonrenewable resource extraction. These scholars demonstrate that exploration activities and the resource price and production path are related [7]; an increase in reserves is generally followed by an increase in production [8,9]. However, as the discovery of further reserves increases to the threshold and exploration activity declines, production also decreases. Additional analyses have been conducted, which Chermak and Patrick [10,11] classified into two primary groups: price path tests and price behavior tests. Price path tests examine if the price of a nonrenewable resource changes according to Hotelling's "r-percent rule" (*i.e.*, whether the price increases at the rate of interest). None of the price path analyses conducted by Barnett and Morse [12], could produce evidence for the theory using actual data [13,14]. However, these tests include strong assumptions that result from simplifications in Hotelling's model. First, technology is assumed to be constant over time, and second, the relation of extraction and production costs to the resource base is not considered.

Large-scale optimization models have been successfully implemented in the copper sector [15,16], as well as in other natural resource industries (e.g., Epstein *et al.* in the forest industry [17] or Baker and Ladson [18], and Dyer *et al.* [19], in the crude oil industry). Based on these systems, decision-makers are able to evaluate alternative operational policies and select those that maximize short-term and long-term profits of the business. However, for most of these models, deterministic inputs in the methodology include the net discounted value, which is predominately used by businesses.

The model utilized in this study builds on Ellis and Halvorsen [20] and Pindyck [21] regarding exploration activity. However, our model has substantially broadened the representation of the gas industry beyond that found in Pindyck's model. In particular, we attempt to address limitations that, in our opinion, made Pindyck's model a less realistic representation for the specific case of indigenous Chinese gas production.

For this reason, beginning with the economic theory of optimal resources exploitation, this paper constructs the exploitation model of resources in combination with the theories of optimal control, net present value and dynamic life cycle. In addition, this study developed a broadened model for the assessment of the effect of technical, cost and policy factors on the optimal exploitation of natural gas resources in China. The contribution of this study is in building the model under dynamic reserves by combining gas field benefits, gas well life cycles and gas field recovery rates into the optimal path model. The results and conclusions will be helpful for the government to formulate policies and for enterprises to make decisions.

3. Optimal Exploitation Model of Natural Gas Resources

3.1. Objective Function

The net present value of total revenue generated from natural gas within the life cycle of natural gas exploitation is presented as follows:

$$V = \sum_{t=0}^{T-1} e^{-rt}[P_t X_t - C_t] \tag{1}$$

where P_t represents the market price of oil and gas resources of year t, X_t represents the exploitation volume of natural gas of year t, e^{-rt} represents the discount coefficient and C_t represents the total cost of natural gas exploitation of year t, including exploration cost, exploitation cost and production cost.

The objective of natural gas exploitation is to maximize the value of exploitation, namely:

$$\max V = \sum_{t=0}^{T-1} e^{-rt}[P_t X_t - C_t] \tag{2}$$

3.2. Constraint Conditions

Because natural gas is non-renewable, reserves will decrease with constant exploitation, thus satisfying:

$$\begin{cases} (Q_{t+1} - Q_t) \times h = -\int_t^{t+\tau} K_t(i) d_i \\ Q_0 = S(0) \\ Q_T = S(T) \end{cases} \quad (t = 0, 1, 2, \cdots\cdots, T-1; i = 0, 1, \cdots, \tau - 1) \tag{3}$$

where T represents the total number of years that gas resources can be exploited; Q_t represents the remaining reserves and state variable of natural gas at the beginning of year t; h represents the recovery rate in the gas field; τ represents the life cycle of gas wells; i denotes the parameter of the period(year) between t and $t+\tau$; $K_t(i)$ represents gas production volume of the t batch of gas wells within the life cycle of τ years; and $\int_t^{t+\tau} K_t(i) d_i$ represents the total exploitation volume of the t batch of gas wells within the life cycle of τ years. $(Q_{t+1} - Q_t) \times h = -\int_t^{t+\tau} K_t(i) d_i$ is a set of differential equations, modifying the transfer law from state t to state $t+1$. $S(0)$ denotes the initial reserves and $S(T)$ the reserves at time T at the end of the period.

In general, exploitation volume X_t is realized through the production volume of different batches of gas wells at time t. Every batch of gas wells experiences an output rising period, a stable period and an output reduction period within a life cycle of τ years; however, the output functions of each batch of gas wells within the life cycle are not necessarily consistent (Figure 1). Let $K_t(i)$ denote the exploitation function of the t batch of gas wells in period i within the life cycle of τ years. The following four inverted "U-shaped" curves demonstrate the schematic diagrams of converting the capacity of different batches of gas wells to output. The production curve is inverted and declines sharply because each well has a short life cycle. Hence, the incremental curve is the accumulative production for the gas fields. In consideration of the functionality of the graph, the schematic diagrams of gas well exhaustion for other years are not plotted.

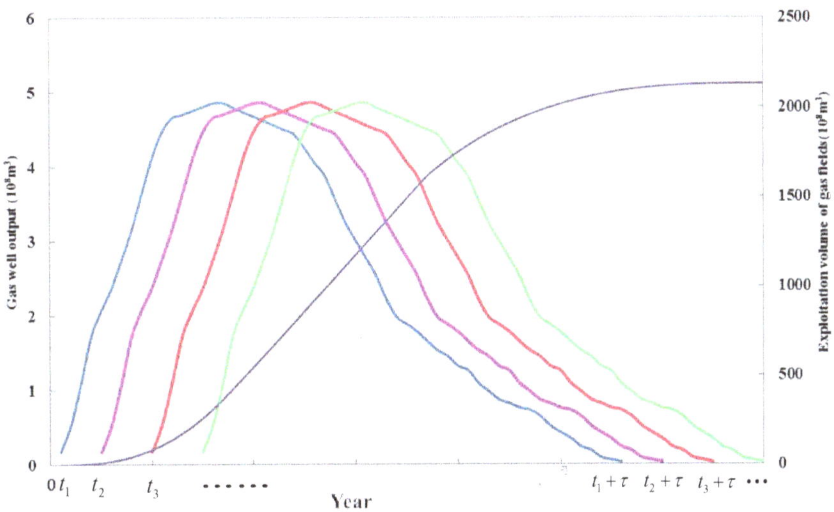

Figure 1. Velocity of converting the capacity of different batches of gas wells to output.

Assuming the life cycle of gas wells is τ, the capacity $\int_t^{t+\tau} K_t(i)d_i$ of the t batch of gas wells utilized in year t will be converted to output between year t and year $t+\tau$. For gas fields, the natural gas exploitation volume of year t is the sum total of the natural gas produced by the $t+\tau$ batch and t batch of gas wells at time t. The conversion of the capacity of different batches of gas wells satisfies the following equation:

$$\begin{cases} \mathrm{X(t)} = \int_0^1 K_t(x)d_x + \int_1^2 K_{t-1}(x)d_x + \cdots + \int_{i-1}^{i} K_{t-i+1}(x)d_x + \cdots + \int_{t-\tau}^{t-\tau+1} K_{t-1}(x)d_i \\ \int_{i-1}^{i} K_t(x)d_x = f(t,i) \end{cases} \quad (4)$$

where $i \in [0,\tau], t \in [0,T]$.

We denote $K_t(i)$ as the exploitation function of the t batch of gas wells at time i within the life cycle of τ years;

$\int_{i-1}^{i} K_t(x)d_x$ denotes the exploitation volume of the t batch of gas wells at time i within the life cycle of τ years.

3.3. Optimal Conditions

The above objective function and constraint conditions constitute the inter-temporal optimized model of natural gas optimal exploitation, namely the optimal exploitation model of dynamic natural gas resources. The model of the reasonable natural gas exploitation scale is transformed into an unconstrained Lagrange extremum problem with $T-1$ constraint conditions, as demonstrated below:

$$\begin{aligned} L &= \sum_{t=0}^{T-1} e^{-rt}[P_t X_t - C] + \sum_{t=0}^{T-1} \{e^{-r(t+1)}\lambda_{t+1}[-\int_t^{t+\tau} K_t(x)d_x + (Q_{t+1}-Q_t)\times h]\} \\ &= \sum_{t=0}^{T-1} e^{-rt}\{[P_t X_t - C] + e^{-r}\lambda_{t+1}[-\sum_{i=1}^{\tau} f(t,i) + (Q_{t+1}-Q_t)\times \mathrm{h}]\} \end{aligned} \tag{5}$$

To calculate the extremum conditions for Equation (5), the Hamilton function is introduced; let:

$$\begin{cases} H = H(X_t, Q, \lambda_{t+1}, t) = P_t X_t - e^{-r}\lambda_{t+1}F(t) \\ F(t) = \sum_{i=1}^{\tau} f(t,i) \end{cases} \tag{6}$$

Substituting Equation (5) into Equation (6) produces the following:

$$L = \sum_{t=0}^{T-1} [H + e^{-r}\lambda_{t+1}(Q_{t+1}-Q_t)\times h]e^{-rt} \tag{7}$$

Conditions necessary to take the extreme value in Equation (7):

$$\begin{cases} \frac{\partial H}{\partial X_t} = P_t - MC - e^{-r}\lambda_{t+1} = 0 \\ \frac{\partial H}{\partial Q_t} = e^{-r(t+1)}\lambda_{t+1} - e^{-rt}\lambda_t = 0 \\ \frac{\partial H}{\partial \lambda_{t+1}} = -\sum_{i=1}^{\tau} f(t,i) + (Q_{t+1}-Q_t)\times h = 0 \end{cases} \quad (t = 0,1,2,\cdots\cdots,T-1) \tag{8}$$

MC represents the marginal cost. The optimal conditions for Equation (5) can be derived from Equations (6)–(8):

$$\begin{cases} P_t - MC - e^{-r}\lambda_{t+1} = 0 \\ e^{-r}\lambda_{t+1} = \lambda_t \\ (Q_{t+1}-Q_t)\times h = -F(t) \end{cases} \quad (t = 0,1,2,\cdots\cdots,T-1) \tag{9}$$

Substitute Equations (8) and (9) into Equation (7); we calculate the following:

$$P_t - MC - e^{rt}\lambda_0 = 0 \ (t = 0,1,2,\cdots\cdots,T-1) \tag{10}$$

It can be determined from the above equation sets that $\lambda_0{}^*$ and optimal path $X_t{}^*$ can be obtained as long as the functional relations between MC, Q_t, X_t, h and $K_t(i)$ in different time periods are available.

4. Optimal Exploitation Path of Natural Gas Resources

4.1. Parameters Setting

Based on the above model analysis, this section selects a gas field in the Ordos Basin as the research target and simulates scenarios to solve for the optimal exploitation scale. The selected gas field is typical, features "three lows" (low porosity, low permeability and low pressure) and represents an onshore gas field in China. This gas field has been exploited for more than ten years and is currently in the output rising period. The analysis of the parameters involved in this study is described in the following text.

4.1.1. Gas Field Exhausting Time T

Because the optimal exhausting rate (an analytical solution with implicit T) involves the setting of time T, the optimal exploitation scale can be solved according to the inter-temporal optimized model of natural gas exploitation scale when T is set. In addition, a gas field has a life cycle. The gas field in the Ordos Basin has a short production period, so its life cycle is divided into three periods (*i.e.*, output rising period, stable period and output reduction period) with t_k, $t_s - t_k$ and $T - t_s$ denoting the corresponding years for each period. In this section, the life cycle of this gas field is denoted as $T = 40$ years, of which $t_k = 12$ years are in the output rising period, $t_s - t_k = 13$ years are in the stable period and $T - t_s = 15$ years are in the output reduction period.

4.1.2. Marginal Exploitation Cost of Natural Gas MC

We consider the fact that estimation of the marginal cost represents the optimization problem of the gas producer. From a modeling perspective, we would prefer the marginal cost to depend on the actual sequence of extraction. This consideration in our formulation is particularly useful for modeling underground mining operations in which production costs tend to increase as extraction progresses. This assumption allows us to formulate the decision-maker's optimization problem using linear programming methods. The exploitation cost of natural gas C refers to the total costs incurred during the development of natural gas, particularly the exploration cost, development cost and production cost. For simplicity, this study assumes that the marginal costs of exploration and development are constant.

In regards to the cost of production, this study divides the exhausting period T of the gas field into the output rising period, the stable period and the output reduction period according to its life cycle. Combining the above hypotheses, the marginal cost of production within period T is determined as demonstrated below:

$$\begin{cases} MCP(t) = at + b & (1 \leqslant t \leqslant t_k) \\ MCP(t) = c & (t_k \leqslant t \leqslant t_s) \\ MCP(t) = dt + e & (t_s \leqslant t \leqslant T) \end{cases} \tag{11}$$

where $MCP(t)$ represents the marginal cost of production at time t and a, b, c, d, e are the coefficients of the marginal cost of production $MCP(t)$ in different time periods.

Marginal costs are determined according to the life cycle. The marginal cost of exploration is $MCE = 0.05$ CNY/m^3; the marginal cost of development is $MCD = 0.05$ CNY/m^3; and the coefficients of the marginal cost of production $MCP(t)$ are $a = 0.02, b = 0.082, c = 0.056, d = 0.004, e = -0.04$.

4.1.3. Discount Rate r

In addition to a risk-free discount rate, the discount rate should consider the rate of risk return; specifically, it should reflect the time value of money and risks associated with the project. Therefore, consideration of capital's time value and the discount of infrastructure is necessary. There are reliable methods of determining the discount rate. Many scholars have determined the discount rate to be between 0.05 and 0.2 [22,23]. In alignment with the current situation and the future development trend of natural gas in China, national long-term loans and treasury bond rate information in related references, the discount rate is denoted as 0.07.

4.1.4. Recovery Rate h

Gas well exploration, exploitation and production processes generally involve converting reserves to capacity and then capacity to output. However, recoverable reserves in gas fields will not necessarily be completely converted to capacity, nor will capacity always be converted to output. Recovery rate refers to the ratio of oil and gas volume extracted from reservoirs with reserves within a certain economic limit and through the use of modern technical skills. The oil and gas recovery rate varies from field to field in China. Taking into account the geological features and natural gas prices in this field, the recovery rate is denoted as 0.6.

4.1.5. Exhausting Rate of Gas Wells τ

The life cycle of different batches of gas wells in the gas field analyzed for this study is: $\tau = 5$ years, where the velocities of converting capacity to reserves in each year are, respectively, 0.1, 0.2, 0.3, 0.3 and 0.1.

4.1.6. Reserves S

The proven recoverable reserves in this gas field is $S(0) = 3000 \times 10^8$ m^3. For simplicity, the closing natural gas reserves are denoted as $S(T) = 0$.

4.2. Optimal Path

The optimal exploitation scale and path of remaining reserves for the gas field under study is calculated by combining the above parameters. As can be noted in Figure 2, for a life cycle of 40 years, the gas field shows an obvious growth period, a stable period and an output reduction period, with the exploitation volume increasing rapidly in the growth period, slowing down in the stable period and decreasing significantly in the output reduction period. Remaining reserves present a reversed S-shaped curve in the life cycle. It can be demonstrated from comparative analysis of the calculation results of the model with the practical data from this gas field that the output error in Period 3 is 26.7%, and the gas produced in a period of 12 years (Periods 1–12) has a relatively lower average error of 9.1%, indicating that the simulation results of the model are reliable (shown in Table 1).

Table 1. Analysis of output error in a gas field model.

Period (Years)	1	2	3	4	5	6	7	8	9	10	11	12
Statistical data (10^8 m^3)	1.2	3.8	7.7	12.1	14.6	16.6	19.2	22	25.3	28.6	32.2	36.1
Practical data (10^8 m^3)	1.4	4	10.5	14.3	13.8	14.2	17.8	22	24	28	31	34
Error (%)	−14.3	−5	−26.7	−15.4	5.8	16.9	7.9	0	5.4	2.1	3.9	6.2

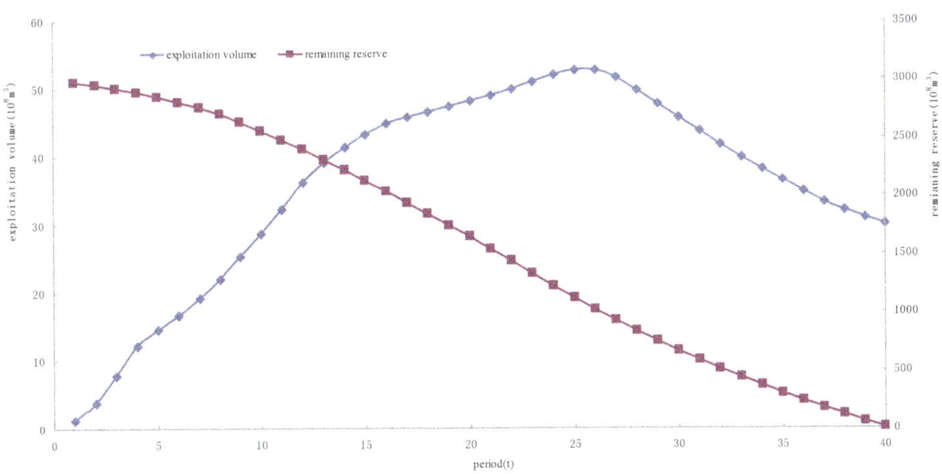

Figure 2. Exploitation scale and remaining reserves in a gas field.

5. Result Analysis and Discussion

Factors, such as the discount rate, recovery rate and price, will inevitably affect the optimal path of natural gas exploitation. This study designates price as the exogenous variable and the discount rate, recovery rate and the speed of converting capacity to output as the endogenous variables. Combining these variables with the above model, this section analyzes the impact of these endogenous variables on the optimal exploitation model of natural gas resources.

5.1. Increase in Discount Rate Accelerating Natural Gas Exploitation

The discount rate reflecting time value is closely related to the bank interest rate included in national macro-regulations. According to the Hotelling rule, the interest rate exerts an important influence on the decision of natural gas exploitation. Here, the natural gas exploitation volume and remaining reserves are simulated when the discount rates are 0.05, 0.07 and 0.12.

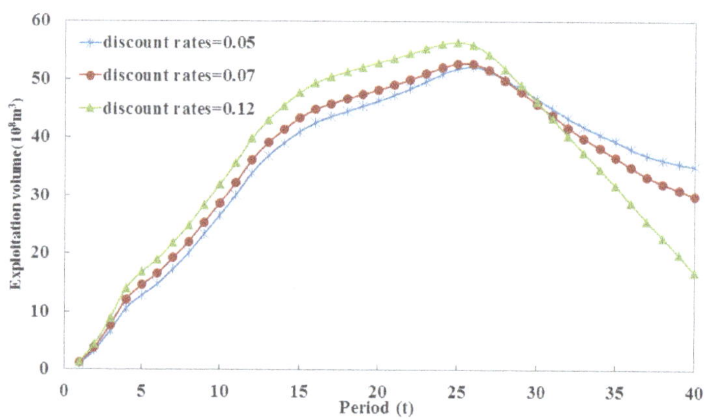

Figure 3. Exploitation volume with different discount rates.

When utilizing different discount rates, the natural gas exploitation curve is generally inconsistent with the remaining reserves curve. To clarify, the exploitation volume can also reflect the output rising period, the stable period and the output reduction period of natural gas, and the remaining reserves curve still presents a reversed S-shaped curve (shown in Figure 3). In the output rising period and stable period, the exploitation volume with a discount rate of 0.12 is obviously higher than one with a discount rate of 0.07. At the end of the stable period, the exploitation volume with a discount rate of 0.05 is similar to one with a discount rate of 0.07. In the output reduction period, the exploitation volume with a discount rate of 0.12 drops fastest, followed by one with a discount rate of 0.07. The three curves intersect at the 29th point of time. In the output rising period and stable period of natural gas exploitation, the exploitation volume will rise with an increase in the discount rate, and investors are more willing to fund natural gas exploitation; while in the output reduction period, the greater the discount rate, the more significant the output reduction is, and investors are less willing to fund natural gas exploitation (shown in Figure 4).

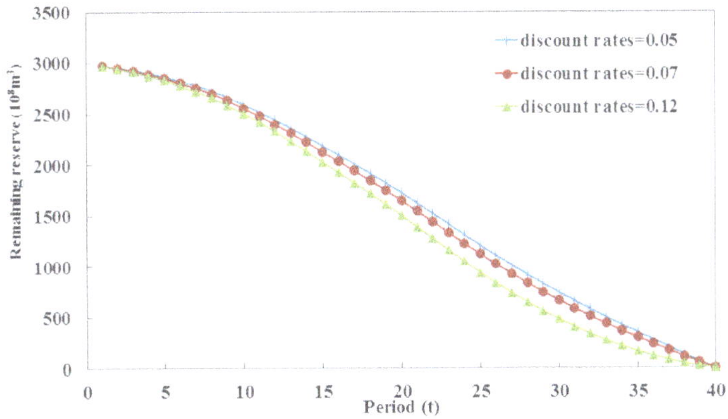

Figure 4. Remaining reserves with different discount rates.

5.2. Increase in Recovery Rate Increases Economically-Recoverable Reserves of Natural Gas

The recovery rate generally affects the recoverable reserves of natural gas, thus affecting the exploitation volume of natural gas in different time periods (shown in Figure 5). The gas field recovery rate in China varies from field to field. Even in different areas of the same gas field, recovery rates are not identical. Exploitation rates of natural gas in the Sichuan Basin, Ordos Basin, Bohai Bay Basin, Songliao Basin, Junggar Basin, Tarim Basin and Qaidam Basin are 40.3%–81.77%, 58.33%–70.07%, 57%–68%, 52%–63.5%, 51.02%–84.48%, 54%–74.08% and 53.83%–75.25%, respectively. Thus, the recovery rates selected in this paper are 0.4, 0.5, 0.6, 0.7 and 0.8. Natural gas with a recovery rate of 0.7 has the same exploitation volume as that with a recovery rate of 0.8, and natural gas with a recovery rate of 0.7 has a higher exploitation volume than that with a recovery rate of 0.6. Through calculation, it is determined that in the range of 0.4–0.7, the recovery rate will increase by 10%, with an average incremental exploitation volume of 3.799 billion cubic meters in yield in the rising period, 7.895 billion cubic meters in the stable period and 15.693 billion cubic meters in the yield reduction period (shown in Figure 6). This indicates that within a certain range, exploitation volume increases as the recovery rate of the gas field increases. When the recovery rate further increases, the exploitation volume of natural gas remains the same.

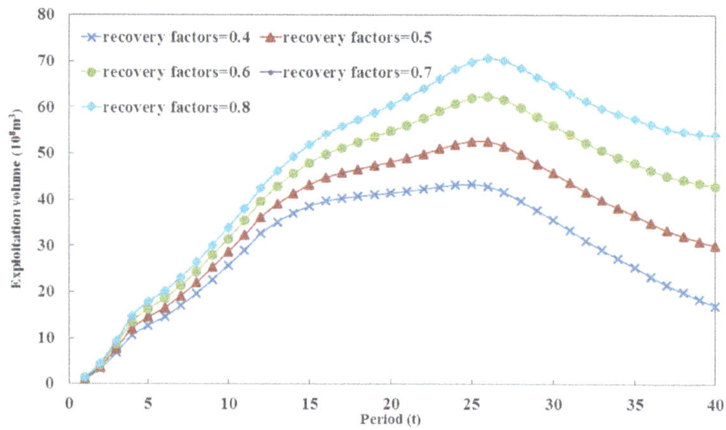

Figure 5. Exploitation volume with different recovery rates.

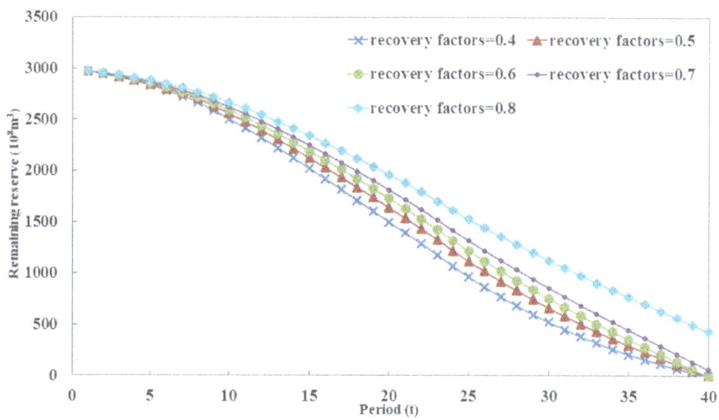

Figure 6. Remaining reserves with different recovery rates.

5.3. Exhausting Rate of Gas Wells Extending the Distribution of Natural Gas in Fields

As stated above, the velocity of converting gas well capacity to output will affect the exploitation scale of natural gas by altering the distribution of every batch of gas wells within the life cycle of the gas field (shown in Figure 7), but annual capacity, or specifically the dynamic reserves, will not change; which explains why the remaining reserves curves of the gas field are entirely consistent with each other (shown in Figure 8). When the life cycle of gas wells is set at three years, the velocities of converting capacity to output are 0.3, 0.4 and 0.3, and in the case of seven years, the corresponding velocities are 0.1, 0.1, 0.2, 0.2, 0.2, 0.1 and 0.1. Overall, the longer it takes for the gas well capacity to be converted to output, or rather the longer the life cycle of gas wells, the further the gas exploitation curve will move to the right. In the output rising period, gas wells with a life cycle of three years have a significant higher exploitation volume than those with a life cycle of five years or seven years. In the stable period, the three exploitation curves essentially overlap each other. In the output reduction period, gas wells with a life cycle of seven years have a significantly higher exploitation volume than those with a life cycle of three years or five years. To clarify, in the output rising period, the faster the gas well capacity is converted to output, the greater the exploitation volume in the gas field, but during the output reduction period, the longer it takes for the gas well capacity to be converted to output, the greater the exploitation volume.

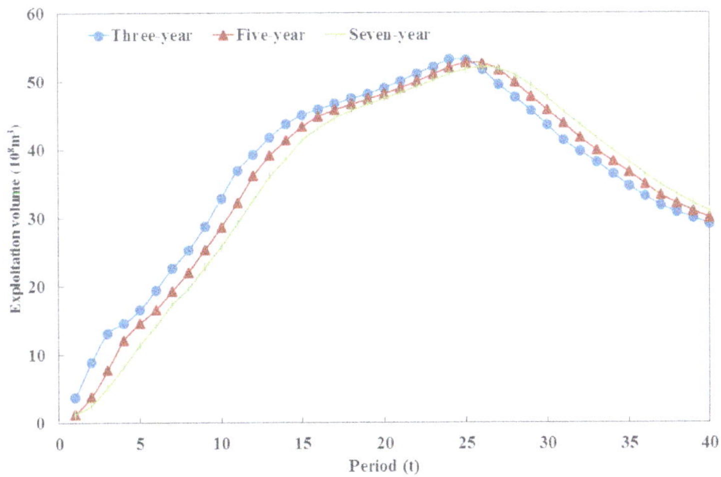

Figure 7. Exploitation volume with different exhausting rates of gas wells.

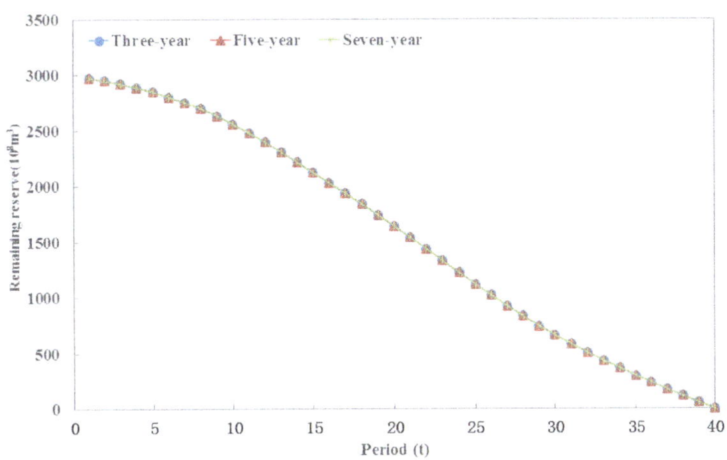

Figure 8. Remaining reserves with different exhausting rates.

5.4. Price Increase Leads to Increased Economically-Recoverable Reserves in Gas Fields

In China, the price of natural gas is regulated by the state, which utilizes a cost-plus pricing mechanism. As a price-taker, enterprises can make corresponding adjustments to allow for price fluctuations. It is quite necessary, therefore, to analyze how price affects the rate of natural gas exploitation.

The initial price of natural gas is determined by the regulation authority in China and is used as an input and an exogenous variable in this study. On the basis of the influence of prices fluctuating in the range of 10%, the path of natural gas exploitation is analyzed. As shown in the Figure 9, when the price increases by 10%, exploitation volume increases in the output rising period and stable period, but decreases in the output reduction period. Otherwise, exploitation volume decreases in the output rising period and the stable period, but increases in the output reduction period. Three curves intersect at the end of the stable period and at the beginning of the output reduction period. Corresponding to the exploitation volume, the remaining reserves decrease rapidly with a higher price in the stable and reduction period (see Figure 10). Due to price increases, the break-even point moves upwards, because the revenue of natural gas exploiting enterprises increases, thus increasing the economic recoverable reserves and then increasing the exploitation volume of natural gas enterprises.

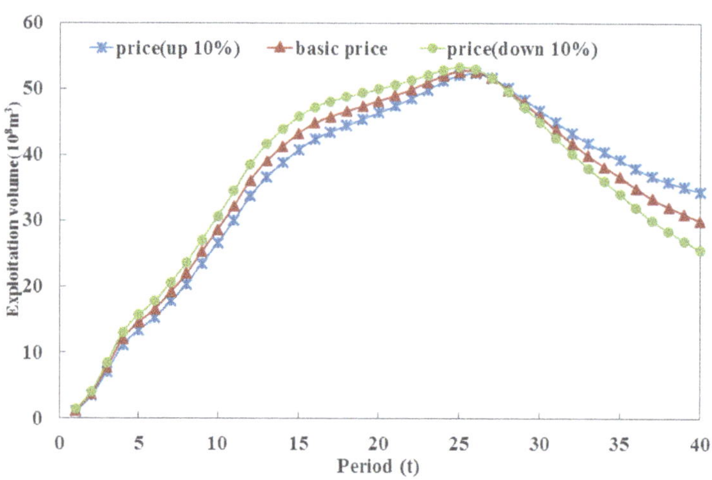

Figure 9. Exploitation volume with different prices.

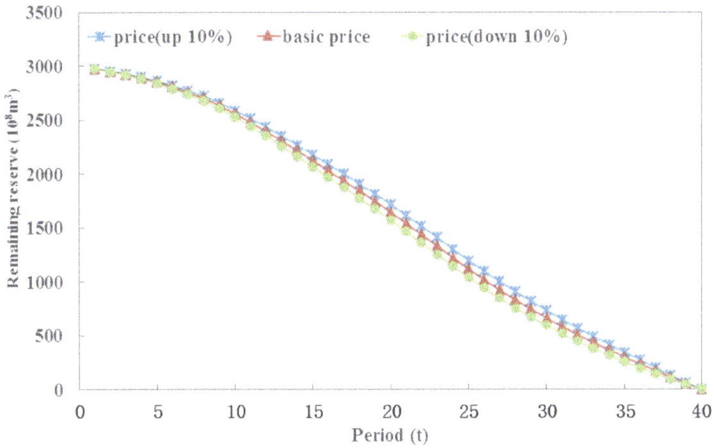

Figure 10. Remaining reserves with different prices.

6. Conclusions and Remarks

Based on current large-scale natural gas exploitation, combining real data with the optimal control theory and using the net present value method, this study builds the optimal exploitation model of natural gas resources to analyze the optimal exploitation path of the gas field within its life cycle using the Ordos Basin as the example. In addition, this study conducts a sensitivity analysis of the factors affecting the path of natural gas exploitation. The following conclusions are drawn from the calculations and analysis.

Upon comparison, the average error between the calculation results of the model and actual data regarding reserves is only 9.1%, which demonstrates that the exploitation path is relatively reasonable at present and suggests a certain degree of practical operability of this model. The gas field used in this analysis is now in the output rising period and will soon enter the stable period. Appropriate adjustments should be made to the optimized model if its exploitation conditions improve.

The discount rate affects the exhausting rate of gas fields. A higher discount rate results in quicker exploitation of natural gas and, therefore, faster exhaustion of gas fields. In addition, investors' decisions regarding natural gas exploitation will be affected by the discount rate; as the discount rate increases, enterprises will invest greater amounts. Thus, the exploitation volume of natural gas shows a significant increase in the output rising period and stable period, increasing the speed of exhaustion for gas fields in the output reduction period.

A larger recovery rate results in an increased exploitation volume of gas fields in each time period and results in greater potential of increasing output in gas fields. The recoverable reserves of gas fields and output-increasing potential will increase as a result of more scientific research data and technological advancements.

Meanwhile, an increase in gas prices may increase the economic recoverable reserves, thus resulting in an increase in the recovery rate of gas fields.

The exhaustion cycle of gas wells affects the velocity of the converting capacity to output and the distribution of the natural gas in the life cycle of gas fields. The faster the conversion, the more output increases in the output rising period and the faster output decreases in the output reduction period.

Price affects the economically-recoverable reserves of gas fields. When prices rise, the economically-recoverable reserves increase, as does the exploitation volume. Because China's natural gas was underpriced for a long period of time due to government control, natural gas prices were expected to increase as a result of pricing reform [24]. It may be reasonable to deduct that natural gas producers' benefit the most from rising prices.

To our knowledge, this is one of the few papers that explicitly analyzes natural gas resources exploitation utilizing an optimizing model for a regulated market. The potential practical value will make it possible to theoretically analyze and calculate the model of optimal resource exploitation rates under uncertain scenarios, which is of great theoretical value and practical significance to the selection of the resource exploitation path.

There are limitations to this analysis. For example, we analyze the effects of increased gas prices of a stated-owned gas producer and ignore the optimal domestic gas pricing level. In addition, we do not consider the constraints of the demand structure in downstream sectors, which shape the behavior of upstream producers. To enhance future natural gas market research, the multi-agent non-linear programming model should be developed to account for the interaction among producers, sellers and end-users.

Acknowledgments: The authors gratefully acknowledge the financial support provided by the Ministry of Education of China under the name of its Philosophy and Social Science Major Issue Research Project "Natural gas production and transportation peak signal extraction and optimization strategies" (No. 11YJC630211).

Author Contributions: Jianzhong Xiao contributed to the design of the article. Xiaolin Wang designed the model and wrote the paper. Ran Wang collected and analyzed the data.

Conflicts of Interest: The authors declare no conflict of interest.

Nomenclature

Abbreviation	
NOC	National Oil Company
Sinopec	China Petroleum and Chemical Corporation
NDRC	National Development and Reform Commission
m^3	cubic meter

Model Symbol Definitions

P	market price of natural gas (CNY/m^3)
C	total cost of natural gas exploitation (CNY/m^3)
t	time series
T	gas field exhaustion time (year)
r	discount rate of capital (%)
h	recover rate of the gas field (%)
τ	exhausting rate of gas wells (%)
S	proven recoverable reserves (m^3)

References

1. Dong, X.; Jie, G.; Mikael, H.; Guanglin, P. Sustainability Assessment of the Natural Gas Industry in China-Using Principal Component Analysis. *Sustainability* **2015**, *7*, 6102–6118.
2. Yan, X.-C. Energy structure optimization and strategic position and role of natural gas in China. *Int. Petr. Econ.* **2010**, *3*, 62–67.
3. Dai, J.; Wu, W.; Fang, C.; Liu, D. Exploration and development of large gas fields in China since 2000. *Nat. Gas Ind. B* **2015**, *2*, 1–8.
4. Hotelling, H. The economics of exhaustible resources. *J. Political Econ.* **1931**, *39*, 137–175.
5. Meadows, D.H.; Goldsmith, E.I.; Meadow, P. *The Limits to Growth*; Earth Island Limited: London, UK, 1972; Volume 381.
6. Solow, R.M. The economics of resources or the resources of economics. In *Classic Papers in Natural Resource Economics*; Palgrave Macmillan: Basingstock, UK, 1974; pp. 257–276.
7. Stiglitz, J.E. Monopoly and the rate of extraction of exhaustible resources. *Am. Econ. Rev.* **1976**, *66*, 655–661.
8. Halvorsen, R.; Smith, T.R. On measuring natural resource scarcity. *J. Political Econ.* **1984**, *92*, 954–964.
9. Halvorsen, R.; Smith, T.R. A test of the theory of exhaustible resources. *Q. J. Econ.* **1991**, *106*, 123–140.
10. Chermak, J.M.; Patrick, R.H. Comparing tests of the theory of exhaustible resources. *Resour. Energy Econ.* **2002**, *24*, 301–325.
11. Chermak, J.M.; Patrick, R.H. A microeconometric test of the theory of exhaustible resources. *J. Environ. Econ. Manag.* **2001**, *42*, 82–103.
12. Barnett, H.J.; Morse, C. The economics of natural resource availability. In *Scarcity and Growth*; Routledge: Abingdon, UK, 2013; Volume 3.
13. Slade, M.E. Trends in natural-resource commodity prices: An analysis of the time domain. *J. Environ. Econ. Manag.* **1982**, *9*, 122–137.
14. Slade, M.E.; Thille, H. Hotelling confronts CAPM: A test of the theory of exhaustible resources. *Can. J. Econ.* **1997**, *30*, 685–708.
15. Mondschein, S.; Schilkrut, A. Optimal investment policies for pollution control in the copper industry. *Interfaces* **1997**, *27*, 69–87.

16. Caldentey, R.; Mondschein, S. Policy model for pollution control in the copper industry, including a model for the sulfuric acid market. *Oper. Res.* **2003**, *51*, 1–16.
17. Epstein, R.; Morales, R.; Seron, J.; Weintraub, A. Use of OR systems in the Chilean forest industries. *Interfaces* **1999**, *43*, 64–79.
18. Baker, T.E.; Ladson, L.S. Successive linear programming at exxon. *Manag. Sci.* **1985**, *31*, 264–274.
19. Dyer, J.S.; Lund, R.L.; Larsen, J.B.; Leone, R.P. A decision support system for prioritizing oil and gas exploration activities. *Oper. Res.* **1990**, *38*, 386–396.
20. Ellis, G.M.; Halvorsen, R. Estimation of market power in a nonrenewable resource industry. *J. Political Econ.* **2002**, *110*, 883–899.
21. Pindyck, R.S. The optimal exploration and production of nonrenewable resources. *J. Political Econ.* **1978**, *86*, 841–861.
22. Hu, M.-C.; Hobbs, B.F. Analysis of multi-pollutant policies for the U.S. power sector under technology and policy uncertainty using MARKAL. *Energy* **2010**, *35*, 5430–5442.
23. Schäfer, A.; Jacoby, H.D. Vehicle technology under CO_2 constraint: A general equilibrium analysis. *Energy Policy* **2006**, *34*, 975–985.
24. Paltsev, S.; Danwei, Z. Natural gas pricing reform in China: Getting closer to a market system? *Energy Policy* **2015**, *86*, 43–56.

Analysis of the Relationship between China's IPPU CO_2 Emissions and the Industrial Economic Growth

Ye Duan, Hailin Mu and Nan Li

Abstract: According to the 2006 IPCC Guidelines for National Greenhouse Gas Inventories (IPCC 2006) and based on the production technology and products, this paper has calculated CO_2 emissions from industrial processes and product use (IPPU), which involves the individual and the summation of five major IPUU CO_2 emissions industrial departments. As there is a classic environmental Kuznets curve between IPPU CO_2 emissions and the economy, this paper discusses the relationship based on the calculation results and the actual situation. The results show that the overall emission level is indeed rising yearly, and that steel and iron alloy manufacturing and nonmetal manufacturing occupy about 80% of the total emissions. The IPPU CO_2 emissions and the corresponding gross industrial output value do not present a classic Kuznets curve in most industrial sectors due to the increasing industrial employed population. The year 2002 appears to be the boundary instead, where prior to 2002, there is a relatively stable function-type growth curve and after 2002, gross industrial output value (*GIOV*) per employed person remained within a certain interval while IPPU CO_2 emissions per employed dipped slightly then increased again. Some, but not all, industrial departments and the combined emissions of per employed person reached maximum values in 2012.

Reprinted from *Sustainability*. Cite as: Duan, Y.; Mu, H.; Li, N. Analysis of the Relationship between China's IPPU CO_2 Emissions and the Industrial Economic Growth. *Sustainability* **2016**, *8*, 426.

1. Introduction

China's economy has made remarkable achievements since the continuous progress of reform and opening-up; however, severe environmental quality deterioration and the predatory exploitation of resources have been associated with the economic growth. The industrial GDP remains around 40% of the total as of today. On the one hand, the rapid development of the industry brings the huge energy consumption and CO_2 emissions. On the other hand, it awakens people's conscienceness about environmental protection that people need higher quality of environment. Therefore, ignoring the effective control of environment quality deterioration will be bound to cause the further contradiction between economic growth and environmental quality, which is not conducive to social stability and sustainable development. Coordinating economic growth alongside improvement in environmental quality may be daunting but is certainly not impossible.

Industrial sector energy consumption creates reaction temperature and reaction conditions, and fossil fuel combustion produces large amounts of CO_2. Controlling energy consumption and reducing the unnecessary waste of energy can slash CO_2 emissions efficiently. Sometimes, carbon elements in raw materials will be released in some industrial production process. Changing the raw materials of production or production technology to make less carbon be discharged into the atmosphere is another obvious solution for CO_2 emissions reduction. This portion of the CO_2 emissions is inextricably linked to the product yield, so if there is a sharp rise in this part of CO_2 emissions, but corresponding department economic levels do not rise, the overcapacity or backward production occurs in some department. The government can impose secondary industry structural adjustment policy according to its assessments of the actual situation, so the relationship between IPPU CO_2 emissions and the economic activity is the focus of this paper. Then, we regard the IPPU CO_2 emissions as the original aspect and the research subject. Finally, this paper puts forward relevant policy suggestions for China's industrial structure adjustment.

Until now, industrial CO_2 emission problems have attracted great attention from society. Few studies have focused on the IPPU CO_2 emission, and even fewer on the relationship between IPPU CO_2 emission and industrial economy activity. Different from other research, an important novelty of this paper is its accounting for China's five major IPPU department CO_2 emission levels as the regression raw data for the first time. The paper serves as a more comprehensive attempt to estimate the IPPU CO_2 emission from 1991 to 2012, including five major IPPU departments. On this basis, the regression relationship between the IPPU CO_2 emissions and the economic level are analyzed for the first time, and the curve shows a realistic significance that the results will contribute to discuss the over-capacity and get some more useful information on the actual situation of China as a result—another innovation point in this paper.

The remainder of this paper is organized as follows. Section 2 provides a literature review focusing on the Carbon Kuznets Curve (CKC) and IPPU CO_2 emission research. Section 3 introduces the accounting method, the foundation of CO_2 Kuznets curve hypothesis and data sources. In Section 4, based on the accounting data and statistic analysis, we present and discuss our results in detail. Section 5 provides conclusions and some policy recommendations for China's industry.

2. Literature Review

Numerous environmental economic scholars have studied the relationship between economic growth and environmental quality. Grossman and Krueger [1], and Shafik [2] analyzed the experience relationship between economic development and environmental quality, and they found that economic development level and pollution emissions presented a morphological characteristics of an inverted "U" type curve, which is called the environmental Kuznets curve. The theory reflects the fact that, in early stages of economic development as income levels rise, environmental quality becomes vestigial, but, at a certain extent of economic development, income level reaches a threshold at which pollution emissions gradually lessen and environmental quality begins to improve.

Alongside the increasing prevalence of climate disasters—and the growing awareness of the dangers of climate change—scholars have begun to focus on CO_2 emissions and their relationship to economic growth. To date, there is no consensus on the existence of the Carbon Kuznets Curve (CKC between economic growth and CO_2 emissions). Due to differences between the selected sample countries (regions), plus the industry and the development phase, the relationship between economic growth and carbon emissions tends to be different. Hettige *et al.* [3], Selden and Song [4], Grossman and Krueger [5] agreed with the traditional inverted "U" CKC conclusion, while some scholars found that the economic growth and carbon emissions relationship type have several shapes, such as "U" shape, "N" shape, inverted "N" shape, "M" shape, monotonous or no specific curve, and other conclusions, such as Hill and Magnani [6], Stem [7], Lantz and Feng [8], Ranjan and Shortle [9], and He and Zhang [10]. Chinese scholars' studies on CO_2 emissions and economic growth are fairly limited and basic, though there have been some valuable contributions to the literature. For instance, Lu [11] established a state space model between CO_2 and GDP per capita, and found that their relationship is not a simple inverted u-shaped curve. By fitting the EKC curve to estimate the turning point of emission levels from up to down, Cai *et al.* [12] estimated and investigated the Chinese inner energy and emissions reduction requirements. They asserted that if greenhouse gases emissions waited for the arrival of the Kuznets inflection point passively, China would not be able to cope with the increasing environmental pressure. After analyzing the relationship between GDP and greenhouse gas emissions of 21 advanced countries, Huang *et al.* [13] finally found that EKC pattern appeared in seven countries. Some scholars denied the existence of the EKC curve. Han and Lu [14] showed that a big difference of the CO_2 Kuznets curve from different research objects (for instance, U-shaped, N-shaped, linear relationship). By applying the panel cointegration error correction model of China's carbon emissions to verify the existence of the Kuznets curve, Zheng and Zhu [15] found that the carbon emissions and economic growth present a long-term and stable inverted U-shaped curve, and further, that the turning point for the GDP per capita is 29,847.29 yuan. Zhang [16] measured the level of the carbon emissions of the manufacturing sector in China and found that no such inverted U-CKC exists.

By employing various approaches, few previous studies have examined sources and reduction potentials of CO_2 emissions from IPPU. Paraschiv *et al.* [17] analyzed the textile industry and sustainable development for the Eastern European Area. Dentoni *et al.* [18] studied environmental sustainability of the alumina industry in Western Europe. Puyvelde *et al.* [19] produced papers investigating emissions and reductions from the cement sector. Sheinbaum *et al.* [20] analyzed energy and CO_2 emission trends of Mexico's iron and steel industry during the period 1970–2006, examining CO_2 emissions related to energy use and production process. Arruda *et al.* [21] analyzed the sustainability in the Brazilian Heavy Construction Industry. Liaskas *et al.* [22] used the algebraic disaggregation method to identify the factors influencing CO_2 emissions generated in the industrial sector of European Union countries. Kim and Worrell [23] and Kirschen *et al.* [24] presented the analysis on energy-related carbon footprint in the iron and steel industrial sector of seven countries and electric arc furnace, respectively.

Zhou *et al.* [25], estimated the carbon footprint of China's ammonia production and analyzed the potential for carbon mitigation in the industry. Dong *et al.* [26] assessed sustainability of the natural gas industry in China using principal component analysis. Liu *et al.* [27] analyzed the industrial carbon footprint calculation of woolen yarn. Liu *et al.* [28] analyzed the regional greenhouse gas (GHG) emission of industrial processes of Shenyang, China. Gao *et al.* [29] explored the impact of carbon cycle in fossil fuel use and industrial production in Henan, China, and estimated the dynamic changes of the IPPU emissions from 2000 to 2009. Chen [30] analyzed embodied CO_2 emissions including industrial process emissions.

From the research above, we found that research has traditionally focused on national level CO_2 emissions and economic growth relationship, fewer on the level of province, and much fewer on specific industry department. Furthermore, studies on industrial CO_2 emissions have been confined to energy-consumption-related emissions, failing to fully account for the emissions caused by industrial processes. With the in-depth study of CO_2 emissions in recent years, IPPU and waste incineration CO_2 emissions have been taken into account as their emissions account for 20% of total emissions. In an effort to fill the knowledge gap in terms of IPPU CO_2 emissions and the related industrial and economic growth, we conducted the present study with a two-fold goal: (1) to determine whether the CO_2 Kuznets curve exists, and if it does, whether it is U-shaped, N-shaped, or grows exponentially; and (2) based on the output data of several industrial production departments in 1991–2012 with industrial CO_2 emissions (IPCC 2006 [31]) as the research object, to comprehensively analyze the relationship between IPPU CO_2 emissions and industrial economic growth. By establishing the relationship between *GIOV* and IPPU CO_2 emissions, the results presented here mark a notable departure from the previous research on Kuznets curves.

3. Methods

3.1. Data Sources

The industrial sectors involved in this paper include the mining industry, chemical industry, metal industry, electronic industry and others listed in IPCC 2006. The industrial product data are obtained from "China Statistical Yearbook [32]", "China Compendium of Statistics 1949–2008 [33]", "China Industrial Economy Statistical Yearbook [34]" and other various industry sectors in the Statistical Yearbook [35,36].

Notations and their explanations used in this paper were presented in Table 1.

Table 1. Notations and explanations used in this paper.

Notations	Explanations
t	IPPU CO_2 emissions industrial department
PC	IPPU CO_2 emissions per employed person
α	The intercept term
$GIOV$	Gross industrial output value
$PGIOV$	Gross industrial output value per employed person
β	The estimated parameter of $PGIOV$
ε	The estimated residuals

3.2. Accounting of IPPU CO_2 Emissions

IPPU CO_2 emissions, as discussed above, are the result of industrial production, namely, from the chemical reaction conditions utilized for the production of raw material. As opposed to fossil fuel combustion, the portion CO_2 produced does not consume energy because the carbon is discharged during production. For example, CO_2 is released as the primary component of calcium carbonate ($CaCO_3$). Emissions are calculated based on the amount of all manner of industrial products' output, multiplied by the corresponding coefficient of carbon emissions. Table 2 lists the CO_2 emission coefficients of all the IPPU departments in IPCC 2006.

The 18 industrial departments we examined here essentially cover all IPPU CO_2 emissions in China. According to the custom of China's industrial departmentalization and the information found in the Statistical Yearbooks, they can be divided into the following five major categories.

Table 2. CO_2 emission coefficient about involved IPPU departments in IPCC 2006.

No.	Department	Coefficient of Carbon Emissions 10^4 t CO_2/10^4 t Product
CO_2a	Steel	65% BOF, 30% EAF, 5% OHF, and the coefficient is 1.46, 0.08, 1.72
CO_2b	Cement	0.55536
CO_2c	Lime	0.683
CO_2d	Calcium carbide	1.154
CO_2e	Glass	0.2
CO_2f	Synthesis ammonia	1.694
CO_2g	Sodium carbonate	0.138
CO_2h	Ferrosilicon	4.38
CO_2i	Manganese-silicon	1.4
CO_2j	Lead(Pb)	0.52
CO_2k	Magnesium(Mg)	3
CO_2l	Zinc(Zn)	1.72
CO_2m	Virgin aluminum	1.69
CO_2n	Lubricating grease	0.2
CO_2o	Urea	0.733
CO_2p	Titanium dioxide	1.4
CO_2q	Ethylene	1.73
CO_2r	Methyl alcohol	0.67

- CO_2a, CO_2h, and CO_2i correspond to the mining and processing of ferrous metal ores and smelting and pressing of ferrous metals, namely steel and iron alloy manufacturing (A1).
- CO_2b and CO_2e correspond to the mining and processing of nonmetal ores and the manufacture of mon-metallic mineral products, namely non-metallic manufacturing (A2).
- CO_2c and CO_2d correspond to the mining of other ores, namely non-metallic mining (A3).
- CO_2f, CO_2g, CO_2n, CO_2o, CO_2p, CO_2q, and CO_2r correspond to the manufacture of raw chemical materials and chemical products, namely the chemical industry (A4).
- CO_2j, CO_2k, CO_2l, and CO_2m correspond to the mining and processing of non-ferrous metal ores and the smelting and pressing of non-ferrous metals, namely the metal industry (A5).

The gross value of industrial output was valued at constant prices in 1990.

3.3. CO_2 *Kuznets Curve Hypothesis*

This paper utilizes the Generalized Least Square method to conduct the regression analysis on CO_2 emissions and economic growth, then uses time series data to verify whether CO_2 emissions and economic growth support the establishment of the emissions Kuznets curve hypothesis. Throughout the predecessors' research of CO_2 emissions and economic growth, the relationship curve may appear in five kinds of shapes. In addition, the variable logarithmic difference approximation is equal to the rate of change approximatively, but the rate of change for economic variable is usually a stable sequence. The application of original data logarithmetics does not impact the co-integration relationship. It can reduce the unwanted fluctuation and ensure the trend linearization, and eliminate the heteroscedasticity. Accordingly, the variable date is transformed into a natural logarithm. This paper adopts the Selden and Song [4] theory model and uses the following simplified model to study the relationship between IPPU carbon emissions and economic growth:

$$\text{Linear}: \ln PC_{it} = \alpha_i + \beta_1 \ln PGDP_{it} + \varepsilon \tag{1}$$

$$\text{Quadratic}: \ln PC_{it} = \alpha + \beta_1 \ln PGDP_{it} + \beta_2 \ln^2 PGDP_{it} + \varepsilon \tag{2}$$

$$\text{Cubic}: \ln PC_{it} = \alpha + \beta_1 \ln PGDP_{it} + \beta_2 \ln^2 PGDP_{it} + \beta_3 \ln^3 PGDP_{it} + \varepsilon \tag{3}$$

The limited scope in this paper is the IPPU department. Certain indices, such as the entire population in China, GDP, and carbon intensity cannot accurately express variable change, so this paper sets employed population in the corresponding industrial sector as the population value, gross industrial output value (*GIOV*) in the corresponding industrial department as the economic data. α means the intercept term, *PC* means the IPPU CO_2 emissions per employed person, *PGIOV* denotes the *GIOV* per employed person. ε is the estimated residuals, β_1, β_2, β_3 denote the first-, second- and third-power estimated parameters of *PGIOV*, *i* represents the IPPU department and *t* represents the time. The relationship between CO_2 emissions and economic growth depends on the final curve form of the estimation model, and curve form depends on β_1, β_2, and β_3 values. Therefore, the original type can be transformed as follows:

$$\ln PC_{it} = \alpha + \beta_1 \ln PGIOV_{it} + \beta_2 \ln^2 PGIOV_{it} + \beta_3 \ln^3 PGIOV_{it} + \varepsilon \tag{4}$$

where *t* represents the time period (1991–2012), and *i* represents the IPPU department (A1 to A5) in this paper. When $\beta_1 > 0$, $\beta_2 = 0$, $\beta_3 = 0$, CO_2 emissions and economic growth curve slopes upward in a straight line. When $\beta_1 < 0$, $\beta_2 = 0$, $\beta_3 = 0$, CO_2 emissions and economic growth curve slopes downward in a straight line. When $\beta_1 > 0$, $\beta_2 < 0$, and $\beta_3 = 0$, CO_2 emissions and economic growth forms an inverted U-shaped curve. When $\beta_1 < 0$, $\beta_2 > 0$, and $\beta_3 = 0$, CO_2 emissions and economic growth forms a regular U-shaped curve. When the $\beta_1 > 0$, $\beta_2 < 0$, and $\beta_3 > 0$, the curve is N-shaped. When $\beta_1 < 0$, $\beta_2 > 0$, and $\beta_3 < 0$, the curve is inverted N-shaped.

Firstly, we consider the existence of the Kuznets curve for individual IPPU departments. Then, we add the five major department categories to determine whether there truly exists a Kuznets curve for China's IPPU CO_2 emissions. The time series is from 1991 to 2012, and the corresponding parameter values are shown in Table 3. Ensuring data stability prior to analysis is a precondition for any reasonable discussion on a co-integration relationship; each sequence we examined is a non-stationary series, so we run ADF unit root tests on six time sequences segmented from the original series. In this paper, the natural logarithm series $\ln PC$, $\ln PGIOV$, $\ln^2 PGIOV$, $\ln^3 PGIOV$ and first order difference sequence should be taken by a unit root test. After getting the cointegration relationship, the generalized least squares method is used to estimate the model. If β_3 was significant, cubic form regression could be considered appropriate. If β_3 was not significant but β_2 was significant, quadratic regressiozhin was appropriate. If neither β_3 nor β_2 were significant, linear regression was appropriate.

We also utilize the D.W. statistic (Durbin-Watson statistic, 0–4) estimation results to judge whether there was any residual autocorrelation sequence: If $0 < \text{D.W.} < D_L$, the first-order autocorrelation disturbance was present (the closer to 0, the greater the correlation will be). If $D_L < \text{D.W.} < D_U$, it was considered impossible to judge whether there was autocorrelation. If $D_U < \text{D.W.} < 4\text{-}D_U$, there was no first-order autocorrelation disturbance; if $4\text{-}D_U < \text{D.W.} < 4\text{-}D_L$, it was impossible to judge whether there was autocorrelation. If $4\text{-}D_L < \text{D.W.} < 4$, there was first-order autocorrelation disturbance (in this case, the closer to four, the greater the correlation). If there was any residual autocorrelation sequence, the lagged variable was necessary to build estimate equations to eliminate the residual autocorrelation sequence. Finally, after necessary diagnostic checks, we were able to use the adjusted R^2 and F values to judge the model fitting effect.

In short, we first use the scatterplot to judge the curve type according to the methods described above. Then, we test the stability of the data for the five IPPU categories, identify the co-integration relationship and use generalized least squares to estimate the model. The analysis and results are shown in Section 4.2.

Table 3. Gross value of industrial output, IPPU CO_2 emissions and industrial employed population.

The Gross Value of Industrial Output at Constant Price in 1990, One Hundred Million Yuan								
	Gross Value of Industrial Output Valued in China	A1	A2	A3	A4	A5	Sum of A1–A5	Sum of A1–A5 Accounted for the Proportion of the Total
1991	25,407.67	1525.04	1040.26	97.69	1527.82	698.42	4889.2	19.24%
1992	28,817.79	1989.28	1226.23	115.15	1606.73	911.90	5849.3	20.30%
1993	35,734.05	3127.20	1751.05	164.43	1740.09	1410.50	8193.3	22.93%
1994	42,702.20	3351.67	1884.13	176.93	2008.06	1539.35	8960.2	20.98%
1995	49,064.82	3541.45	2004.71	188.25	2534.17	1639.18	9907.8	20.19%
1996	50,487.70	3476.70	2090.92	196.35	2620.33	1648.03	10,032.3	19.87%
1997	50,336.24	3395.53	2082.55	195.56	2502.42	1638.93	9814.99	19.50%
1998	48,272.45	3165.72	2011.75	188.91	2324.75	1538.35	9229.48	19.12%
1999	47,113.91	3031.52	1965.48	184.57	2243.38	1470.29	8895.24	18.88%
2000	48,433.10	3125.46	1957.61	183.83	2265.81	1501.79	9034.51	18.65%
2001	47,803.47	3081.71	1900.84	178.50	2200.10	1480.77	8841.92	18.50%
2002	46,751.80	3006.82	1855.98	174.29	2148.18	1444.79	8630.06	18.46%
2003	47,827.09	3211.28	1898.67	178.30	2197.59	1543.03	9028.87	18.88%
2004	50,744.54	3849.41	1966.44	188.71	2422.18	1830.28	10,257.0	20.21%
2005	53,231.02	4041.59	1994.89	206.35	2627.58	1973.31	10,843.7	20.37%
2006	54,827.95	3881.47	2026.83	211.51	2638.10	2021.43	10,779.3	19.66%
2007	56,527.62	4192.59	2056.89	218.07	2738.34	2208.01	11,413.9	20.19%
2008	60,428.03	5044.22	2225.95	242.01	3044.76	2547.36	13,104.3	21.69%
2009	57,164.91	4207.09	2218.45	236.79	2681.52	2171.49	11,515.4	20.14%
2010	60,308.98	4538.66	2274.58	251.94	2896.04	2401.74	12,363.0	20.50%
2011	63,943.20	4990.76	2438.76	274.91	3178.75	2674.49	13,557.7	21.20%
2012	62,856.17	4460.67	2417.50	284.25	3051.60	2458.34	12,672.7	20.16%

The IPPU CO_2 Emissions, Ten Thousand Tons						
	Total Emissions	A1	A2	A3	A4	A5
1991	28,932.76	6255.34	10,608.97	8066.38	3705.85	296.22
1992	32,793.39	7403.00	12,931.85	8187.93	3922.90	347.71
1993	35,702.32	8517.35	14,741.26	8265.23	3795.12	383.35
1994	38,973.08	9291.05	16,763.39	8330.01	4151.74	436.89
1995	41,984.70	9929.32	18,470.31	8417.36	4677.05	490.66
1996	44,110.32	10,323.50	19,519.43	8405.51	5307.18	554.70
1997	45,423.10	11,017.76	20,132.61	8463.07	5159.83	649.83
1998	47,296.88	11,807.87	21,003.49	8408.29	5330.88	746.35
1999	50,568.35	13,296.09	22,098.14	8592.42	5773.38	808.32
2000	54,336.91	14,419.78	24,386.67	8838.46	5765.77	926.23
2001	60,625.81	17,573.42	27,009.09	9100.93	5882.81	1059.56
2002	67,319.05	20,985.85	29,238.13	9508.61	6328.02	1258.43
2003	79,354.04	26,314.32	34,753.22	10,046.85	6733.92	1505.72
2004	93,941.46	34,932.74	39,289.89	10,689.67	7244.04	1785.12
2005	105,843.67	41,318.73	42,608.91	11,592.52	8285.95	2037.56
2006	124,577.68	51,431.70	48,948.68	12,670.03	9093.12	2434.16
2007	138,898.13	62,057.72	49,995.35	13,913.11	9859.33	3072.61
2008	146,197.04	63,801.32	54,858.13	14,815.03	9465.80	3256.75
2009	163,610.34	76,558.84	57,944.49	16,187.72	9578.13	3341.15
2010	184,558.60	87,784.08	64,639.79	17,523.62	10,566.39	4044.72
2011	204,635.74	97,243.07	71,953.24	19,412.64	11,659.41	4367.38
2012	214,239.57	104,330.24	71,755.58	21,336.45	12,107.13	4710.17

Table 3. *Cont.*

	The Industrial Employed Population, 10,000 persons							
	The Industrial Employed Population in China	A1	A2	A3	A4	A5	Sum of A1–A5	Sum of A1–A5 Accounted for the Proportion of the Total
1991	6551.00	334.00	461.00	113.00	398.00	161.00	1467.00	22.39%
1992	6621.00	344.00	464.00	112.00	408.00	160.00	1488.00	22.47%
1993	6626.00	361.00	456.00	112.00	389.00	145.00	1463.00	22.08%
1994	6580.00	368.00	463.00	112.00	405.00	163.00	1511.00	22.96%
1995	6610.00	368.00	488.00	116.00	412.00	163.00	1547.00	23.40%
1996	6450.00	358.00	465.00	116.00	407.00	161.00	1507.00	23.36%
1997	6215.00	341.00	441.00	112.00	392.00	152.00	1438.00	23.14%
1998	4753.00	272.00	327.00	90.00	304.00	124.00	1117.00	23.50%
1999	4428.00	258.00	301.00	86.00	282.00	120.00	1047.00	23.64%
2000	4102.00	237.00	273.00	74.00	254.00	115.00	953.00	23.23%
2001	3838.00	218.00	248.00	66.00	230.00	111.00	873.00	22.75%
2002	3729.00	205.00	234.00	60.00	217.00	106.00	822.00	22.04%
2003	5748.57	283.30	441.83	183.40	311.33	147.97	1367.83	23.79%
2004	6098.62	290.63	452.67	191.07	315.66	155.22	1405.25	23.04%
2005	6895.96	328.08	461.14	222.76	339.99	172.61	1524.58	22.11%
2006	7358.43	341.40	470.59	238.68	357.78	182.13	1590.58	21.62%
2007	7875.20	353.57	495.03	265.06	380.28	211.38	1705.32	21.65%
2008	8837.63	375.02	552.96	317.35	429.64	238.71	1913.68	21.65%
2009	8831.22	380.50	564.02	337.95	440.49	227.68	1950.64	22.09%
2010	9544.71	412.67	601.15	369.46	474.14	246.99	2104.41	22.05%
2011	9167.29	405.12	570.56	360.90	454.86	245.99	2037.43	22.22%
2012	9095.21	393.33	572.17	346.42	449.78	239.84	2001.54	22.01%

4. Results and Discussions

4.1. The Analysis of China's IPPU CO_2 Emissions from 1991 to 2012

Five major IPPU industrial departments' CO_2 emissions and gross value of industrial output are shown in Table 3. In addition, due to the consideration of the impact of industrial employed population scale on IPPU CO_2 emissions and economy, Table 3 also illustrates the employed population scale of each industrial department. It is important to note that the line chart in Figure 1 represents the gross value of industrial output at a constant price in 1990 (blue line) and industrial employed population (red line), the units are one hundred million yuan and ten thousand people.

Some conclusions can be drawn from Table 3 and Figure 1. In terms of the overall scale, the proportion of gross industrial output value and industrial employed population of the five major IPPU CO_2 emissions departments is about 20% of the total, and remained stable for a generous portion of the time period. It is also determined by each IPPU CO_2 emission department's characteristic: they represent favorable industrial bases and facilities overall, both in basic manufacturing and mining departments, as far as ensuring stable development of the national economy and livelihood, and more stable product supply and marketing demand; thus, the proportion remains around 20% steadily.

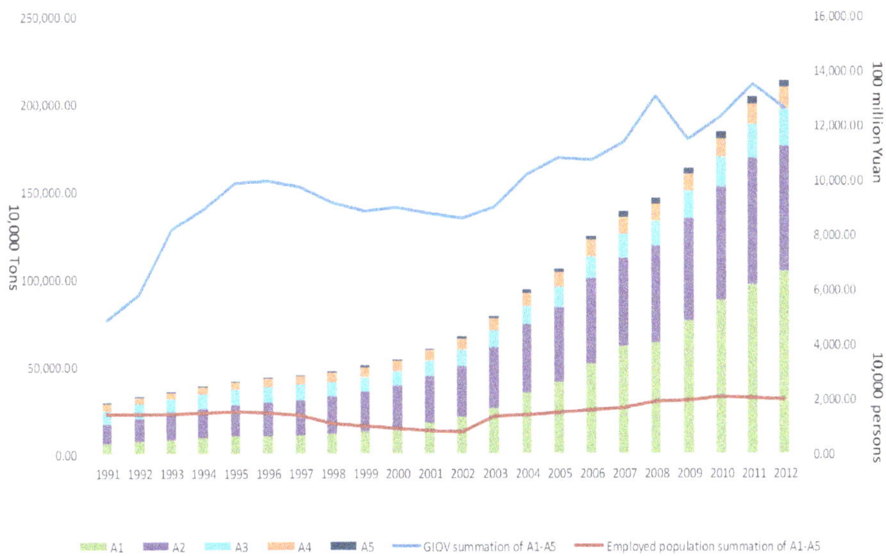

Figure 1. IPPU CO_2 emissions from A1–A5, Gross value of industrial output and industrial employed population.

From the overall view, CO_2 emissions are determined by the product yield, so with the increase of product output, CO_2 emissions rose and reached their maximum in 2012, 2142.3957 million tons, while the industrial department employed population and gross industrial output value firstly began to decline from 1996 to 2002 steadily and then increased after 2002. This is because, since 1998, China's accumulation of deep-seated economic issues became problematic, and layoffs and unemployment became common social phenomena in China's state-owned enterprises. In the middle of the 1990s, China experienced an economic recession characterized by lack of external or internal demand. The stagnation of the whole industry chain made all related enterprises face a very embarrassing situation. Large- and medium-sized, state-owned enterprises going bankrupt reduced the *GIOV* during this period; both reached their nadir in 2002, at 8.22 million (the total employed population in 2002 is 37.29 million) and 863.006 billion (the total China's *GIOV* in 2002 is 4675.18 billion yuan), respectively. After 2002, with China's accession to the WTO (World Trade Organization), a massive boom in domestic and foreign demand strengthened the economy and saved many state-owned enterprises from bankruptcy. Particularly, large state-owned enterprises like iron and steel producers began to make the transition from the planned economy to the market economy, ultimately rising to meet renewed external and internal demand. The industrial worker population and *GIOV* rose again; worker population peaked in 2010 at 21.0441 million (compared to the total population of China, 95.4471 million) and *GIOV* in 2011 at 1355.766 billion yuan (compared to the total gross value of China, 6,394.32 billion yuan), but, in 2011 and 2012, both values leveled off as the growth of the global economy slowed; this weakened external demand considerably, causing declining domestic infrastructure investment and weakening China's real estate market regulations. These reasons sharply slowed down the speed of China's economic growth, especially in manufacturing. Manufacturing's employed demand growth slowed markedly while the construction, resident service and other services employed demand grew rapidly. As a result, the employment pressure will be more and more aggravated. In the next few years, China's labor supply and structure will continue to change, the overall workforce will usher in a new turning point in a few years, and the new labor force and employed population will continue to decline among young adults. In the next few years, China's labor supply and structure will certainly continue to change. New turning points in the country's overall workforce are inevitable, including reducing the number of young adults employed in the industrial field. As the economy slows down and reaches stagnation, the employed population size will be stable.

From the view of the five major IPPU CO_2 emission industrial departments, steel and iron alloy manufacturing, and non-metallic manufacturing accounted for more than 80% of total emissions, and reached 82.2% in 2012. With the increasing production of steel and cement, glass, CO_2 emissions will continue to increase steadily over these years, while the rest of the industry growth rates are slower than A1 and A2. Taking the five major IPPU CO_2 emission industrial departments as a whole, the steel and iron alloy manufacturing and chemical industry occupy nearly 60% of gross industrial output value, and non-metal and metal manufacturing industry occupies 20%. Non-metallic mining comprises the smallest portion at only about 2%. The non-metallic manufacturing industry, interestingly, shows the greatest proportion of the worker population at nearly 30%, while steel and iron alloy manufacturing and the metal industry comprises 19% and 11% of the worker population, respectively. The non-metallic mining worker population also presents a rising trend, from 7.7% in 1991 to 17.3% in 2012, while the chemical industry fell about 5% in 2012 in comparison to that of the 1990s. Due to the relative stability of China's industrial development in recent years, the employed population structure also tends to be stable, both as a reflection of domestic and foreign demand characteristics. Due to the relative stability of China's industrial development in recent years, the employed population structure also tends to be stable, both as a reflection of domestic and foreign demand characteristics. As China's cement, glass, and iron alloy manufacturing industries are national leaders in production (and accordingly are top-priority infrastructure investment targets, demand-creators, and policy-drivers), the corresponding *GIOV* and employed population stayed at high levels toward the end of the time period we examined. The chemical industry and metal industry are pillars of China's industrial infrastructure and effectively serve as the foundational livelihood of a massive employed population and of China's economic and social stability—in short, no changes to this infrastructure can be expected in the short term. Non-metallic mining can be considered as a bottleneck industry, conversely. Due to its high dependence on natural resources, this industry tends to be more exploitative (*i.e.*, its *GIOV*, in effect, inextricably corresponds to the exploitation and industrial pretreatment of its products.). Obviously, making any substantial increase in industrial output of this kind of resource-intensive industry is impossible in terms of environmental sustainability.

4.2. The Verification of IPPU CO_2 Kuznets Curve

Upon testing, the time sequence series are consistent with cointegration test precondition. The result of the unit root test shows that the IPPU CO_2 emissions of per employed person time series and the *GIOV* per employed person time series are both non-reposeful time series and integrated series of the first order. Then, taking the variable cointegration with the Johanse–Juselius test. The examination shows that the corresponding variable co-integration relationship exists as well. Therefore, scatter plot description and regression analysis can proceed.

Figures 2–7 shows the scatterplots of five major IPPU CO_2 emissions departments as well as their summation. It can reflect the relationship between IPPU CO_2 emissions per employed person and *GIOV* per employed person, as well as the change trend in both. The abscissa represents *GIOV* per employed person and the unit is ten thousand yuan. The ordinate represents the IPPU CO_2 emissions per employed person, and the unit is tons. Natural logarithm processing has been done to all data.

Unlike those drawn in previous studies, the scatter do not present any classic inverted U-shaped, N-shaped and other Kuznets curve types. Except non-metallic mining's scatterplot presenting a function type growth curve, the rest of the IPPU department data are divisible into two parts marked with red and black splashes, respectively. The black scattered data represents 1991–2002 values which presented a function type growth curve and the red scattered data represents 2002–2012 values, which created a gathering type phenomenon, and all red scattered data are upper left of the black curve, namely *GIOV* per employed person value declined after 2002.

We consider non-metallic mining as a regression example in Table 4 because β_3 is not significant in cubic form, while β_2 is significant, so quadratic regression is adopted. Then, we checked the D.W. value to judge whether the regression residuals show series autocorrelation.

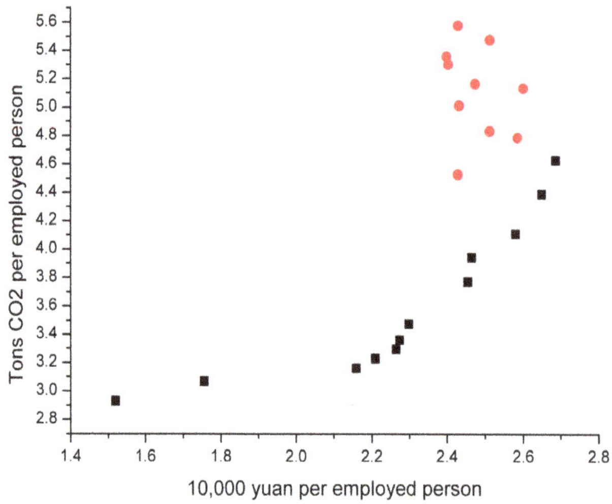

Figure 2. Scatterplots of CO_2 emissions and *GIOV* in steel and iron alloy manufacturing.

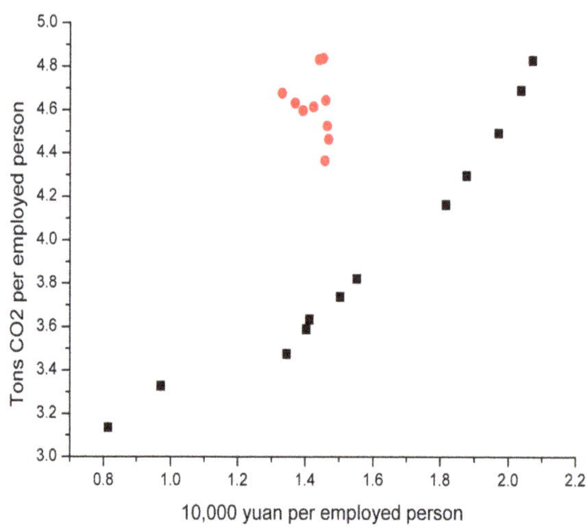

Figure 3. Scatterplots of CO_2 emissions and *GIOV* in non-metallic manufacturing.

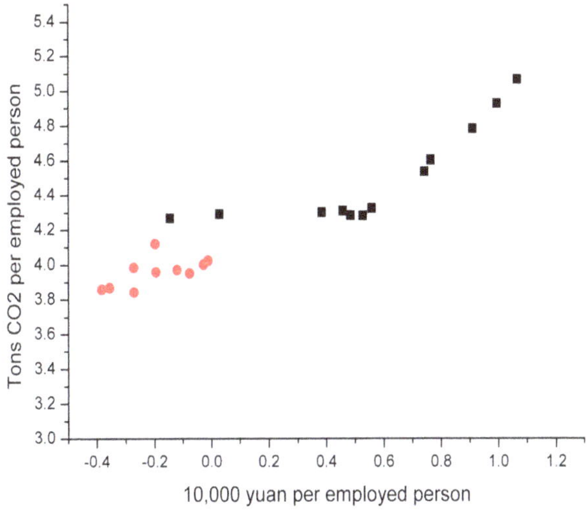

Figure 4. Scatterplots of CO_2 emissions and *GIOV* in non-metallic mining.

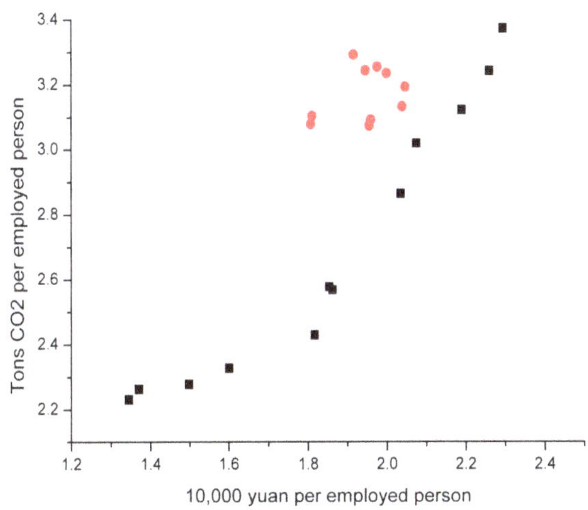

Figure 5. Scatterplots of CO_2 emissions and *GIOV* in the chemical industry.

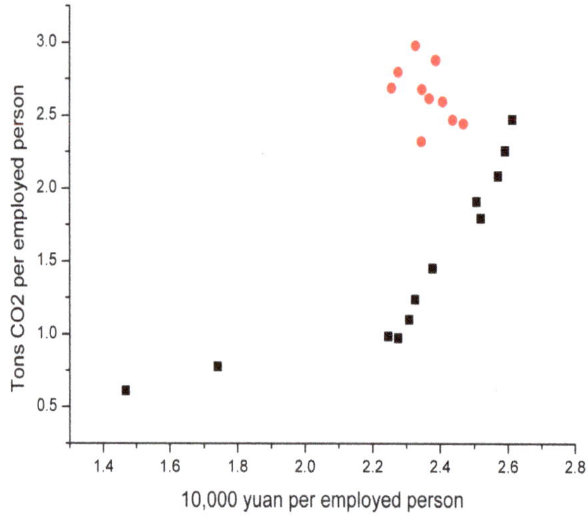

Figure 6. Scatterplots of CO_2 emissions and *GIOV* in the metal industry.

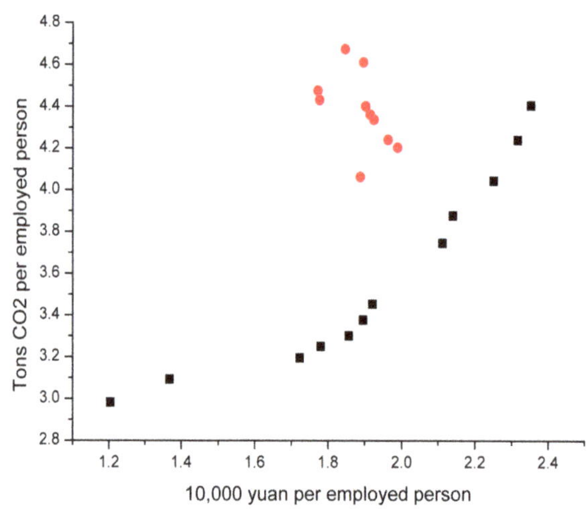

Figure 7. Scatterplots of CO_2 emissions and *GIOV* in the sum of the five major IPPU CO_2 emission departments.

Table 4. Cubic form and quadratic form regression in non-metallic mining.

Variable	Coefficient	Std. Error	*t*-Statistic	Prob.
		Cubic Form		
C	4.1099992	0.035799	114.8081	0.0000
X	0.412677	0.087479	4.717469	0.0002
X^2	−0.386025	0.329525	−1.171459	0.2567
X^3	0.801438	0.295341	2.713604	0.0142
		Quadratic Form		
C	4.046176	0.031185	129.7469	0.0000
X	0.406503	0.101038	4.023275	0.0007
X^2	0.444054	0.141575	3.136520	0.0054

According to the result, D.W. = 0.498941 < D_L = 0.9711, which indeed demonstrates autocorrelation between CO_2 emissions and *GIOV*. Then, we added AR (AR(1) means the random disturbance is the first-order autoregressive form of serial correlation) to the estimating equations to eliminate autocorrelation sequence via the generalized least square method and obtain the results list in Table 5.

Table 5. Regression result using the generalized least square in non-metallic mining.

Variable	Coefficient	Std. Error	*t*-Statistic	Prob.
C	4.004083	0.041465	96.56548	0.0000
X	0.382922	0.135533	2.825304	0.0117
X^2	0.533774	0.153956	3.467047	0.0029
AR(1)	0.607073	0.170561	3.559266	0.0024
R^2	**Adjusted R^2**	**S.D. Dependent Var**	***F*-Statistic**	**D.W.**
0.972655	0.967829	0.356543	201.5612	1.422968

The final result (in which we took four decimal places, the same as the following regression analysis results):

$$\ln PC_{A3} = 4.0041 + 0.3829 \ln PGIOV_{A3} + 0.5338 \ln^2 PGIOV_{A3} + \varepsilon \quad (5)$$

The following regression analysis results can be obtained in Table 6 with the Generalized Least Squares. According to the results, estimate equation and the analysis of the results are shown as follows:

Table 6. Regression analysis results of the IPPU CO_2 emissions departments.

		Steel and Iron Alloy Manufacturing		
Variable	**Coefficient**	**Std. Error**	***t* -Statistic**	**Prob.**
C	10.92199	1.030504	10.59869	0.0000
X	−8.609946	0.914439	59.415545	0.0000
X^2	2.325560	0.201424	11.54560	0.0000
AR(1)	50.335305	0.283262	51.183727	0.2752
R^2	**Adjusted** R^2	**S.D. Dependent Var**	***F*-Statistic**	**D.W.**
0.990074	0.985820	0.529344	232.7395	2.247248
		Non-Metallic Manufacturing		
Variable	**Coefficient**	**Std. Error**	***t*-Statistic**	**Prob.**
C	5.821201	0.408260	14.258555	0.0000
X	−3.665195	0.462456	−7.925494	0.0001
X^2	1.528215	0.134834	11.33406	0.0000
AR(1)	0.516023	0.059599	8.658205	0.0001
R^2	**Adjusted** R^2	**S.D. Dependent Var**	***F*-Statistic**	**D.W.**
0.997692	0.996703	0.514686	1008.591	1.807023
		Non-Metallic Mining		
Variable	**Coefficient**	**Std. Error**	***t*-Statistic**	**Prob.**
C	4.004083	0.041465	96.56548	0.0000
X	0.382922	0.135533	2.825304	0.0117
X^2	0.533774	0.153956	3.467047	0.0029
AR(1)	0.607073	0.170561	3.559266	0.0024
R^2	**Adjusted** R^2	**S.D. Dependent Var**	***F*-Statistic**	**D.W.**
0.972655	0.967829	0.356543	201.5612	1.422968
		Chemical Industry		
Variable	**Coefficient**	**Std. Error**	***t*-Statistic**	**Prob.**
C	4.651413	0.853222	5.451585	0.0010
X	−3.519916	0.934448	−3.766841	0.0070
X^2	1.291968	0.251070	5.145845	0.0013
AR(1)	0.117817	0.378453	0.311313	0.7646
R^2	**Adjusted** R^2	**S.D. Dependent Var**	***F*-Statistic**	**D.W.**
0.985567	0.979382	0.406903	159.3356	1.907908
		Metal Industry		
Variable	**Coefficient**	**Std. Error**	***t*-Statistic**	**Prob.**
C	12.78229	1.177808	10.85261	0.0000
X	−13.05701	1.074519	−12.15149	0.0000
X^2	3.472181	0.242660	14.30886	0.0000
AR(1)	−0.415051	0.204208	−2.032488	0.0816
R^2	**Adjusted** R^2	**S.D. Dependent Var**	***F*-Statistic**	**D.W.**
0.989575	0.985107	0.581921	221.4871	2.100003

Table 6. Regression analysis results of the IPPU CO_2 emissions departments.

The Sum of the 5 Major IPPU CO_2 Emission Departments				
Variable	**Coefficient**	**Std. Error**	***t*-Statistic**	**Prob.**
C	6.198014	0.408532	15.17141	0.0000
X	−4.404028	0.429709	−10.24887	0.0000
X^2	1.540928	0.111527	13.81664	0.0000
AR(1)	−0.296303	0.302890	−0.978251	0.3605
R^2	**Adjusted** R^2	**S.D. Dependent Var**	***F*-Statistic**	**D.W.**
0.996164	0.994494	0.452141	603.1067	2.482855

After obtaining the coefficient and the related parameters in the equation, due to the uncertainty of initial definition, diagnostic tests (tests on residuals, specification and stability tests and so on) of regression equations should be checked. In this paper, White heteroscedasticity testing, Chow breakpoint test, and Granger causality test are adopted from different perspectives. Consider the chemical industry as a regression example, and the results are as follows:

From the result in Table 7, *F*-statistic and the corresponding probability indicate that there is no heteroscedasticity in the residual and regression coefficient keeping stable at different times, PC_{A4} does not Granger Cause $PGIOV_{A4}$ and $PGIOV_{A4}$ does not Granger Cause PC_{A4}.

Table 7. White Heteroskedasticity Test, Chow Breakpoint Test, Granger Causality Test in Chemical Industry.

White Heteroskedasticity Test				
***F*-Statistic**	**Prob.**		**Obs** R^2	**Prob.**
1.256214	0.3208		3.810624	0.2827
Chow Breakpoint Test: 1997 (Other Hypothetical Breakpoint Dates Have the Same Conclusion)				
***F*-Statistic**	0.851206		**Prob.**	0.5087
Granger Causality Test, Lags = 2				
PC_{A4} does not Granger Cause $PGIOV_{A4}$	**F-Statistic**	0.36198	**Prob.**	0.7132
$PGIOV_{A4}$ does not Granger Cause PC_{A4}		1.07885		0.4078

Similarly, other industry departments and the total industrial department also carry out the diagnostic test and get the same conclusion. Because of the limitation of length, there is no more tautology here. Finally, the estimate equations are obtained in Section 4.2.1 to Section 4.2.6.

4.2.1. Steel and Iron Alloy Manufacturing

1991–2002

$\ln PC_{A1} = 10.9220 - 8.6099 \ln PGIOV_{A1} + 2.3256 \ln^2 PGIOV_{A1} + \varepsilon$

In 2002, $PGIOV_{A1}$, PC_{A1} reached maximum, $\ln PGIOV_{A1} = 2.686$, $\ln PC_{A1} = 4.629$.

2003–2012

$\ln PGIOV_{A1}$ stays within the interval (2.398, 2.599). $\ln PC_{A1}$ began to rise resiliently in recent years after a slight decline, and reached the maximum, 5.581, in 2012.

4.2.2. Non-Metallic Manufacturing

1991–2002

$\ln PC_{A2} = 5.8212 - 3.6652 \ln PGIOV_{A2} + 1.5282 \ln^2 PGIOV_{A2} + \varepsilon$

In 2002, $PGIOV_{A2}$, PC_{A2} reached maximum, $\ln PGIOV_{A2} = 2.071$, $\ln PC_{A2} = 4.828$.

2003–2012

$\ln PGIOV_{A2}$ stays within the interval (1.331, 1.469). $\ln PC_{A2}$ began to rise resiliently in recent years after a slight decline, and reached its maximum, 4.837, in 2011. The value was 4.832 in 2012.

4.2.3. Non-Metallic Mining

1991–2012

$\ln PC_{A3} = 4.0041 + 0.3829 \ln PGIOV_{A3} + 0.5338 \ln^2 PGIOV_{A3} + \varepsilon$

In 2002, $PGIOV_{A3}$, PC_{A3} reached the maximum of the whole time series from 1991 to 2012, $\ln PGIOV_{A3} = 1.066$, $\ln PC_{A3} = 5.066$. In fact, compared to the value in 2002, $\ln PGIOV_{A3}$ and $\ln PC_{A3}$ in 2003–2012 both showed a greater degree of decline and did not follow the increasing trend pre-2002, though the data do fall onto the overall fitting curve coincidentally.

4.2.4. Chemical Industry

1991–2002

$\ln PC_{A4} = 4.6514 - 3.5199 \ln PGIOV_{A4} + 1.2920 \ln^2 PGIOV_{A4} + \varepsilon$

In 2002, $PGIOV_{A4}$, PC_{A4} reached maximum, $\ln PGIOV_{A4} = 2.292$, $\ln PC_{A4} = 3.373$.

2003–2012

$\ln PGIOV_{A4}$ stays within the interval (1.806, 2.045). $\ln PC_{A4}$ began to rise resiliently in recent years after a slight decline, and reached its maximum, 3.293, in 2012.

4.2.5. Metal Industry

1991–2002

$\ln PC_{A5} = 12.7823 - 13.0570\ln PGIOV_{A5} + 3.4722\ln^2 PGIOV_{A5} + \varepsilon$

In 2002, $PGIOV_{A5}$, PC_{A5} reached maximum, $\ln PGIOV_{A5} = 2.612$, $\ln PC_{A5} = 2.474$.

2003–2012

$\ln PGIOV_{A5}$ stays within the interval (2.255, 2.467). $\ln PC_{A5}$ began to rise resiliently in recent years after a slight decline, and reached its maximum, 2.977, in 2012.

4.2.6. The Sum of the Five Major IPPU CO_2 Emission Departments

1991–2002

$\ln PC_A = 6.1980 - 4.4040\ln PGIOV_A + 1.5409\ln^2 PGIOV_A + \varepsilon$

In 2002, $PGIOV_A$, PC_A reached its maximum, $\ln PGIOV_A = 2.351$, $\ln PC_A = 4.405$.

2003–2012

$\ln PGIOV_A$ stays within the interval (1.771, 1.988). $\ln PC_A$ began to rise resiliently in recent years after a slight decline, and reached its maximum, 4.673, in 2012.

Compared with the previous studies on industry EKC, the source of basic data and curve shape are not the same for several situations, but the vast majority of the carbon emission data are from the energy consumption [11,12,15,16], there is almost no carbon emissions EKC analysis caused by industrial production process. On the other hand, from the point of curve shape, although the curve shapes exist in a variety of forms [3–10,14–16], the two obvious parts of the scatters have never appeared before. This paper focused on the relationship between industry CO_2 emissions and economic activity, and CO_2 emissions are accounted for in terms of IPPU rather than energy consumption. Though the surface result is similar with the research [10,16] that an inverted-U EKC does not exist, due to the IPPU CO_2 data the first time, our novel use of IPPU CO_2 data made our scatter and regression analysis quite different from those of previous researchers. Because of the data result accounting and the curve fitting situation, a new relationship between IPPU CO_2 emission and *GIOV* is put forward for the first time in this paper. The reasons and discussions are as follows.

Prior to 2002, reduction in domestic and foreign demand caused either stagnation or decline in China's manufacturing sector overall (including the steel, cement, glass, and chemical industries) as well as a decline in the *GIOV* as prices remained constant and employed population declined, while the decreasing amount is far from the range of laid-off and unemployed workers. Several of China's state-owned enterprises reduced surplus staff during this time in effort to prevent bankruptcy, maintain (low) production levels, and avoid collapse exponentially of the employed population. This caused the ascendant trend in the *GIOV* per employed curve.

After 2002, with China's accession to the WTO, the increasing domestic and foreign demand and widespread new construction of domestic infrastructures required a larger workforce; accordingly, the population of employed persons increased yearly from this point forward. The economic growth rate also tended to be stable in those years, though it grew at a slower rate than that of the worker population. To this effect, the *GIOV* per worker data moved to the left of the black curve after 2002. After 2010, China's labor force structure was fairly stable, and the number of workers was around 90 million. As-affected by a sluggish global economy, weakened external demand, slowed domestic infrastructure investment, new real estate market regulations, and other factors, China's economic growth then slowed down (especially in terms of manufacturing growth rate). The *GIOV* stalled once again while the *GIOV* per worker remained unchanged, dropping below the maximum in 2002.

The production increase is related to domestic and foreign demand directly working on increasing CO_2 emissions; likewise, emissions decline when production declines. After 2002, IPPU CO_2 emissions per employed person in all departments presented a slight decline in 2003, followed by an overall increase until reaching maxima in 2011 and 2012. Among the separate curves for individual department, steel and iron alloy manufacturing and metal industry IPPU CO_2 emissions per employed person reached its maximum in 1991–2012, while the remaining departments' extremum values were slightly smaller than that of the 2002 maximum.

GIOV is affected by the apparent output consumption data accounted in this paper by IPPU CO_2 emission, and there was a greater distance between data points in 2003 to 2012 and the fitting curve in 1991–2012 that reflects the greater difference between product output and consumption. Take the iron and steel industry for example, before 2003, China's steel industry was in short supply and domestic production did not meet the needs of economic development, but, after 2003, with the capacity of expanding and enterprise's production enthusiasm, the iron and steel industry appeared in the steel stocks. As the economy gradually stabilized over these years, the overcapacity problem got worse noticeably. In 2012, the difference between demand and supply reached its maximum, nearly 300 million tons, and its reflective phenomenon reflects the scatter performance.

In fact, the calcium carbide and lime mining also complies with the principle of the above. In 2002, $\ln PC_{A3}$ and $\ln PGIOV_{A3}$ reached the maximum, then it dipped to a low level.

The scatter we created for the summation of all five major IPPU CO_2 emission departments was similar to that of individual manufacturing sectors (apart from that of non-metallic mining, which was quite a bit smaller and had fewer effects on the whole). In 2002, the *GIOV* per employed person value reached its maximum and then began to decline, as discussed above. The IPPU CO_2 emissions per employed person rose slightly after a slight decline in 2002, then reached its maximum in 2012.

According to the classic Kuznets curve theory, carbon emissions and economic growth present the inverted U-shaped curve. However, in this paper, the results show a weaker relationship between IPPU CO_2 emissions per worker and *GIOV* per worker (*i.e.*, two different curve shapes) in terms of IPPU CO_2 emissions. These scatterplots may provide a valuable reference for future policy adjustments made to the industrial sector as well.

Especially the data since 2003, which moved to the left above the curve in 1991-2002, the per capita CO_2 emissions presents an increasing trend in recent years and reached the maximum in 2011 or 2012. This reflects a serious circumstance: due to small changes in *GIOV* and the employed population, China's iron and steel, cement and other industries demand have tended to become saturated and even reduced. While the production is still increasing, the problem of excess production capacity has appeared in China in recent years, and a worsening overcapacity situation has been alarming the government.

5. Conclusions

Based on IPPU CO_2 emissions calculating equations in IPCC 2006, this paper preliminarily calculated and established a time series for China's IPPU CO_2 emissions in 1991-2012. Combined with the *GIOV* and employed population, this paper analyzed the relationship between IPPU CO_2 emissions level and industrial economic development, and reached some interesting conclusions.

The overall trend shows that CO_2 emissions are determined by the product yield. CO_2 emissions increased until reaching a maximum in 2012; the IPPU CO_2 emissions related to the employed population and *GIOV* first steadily declined from 1996 to 2002, then increased from 2002 to the end of the observed time period. Steel and iron alloy manufacturing and non-metallic manufacturing CO_2 emissions accounted for more than 80% of total emissions, and up to 82.2% in 2012. The steel and iron alloy manufacturing and chemical industry occupied nearly 60% of *GIOV*, while non-metal and metal manufacturing industries occupied 20%, and non-metallic mining are only 2%. The non-metallic manufacturing industry had the greatest proportion of the employed population, however, at nearly 30%.

The curves we drew to represent the relationship between *GIOV* and IPPU CO_2 emissions marked a departure from Kuznets curves in previous studies: The scatterplots did not present any classic inverted U-shaped or N-shaped curves. Most of the industrial department (except non-metallic mining) and the sum of all sectors presented two-part curves. Before 2002, the data formed a quadratic or a cubic function curve, and after 2003, *GIOV* per employed person remained within a certain range while IPPU CO_2 emissions per worker fell slightly, then continued to rise gradually.

According to the current situation of economic and employment, it is impossible for large-scale industrial economy to grow and the employed population to change. Therefore, decreasing CO_2 emission intensity and per capita indicators should come directly from the CO_2 emission source—IPPU. The government should design policy measures from the perspective of the removal of excess capacity and elimination of backward production capacity, namely, reducing new production capacity, repeated use of products whenever possible, and replacing outdated equipment to reduce carbon emissions. Certain processes, such as those practiced in the non-metallic mining industry, metal industry, and chemical industry are unitary and stable and they create only relatively low IPPU CO_2 emissions, while steel and cement manufacturing are primary culprits of high IPPU CO_2 emissions as well as industries that are traditionally vulnerable to overcapacity. It is absolutely imperative to cut down the CO_2 emissions from these two industries.

The most direct path for the cement industry is to reduce production, use up current inventory and extend the service life of cement materials. On one hand, carbonate decomposition can produce a massive amount of CO_2 emissions; thus, blasting furnace slag, fly ash and other industrial waste instead of carbonate can be used to reduce IPPU CO_2 significantly. In addition, mixed material and mineralizing agent can be applied to the cement clinker production process as well. On the other hand, to improve the cement performance, which would improve the material's service life, would reduce demand for cement as well. There has been a similar situation in steel and iron alloy products, and some areas' production has already met the needs of the next 10 years. Reducing production, using up inventory, and smelting scrap steel are the most direct paths to sustainability. A transition to the electric arc furnace could substantially reduce the IPPU values of the steel and iron alloy industries. Though most open-hearth steelmaking production has been eliminated, there is much work yet to be done to make these industries environmentally sustainable. Because the current cement industry and iron and steel industry have gone into a downturn, and there has been no mass production or demand of steel and cement possible in recent years due to the depressed market, it is a golden age to make the scatter return to more normal levels by digesting inventory and adjusting the technology structure. If steel and cement industries do not take advantage of the depressed market situation now and do not adjust structure quickly, the future to reduce the IPPU CO_2 emission will be a more serious issue while the rapid economic development appears once again.

Though current IPPU carbon emissions and corresponding *GIOV* do not show a Kuznets curve, they do indeed show rising trends. The top five industries examined in this study comprise about 20% of China's total *GIOV*, reflecting that they are pillars of the country's economy and central to the livelihood of millions of workers, meaning that substantive changes to infrastructure in the short term are effectually impossible. Instead, promoting investment in industrial capital and equipment is imperative to the technological improvement necessary for environmental sustainability. Improving R & D efficiency to raise product quality, eliminating production of low output value materials, and minimizing energy consumption are also effective paths to a two-fold goal of ensuring valuable products and a healthy economy while ensuring the stable livelihood of the workforce. In effect, sustainability measures can improve the value of *GIOV* per employed person in China, ushering China into a new era of economic and environmental sustainability.

Acknowledgments: The authors gratefully acknowledge the financial support from the National Natural Science Foundation of China (71273039). This research has also been supported by the Key Laboratory of Ocean Energy Utilization and Energy Conservation of Ministry of Education, the Fundamental Research Funds for the Central Universities (DUT14RC(3)151) and the China Postdoctoral Science Foundation (2015M571309).

Author Contributions: Both authors contributed equally to this work. In particular, Hailin Mu and Ye Duan had the original idea for the study, and both coauthors conceived of and designed the methodology. Ye Duan drafted the manuscript, which was revised by Hailin Mu and Nan Li. All authors have read and approved the final manuscript.

Conflicts of Interest: The authors declare no conflict of interest.

References

1. Grossman, G.M.; Krueger, A.B. *Environmental Impacts of a North American Free Trade Agreement*; Working Paper No. 3914; National Bureau of Economic Research: Cambridge, UK, 1991; pp. 1–37.
2. Shafik, N. Economic development and environmental quality: An econometric analysis. *Environ. Econ.* **1994**, *46*, 757–773.
3. Hettige, H.; Lucas, R.E.B.; Wheeler, D. The toxic intensity of industrial production; global patterns, trends, and trade policy. *Am. Econ. Rev.* **1992**, *82*, 478–481.
4. Selden, T.M.; Song, D. Environmental quality and development: Is there a Kuznets curve for air pollution emissions? *J. Environ. Econ. Manag.* **1994**, *27*, 147–162.
5. Grossman, G.M.; Krueger, A.B. Economic growth and the environment. *Q. J. Econ.* **1995**, *110*, 353–377.
6. Hill, R.J.; Magnani, E. An exploration of the conceptual and empirical basis of the environmental Kuznets curve. *Aust. Econ. Pap.* **2002**, *42*, 239–254.
7. Stem, D.L. The rise and fall of the environmental Kuznets curve. *World Dev.* **2004**, *32*, 1419–1439.
8. Lantz, V.; Feng, Q. Assessing Income, Population and technology impacts on CO_2 emissions in Canada: Where's the EKC? *Ecol. Econ.* **2006**, *57*, 229–238.
9. Ranjan, R.; Shortle, J. The environmental Kuznets curve when the environment exhibits hysteresis. *Ecol. Econ.* **2007**, *64*, 204–215.

10. He, X.; Zhang, Y. Influence Factors and Environmental Kuznets Curve Relink Effect of Chinese Industry's Carbon Dioxide Emission—Empirical Research Based on STIRPAT Model with Industrial Dynamic Panel Data. *China Ind. Econ.* **2012**, *1*, 26–35.
11. Lu, H. An Analysis on China's Economy Development and on State space Model of Environment—Take Air Pollution as an Example. *Study Financ. Econ.* **2000**, *26*, 53–59.
12. Cai, F.; Du, Y.; Wang, M. The Political Economy of Emission in China: Will a Low Carbon Growth Be Incentive Compatible in Next Decade and Beyond? *Econ. Res. J.* **2008**, *6*, 4–11.
13. Huang, W.M.; Lee, G.W.M.; Wu, C.C. GHG Emissions, GDP Growth and the Kyoto Protocol: A Revisit of Environmental Kuznets Curve Hypothesis. *Energy Policy* **2008**, *36*, 239–247.
14. Han, Y.; Lu, Y. The Relationship between Economic Growth and Environmental Quality: An Empirical Test on the Environmental Kuznets Curve of CO_2. *Econ. Theory Bus. Manag.* **2009**, *3*, 5–12.
15. Zheng, L.; Zhu, Q. Study on Existence of Environmental Kuznets Curve of Carbon Emissions in China. *Stat. Res.* **2012**, *29*, 58–65.
16. Zhang, M. Measurement of Carbon Emission in the Manufacturing Segmented Industry of China-Giving Discussion of the Existence of EKC in the Manufacturing Industry. *Soft Sci.* **2015**, *29*, 113–116.
17. Dorel, P.; Cristiana, T.; Radu, P. The Textile Industry and Sustainable Development: A Holt-Winters Forecasting Investigation for the Eastern European Area. *Sustainability* **2015**, *7*, 1280–1291.
18. Valentina, D.; Battista, G.; Giorgio, M. Environmental Sustainability of the Alumina Industry in Western Europe. *Sustainability* **2014**, *6*, 9477–9493.
19. Van Puyvelde, D. CCS opportunities in the Australian Industrial Processes sector. *Energy Procedia* **2009**, *1*, 109–116.
20. Sheinbaum, C.; Ozawa, L.; Castillo, D. Using logarithmic mean divisia index to analyze changes in energy use and carbon dioxide emissions in Mexico's iron and steel industry. *Energy Econ.* **2010**, *32*, 1337–1344.
21. Arruda, L.R.; de Jesus Lameira, V.; Quelhas, O.L.G.; Pereira, F.N. Sustainability in the Brazilian Heavy Construction Industry: An Analysis of Organizational Practices. *Sustainability* **2013**, *5*, 4312–4328.
22. Liaskas, K.; Mavrotas, G.; Mandaraka, M.; Diakoulaki, D. Decomposition of industrial CO_2 emissions: The case of European Union. *Energy Econ.* **2000**, *22*, 383–394.
23. Kim, Y.; Worrell, E. International comparison of CO_2 emission trends in the iron and steel industry. *Energy Policy* **2002**, *30*, 827–838.
24. Kirschen, M.; Risonarta, V.; Pfeifer, H. Energy efficiency and the influence of gas burners to the energy related carbon dioxide emissions of electric arc furnaces in steel industry. *Energy* **2009**, *34*, 1065–1072.
25. Zhou, W.; Zhu, B.; Li, Q.; Ma, T.; Hu, S.; Griffy-Brown, C. CO_2 emissions and mitigation potential in China's ammonia industry. *Energy Policy* **2010**, *38*, 3701–3709.
26. Dong, X.; Guo, J.; Höök, M.; Pi, G. Sustainability Assessment of the Natural Gas Industry in China Using Principal Component Analysis. *Sustainability* **2015**, *7*, 6102–6118.

27. Liu, W.; Wu, X.; Ding, X. Industrial carbon footprint calculation of woolen yarn. *Wool Text. J.* **2015**, *2*, 57–61.
28. Liu, Z.; Dong, H.; Geng, Y.; Lu, C.; Ren, W. Insights into the Regional Greenhouse Gas (GHG) Emission of Industrial Processes: A Case Study of Shenyang, China. *Sustainability* **2014**, *6*, 3669–3685.
29. Gao, J.; Wang, Y.; Li, Q. Estimation on Carbon Emissions from Fossil Fuel and Industrial Production from 2000 to 2009 in Henan Province. *Territ. Nat. Resour. Study.* **2011**, *5*, 48–50.
30. Chen, H. Analysis on Embodied CO_2 Emissions Including Industrial Process Emissions. *China Popul. Resour. Environ.* **2009**, *19*, 25–30.
31. IPCC. *IPCC Guidelines for National Greenhouse Gas Inventories*; United Kingdom Meteorological Office: Bracknell, UK, 2006.
32. CSY. *China Statistical Yearbook*; National Bureau of Statistics of the People's Republic of China: Beijing, China, 1991–2012.
33. NBS. *China Compendium of Statistics 1949–2008*; National Bureau of Statistics of the People's Republic of China: Beijing, China, 2010.
34. CIESY. *China Industrial Economy Statistical Yearbook, 1991–2012*; National Bureau of Statistics of the People's Republic of China: Beijing, China, 1991–2012.
35. CCIY. *China Chemical Industrial Yearbook, 1991–2012*; National Bureau of Statistics of the People's Republic of China: Beijing, China, 1991–2012.
36. NBS. *The Yearbook of Nonferrous Metals Industry of China, 1991–2012*; National Bureau of Statistics of the People's Republic of China: Beijing, China, 1991–2012.

Assessment and Decomposition of Total Factor Energy Efficiency: An Evidence Based on Energy Shadow Price in China

Peihao Lai, Minzhe Du, Bing Wang and Ziyue Chen

Abstract: By adopting an energy-input based directional distance function, we calculated the shadow price of four types of energy (*i.e.*, coal, oil, gas and electricity) among 30 areas in China from 1998 to 2012. Moreover, a macro-energy efficiency index in China was estimated and divided into intra-provincial technical efficiency, allocation efficiency of energy input structure and inter-provincial energy allocation efficiency. It shows that total energy efficiency has decreased in recent years, where intra-provincial energy technical efficiency drops markedly and extensive mode of energy consumption rises. However, energy structure and allocation improves slowly. Meanwhile, lacking an integrated energy market leads to the loss of energy efficiency. Further improvement of market allocation and structure adjustment play a pivotal role in the increase of energy efficiency.

Reprinted from *Sustainability*. Cite as: Lai, P.; Du, M.; Wang, B.; Chen, Z. Assessment and Decomposition of Total Factor Energy Efficiency: An Evidence Based on Energy Shadow Price in China. *Sustainability* **2016**, *8*, 408.

1. Introduction

Since the reform and opening up, extensive economic growth mode with high input and high energy consumption has made contributions to the boom of the Chinese economy with the side effect of energy and environmental problems. With the development of urbanization and industrialization, the contradiction of energy supply and demand would be evident in the long term due to the growth of energy consumption. For example, the energy consumption reached 42.6 trillion Ton of standard Coal Equivalent (Tce) in China in 2014. Saving energy and reducing consumption is necessary for sustainable development, owing to the non-renewability and scarcity of fossil fuels such as coal and gas. Therefore, in China's 12th Five-Year Plan, there are some main targets proposed. The first target is to reduce the energy consumption in GDP by 16% compared with that in 2010 by the control of energy consumption intensity and total energy. The second one is to optimize energy structure by lowering the reliance on fossil fuels. The third target is to promote marketization of energy price by reforming energy pricing mechanism. To realize these aims, we need to answer these three questions: (1) Is the energy pricing reasonable in China? (2) Which level is the total macro-energy efficiency at in China? (3) Is the promotion of energy marketization and optimization of energy structure effective in China?

Compared with advanced level in the world, total energy efficiency in China is not high enough for low energy using efficiency. By evaluating total factor energy efficiency of different economies, China is found as the economy with the lowest value in 17 APEC countries or areas [1]. Even though total factor energy efficiency has greatly increased in China in recent years, and narrowed the gap with that of developed countries, China still has large quantities of energy consumption, leading to potentiality of energy-saving and emission-reduction [2]. Increasing numbers of scholars focus on energy inefficiency and technique inefficiency for the reason of low energy efficiency in China [3]. Firstly, as for the energy resources in China, the proportion of coal is too large while the production of clean energy is not enough, resulting in unreasonable energy input structure. This part of loss of the energy efficiency can be remedied by upgrading energy structure, aiming at restricting carbon emission intensity [4,5]. The other reason is energy market distortions resulted from government intervention. During the economic transformation, lag of energy marketization causes energy price distortion, which further restrains the improvement of energy efficiency in China [6,7]. Therefore, some measures, such as energy marketization, reduction of government intervention and optimization of energy allocation by market, contribute to the realization of energy saving and emission reduction.

In this paper, a macro-energy efficiency index in China (H) is established for quantized analyses of the impact of factors, such as energy structure and allocation on energy efficiency. The index is then decomposed into intra-provincial technical efficiency (ATE), allocation efficiency of energy input structure efficiency (AAE) and inter-provincial energy allocation efficiency (RE). The first index is the efficiency loss caused by technical inefficiency, while the other two are allocation inefficiency caused by market distortions. The result of the research would offer theoretical support for the assessment of energy structure adjustment and promotion of the market reform of energy prices.

The rest of the paper is organized as follows: Section 2 reviews the total factor energy efficiency and shadow price. Section 3 introduces energy-input based directional distance function (DDF) and estimation of energy shadow price and constructs macro-energy efficiency indices as well as corresponding deposition model. Section 4 is about data processing and analysis. Section 5 is an empirical analysis of a case. The last section is the conclusion.

2. Literature Review

To evaluate the energy efficiency in China accurately, we are required to select the indices of energy efficiency reasonably, which can mainly be divided into single factor energy efficiency and total factor energy efficiency. With the feature of easy comparison and simple calculation, single factor energy efficiency is introduced mainly for analyzing the relationship of single energy input and output. Early research mostly applied this method to evaluate energy efficiency. For example, energy consumption of the Chinese industrial sector in the 1990s was decomposed by an improved Laspeyres Index [8]. The research found that technical effect is the leading factor of the changes in industrial energy consumption. This method is criticized by many researchers because in the single factor energy efficiency, the contribution of labor and capital to output is neglected, and then the substitution effect among different production factors is not taken into account.

Therefore, Hu and Wang [9] proposed the concept of total factor energy efficiency originally, which brought production factors such as labor and capital into efficiency analysis. By considering the substitution effect between energy and other production factors, real production process could be well simulated. Thus, this method became the main evaluation method for energy efficiency. By window analysis of DEA method, total factor energy efficiency of 23 developing countries from 1980 to 2005 were tested before analyzing the relation between total factor energy efficiency and incomes based on Tobit model [4]. Meanwhile, a parametric meta-frontier approach based on the Shephard energy distance was introduced, dividing 30 provinces in China into three technical frontiers for analyzing efficiency differences due to technology gap among areas [10].

Some researches [11–13] discussed multiple inputs and outputs based on the model of multiple input and single output. That means they combined undesirable outputs such as pollutant during production process into output system, which better simulated the byproducts in actual production process such as carbon dioxide, sulfur dioxide and other industrial wastes. For example, Fujii *et al.* [11] employed a DDF that can handle multiple inputs and outputs to compute the change of TFP. Then it was found that environmentally sensitive productivity (ESPs) of China's iron and steel industry had continuously improved, even in the period when the conventional economic productivity (CEP) declined in the 1990s. Moreover, Li *et al.* [12] brought byproducts (carbon dioxide and sulfur dioxide) into the production process, establishing an environmental total factor energy efficiency index. According to SBM model, energy efficiency is commonly lower than that regardless of bad output, which means energy efficiency in China is probable to be over-estimated. Moreover, Li *et al.* [13] introduced a meta-frontier framework with the improved DDF to measure the meta-frontier energy efficiency with carbon dioxide emissions. The research indicated that the main reason for low energy efficiency in China is managerial failure in the east and west areas, and technological differences in middle part.

Recently, more and more scholars [14–17] pay attention to the impact of the energy input structure and shadow price of total factor energy efficiency, resulting in incapability of estimating energy allocation efficiency, which was first analyzed by Ouyang and Sun [14] as well as Sheng [15,16]. As we know, underestimation of energy price leads to excessive energy consumption during production process. For example, Ouyang *et al.* [14] pointed out that factor prices of capital, labor and energy are distorted in China due to government regulations. Moreover, energy prices is relatively low compared with capital price, while is relatively high compared to labor price. Meanwhile, energy type would affect the energy allocation significantly [16,17]. Kumar *et al.* [16] estimated the Morishima elasticity of substitution between different fossil fuels and renewable resources, and found that the types of energy would make different contributions to total factor energy efficiency. All these factors would have further influence on total factor energy efficiency.

Energy shadow price reflects the marginal cost of production or marginal output, which is the price when production process gets best allocation. The estimation of shadow price can offer theoretical support for policy decision. In recent years, shadow price is widely used in the area of estimation of pollutant price [18–21]. For example, Kaneko *et al.* [18] used a non-parametric DDF approach to calculate the shadow prices of sulfur dioxide, and then estimated the regional pollution abatement cost under the financial allocation strategy in China. Moreover, Ishinabe *et al.* [19] evaluated the shadow price of greenhouse gas (GHG) emissions for 1024 international companies. Meanwhile, Molinos-Senante *et al.* [19] analyzed the carbon dioxide emission efficiency and shadow prices of sewage plant based on a parametric quadratic DDF and got the conclusion that shadow price mechanism of carbon dioxide contributes to emission reduction better than command-and-control regulation.

Among these studies, the methods to estimate the shadow price differ from one another, depending on the respective needs. For example, Sheng *et al.* [15] estimated energy shadow prices by non-parametric approach. With DEA to evaluate energy DDF, the result showed that both national and regional energy shadow prices are higher than market price. Further research (Sheng *et al* [16]) found that the situation of total factor energy inefficiency in China has improved during the 15 years. The similar method has been adopted by Kaneko *et al.* [18] and Li *et al.* [13]. Although the non-parametric DEA has the advantage of envelope characteristics, it shows the limitation of excluding statistical noise. Based on the stochastic frontier analysis (SFA), Ouyang *et al.* [14] analyzed the factor allocative efficiency of China's industrial sector, and estimated the energy savings potential from the perspective of allocative inefficiency. Comparing with DEA, SFA has the limitation of envelopment, but it is convenient to do parametric test. Another method to estimate shadow price is parametric linear programming (PLP). It combines the envelope feature of DEA with parameters nature of SFA, which is widely used to estimate DDF [17,19]. For instance, Kumar *et al.* [17] applied quadratic function to approximate the distance function, and estimated the Morishima elasticity of substitution between different fossil fuels and renewable resources.

In this paper, we are extending the work of Ouyang and Sun [14], Sheng *et al.* [16] and Kumar *et al.* [17] to include energy structure. Firstly, an energy-input DDF is established to evaluate total factor energy efficiency. Next ,we break down the shadow price into four sources of energy. Similar to Kumar *et al.* [17], by adopting PLP, we set up a macro-energy index in China and divide energy inefficiency into three parts, energy technical inefficiency in each area, energy efficiency loss in each area due to irrational input structure and allocation inefficiency due to energy misallocation nationwide.

3. Methodology

3.1. Energy Efficiency and Directional Distance Function

Firstly, we assume that in each province there are J types of energy input, $e = (e_1, e_2, \cdots, e_J) \in R_J^+$, N types of other input, $x = (x_1, x_2, \cdots x_N) \in R_N^+$, and M kinds of output, $y = (y_1, y_2, \cdots y_M) \in R_M^+$. Therefore, energy technology is represented by production possibility set T.

$$T = \{(x,e,y) : (x,e)\ can\ \ produce\ \ y\} \tag{1}$$

where T, as a set containing all possible input–output vectors, is constructed for describing technical efficiency of production. According to production theory, T is assumed as a closed finite set. Besides, both inputs and outputs should have the property of strong disposability. That means, if $(x', e') \geqslant (x, e)$ and $y' \leqslant y$, then $(x', e', y') \in T$.

Aiming at remanding the neglect of definition and measurement of energy efficiency, Zhou *et al.* [22] defined the energy-input based Shephard Distance Function.

$$D_I(x,e,y) = sub\,\{\alpha | (x, e/\alpha, y) \in T\} \tag{2}$$

According to production theory, the energy-input based Shephard Distance Function has two features. One is $D_I(x,e,y) \leqslant 1$. The other is that $D_I(x,e,y)$ is a homogeneous linear function of energy input. Meanwhile, energy-input based directional distance function reflects the maximal cuttable ratio of energy input when technique and other factor inputs (*i.e.*, labor, capital, *etc.*) remain [23]. Therefore, $e/D_I(x,e,y)$, as the energy input when the maximal energy efficiency occurs in the area, is the optimal energy input theoretically.

Assuming that g_e is direction vector and $g_e \neq 0$, the energy-input based directional distance function is

$$\overrightarrow{D}_I(x,e,y;g_e) = \max\{\beta : (x, e - \beta g_e, y) \in T\} \quad (3)$$

Corresponding to the homogeneity of the energy-input based Shephard Distance Function, the feature of the energy-input based directional distance function is

$$\overrightarrow{D}_I(x, e - \beta g_e, y; g_e) = \overrightarrow{D}_I(x,e,y) - \beta \quad (4)$$

Let $g_e = e$, the relation between directional distance function and the energy-input based Shephard Distance Function could be calculated.

$$\begin{aligned}\overrightarrow{D}_I(x,e,y;e) &= \sup\{\beta : D_I(x, e - \beta e, y) \leqslant 1\} \\ &= \sup\{\beta : (1-\beta) D_I(x,e,y) \leqslant 1\} \\ &= \sup\{\beta : \beta \leqslant 1 - \frac{1}{D_I(x,e,y)}\} \\ &= 1 - \frac{1}{D_I(x,e,y)}\end{aligned} \quad (5)$$

Directional input distance function should satisfy the following constraints as well.

$$\text{If } e' \geqslant e, \text{then } \overrightarrow{D}_I(x,e',y) \geqslant \overrightarrow{D}_I(x,e,y) \quad (6)$$

$$\text{If } \overrightarrow{D}_I(x,e,y;g_e) \leqslant 1, \text{and } \alpha \geqslant 1, \text{then } \overrightarrow{D}_I(x,\alpha e,y) \leqslant 1 \quad (7)$$

3.2. Energy Shadow price Evaluation

Shadow price of energy reflects both the scarcity and marginal use value of energy. Moreover, the scarcity of energy is pivotal for working out reasonable energy price, regulating the supply–demand relation of energy market and promoting enterprises to raise the energy using efficiency. As a result, decomposition of energy efficiency is analyzed by shadow price in this paper.

Dual relation between maximized profit function and energy-input based directional distance function should be considered for calculating the shadow price of energy input. We assume that price of J types of energy inputs are $p = (p_1, p_2, \cdots, p_J) \in R_J^+$, price of M types of other input factors are $q = (q_1, q_2, \cdots, q_M) \in R_M^+$, price of N types of outputs are $w = (w_1, w_2, \cdots, w_N) \in R_N^+$. Then the profit function is

$$\pi(q,p,w) = \max_{x,e,y}\{wy - qx - pe : e \in T\} \quad (8)$$

The profit function means that producer would earn the maximal profit when the output is given and all the factors except energy remain. Because $\overrightarrow{D}_I(x,e,y;g_e) \geqslant 1$ be equivalent to $e \in T$, profit function could be converted as

$$\pi(q,p,w) = \max_{x,e,y}\{wy - qx - pe : \vec{D}_I(x,e,y;g_e) \geqslant 1\} \quad (9)$$

It is feasible to reduce energy input along the direction g. Thus profit, function could be converted as

$$\pi(q,p,w) \geqslant (wy - qx - pe) + p \times \vec{D}_I(x,e,y;g_e) \times g_e \quad (10)$$

The surplus receipt is the value of conserved energy. After transposition,

$$\vec{D}_I(x,e,y;g_e) \leqslant \frac{\pi(q,p,w) - (wy - qx - pe)}{pg_e} \quad (11)$$

Therefore, the defined directional distance function is

$$\vec{D}_I(x,e,y;g_e) = \min_p \left\{ \frac{\pi(q,p,w) - (wy - qx - pe)}{pg_e} \right\} \quad (12)$$

Envelope theorem is applied in Equation (12). Then the shadow price model is

$$\frac{\partial \vec{D}_I(x,e,y;g_e)}{\partial e} = \frac{p}{pg_e} \geqslant 0 \quad (13)$$

$$\frac{\partial \vec{D}_I(x,e,y;g_e)}{\partial y} = \frac{-w}{pg_e} \leqslant 0 \quad (14)$$

Therefore, if the price of the jth output is w_i, the shadow price of the ith type of energy could be calculated.

$$\frac{p_i}{w_j} = \frac{\partial \vec{D}_I(x,e,y;g_e)/\partial(e_i)}{\partial \vec{D}_I(x,e,y;g_e)/\partial(y_j)} \quad (15)$$

3.3. Parametric Forms of Directional Input Distance Function

Translog function is commonly introduced for parameterizing Shephard Distance Function but not directional distance function because the form of translog function cannot be limited to satisfy the transfer properties. Moreover, quadratic function is the second-order approximate of an unknown distance function, which could perfectly satisfy the properties of directional distance function [24]. Therefore, energy efficiency is estimated by quadratic directional distance function.

$$\overrightarrow{D}_I^t(x_i^t, e_i^t, y_i^t; g_e) = \alpha + \sum_{n=1}^{N} \beta_n x_{nk}^t + \sum_{m=1}^{M} \alpha_m y_{nk}^t + \sum_{j=1}^{J} \gamma_j e_{jk}^t$$
$$+ \frac{1}{2}\sum_{n=1}^{N}\sum_{n=1}^{N} \beta_{nn'} x_{nk}^t x_{nk}^t + \frac{1}{2}\sum_{m=1}^{M}\sum_{m=1}^{M} \alpha_{mm'} y_{mk}^t y_{mk}^t + \frac{1}{2}\sum_{j=1}^{J}\sum_{j=1}^{J} \gamma_{jj'} e_{jk}^t e_{jk}^t \tag{16}$$
$$+ \sum_{n=1}^{N}\sum_{m=1}^{M} \delta_{nm} x_{nk}^t y_{mk}^t + \sum_{n=1}^{N}\sum_{j=1}^{J} \eta_{nj} x_{nk}^t e_{jk}^t + \sum_{m=1}^{M}\sum_{j=1}^{J} \mu_{mj} y_{mk}^t e_{jk}^t$$

where $\alpha = \alpha_0 + \sum_{k=1}^{K} D_{Sk} S_k + \sum_{t=1}^{T} Dt_t Time_t$.

Parameter linear program was used for the estimation of quadratic directional distance function in this paper according to Yuan [24] and Wang *et al.* [25].

$$\max \sum_{t=1}^{T}\sum_{k=1}^{K} \left[\overrightarrow{D}_I^t(x_k^t, e_k^t, y_k^t, g_e) - 0 \right] \tag{17}$$

S. t.

(1) Energy-input based directional distance function

$$\overrightarrow{D}_I^t(x_k^t, e_k^t, y_k^t; g_e) \leqslant 1 \tag{18}$$

(2) Monotonicity in inputs

$$\frac{\partial \overrightarrow{D}_I^t(x_k^t, e_k^t, y_k^t; g_e)}{\partial x_i} \geqslant 0 \tag{19}$$

$$\frac{\partial \overrightarrow{D}_I^t(x_k^t, e_k^t, y_k^t; g_e)}{\partial e_i} \geqslant 0 \tag{20}$$

(3) Monotonicity in outputs

$$\frac{\partial \overrightarrow{D}_I^t(x_k^t, e_k^t, y_k^t; g_e)}{\partial y_i} \leqslant 0 \tag{21}$$

(4) Homogeneous linear restriction

$$\begin{aligned}
&\sum_{j=1}^{J} \gamma_j = 1; \\
&\sum_{m=1}^{M} \alpha_{mm'} = \sum_{j=1}^{J} \mu_{mj}, m = 1, \cdots M \\
&\sum_{j'=1}^{J} \gamma_{jj'} = \sum_{m=1}^{M} \mu_{mj}, j = 1, \cdots J \\
&\sum_{m=1}^{M} \delta_{nm'} = \sum_{j=1}^{J} \eta_{nj}, n = 1, \cdots N
\end{aligned} \tag{22}$$

(5) Symmetry of quadratic form

$$\begin{aligned}
&\beta_{nn'} = \beta_{n'n}, n \neq n'; \\
&\alpha_{mm'} = \alpha_{m'm}, m \neq m'; \\
&\gamma_{jj'} = \gamma_{j'j}, j \neq j';
\end{aligned} \tag{23}$$

Therefore, shadow price of the *i*-th type of energy is

$$p_i = \frac{\partial \overrightarrow{D}_I^t(x_k^t, e_k^t, y_k^t; g_e)/\partial(e_i)}{\partial \overrightarrow{D}_I^t(x_k^t, e_k^t, y_k^t; g_e)/\partial(y)} w \tag{24}$$

Energy efficiency of the corresponded area is

$$F(x, e, y) = \frac{1}{D_I(x, e, y)} = 1 - \beta \tag{25}$$

3.4. Establishment and Decomposition of Overall Energy Efficiency

Referring to the research of Li and Ng [26], a national index was constructed and called national energy efficiency. To build the index, the first step is to construct a virtual area, where the input and output of the area equals to national average. That is to say, $x^* = \sum_{i=1}^{I} x_i/I = x^0/I$ and $e^* = \sum_{i=1}^{I} e_i/I = e^0/I$ are national input, while $y^* = \sum_{i=1}^{I} y_i/I = y^0/I$ is national output. When T is convex, national energy efficiency equals to the energy efficiency of the virtual area. Therefore, the national energy efficiency is: $H(x^0, e^0, y^0) = 1 - \beta^*$.

According to Li and Ng [26], the first factor equals to the sum of the minimal energy shadow input of all areas, $R^{TE} = \sum_{i=1}^{I} F(x_i, e_i, y_i) \times (p^* \times e_i)$, divided by the shadow price of total real energy input, $R^0 = p^* \times e^0$. The factor equals to the weighted average of technical efficiency index of each area as well.

$$ATE = \frac{R^{TE}}{R^0} = \frac{\sum_{i=1}^{I} F(x_i, e_i, y_i) \times p^* \times e_i}{p^* \times e^0} = \sum_{i=1}^{I} \tau_i F(x_i, e_i, y_i) \tag{26}$$

where p^* is the shadow prices of minimal energy shadow input in China.

The second factor equals to the minimal real energy shadow input of each area, $R^{AE} = \sum_{i=1}^{I} F(x_i, e_i, y_i) \times p_i \times e_i$, divided by the sum of the minimal energy shadow input of all areas. Therefore, misallocation of available energy due to the price, which is called allocation efficiency of energy input structure, would lead to efficiency loss of energy structure.

$$AAE = \frac{R^{AE}}{R^{TE}} = \frac{\sum_{i=1}^{I} tr_i}{R^{TE}} \tag{27}$$

where $tr_i = \min\{p_i \times e_i : (x_i, e_i, y_i) \in T\}$.

The third factor is the efficiency loss due to energy misallocation among areas. For example, when area A is under the condition of increasing returns to scale, while area B is under the condition of decreasing returns to scale, energy flows from area A to B would lead to the raise of total output. By eliminating the effect of the first two factors from total energy efficiency, the impact of misallocation is

$$RE = \frac{H(x^*, e^*, y^*) \times (p^* \times e^0)}{R^{AE}} \tag{28}$$

where RE is inter-provincial energy allocation efficiency index. Based on the definitions above, national energy efficiency could be decomposed as

$$H(x^0, e^0, y^0) = ATE \times AAE \times RE \tag{29}$$

4. Data

After dropping the data of Hong Kong, Macau, Taiwan and Tibet, data of 30 areas from 1998 to 2012 are selected to make an input–output panel for empirical study. All data come from China Statistical Yearbook [27] and China Energy Statistical Yearbook [28].

The variables include the inputs (labor, capital stock and energy consumption), and an output (gross regional product). In previous studies, labor force was widely used as the index of labor input. However, human capital is a better index on measuring the contribution of labor during production [29]. Thus, the index in this paper is the number of the employees weighted by average education attainment of different areas at the end of the year. As for the capital stock, we introduce the permanent inventory method to estimate the capital input. According to the research [30], investment of fixed capital is selected as the investment index of the year. By constructing the price index of investment in the fixed assets from 1952 to 2010, the real investment prices of areas are calculated. The model for capital stock accounting is

$$K_t = I_t + (1 - \delta)K_{t-1} \tag{30}$$

where K_t is the stock of capital in period t, δ is depreciation rate, and I_t is the amount of investment in period t.

Data of merely energy consumption of each area come from China Energy Statistical Yearbook [28]. After categorizing the data into four main types, including coal, oil, gas and electricity (*i.e.*, hydropower and nuclear power) based on energy balance sheet of each area, the measurement unit is converted to 1000 Tce by heat quantity. Specifically, data of Hainan in 2002 and data of Ningxia from 2000 to 2002 are estimated based on other data of the area. Gross regional product of each area is selected as the output in this paper. To eliminate the effect of price factor, we use gross regional product deflator. All nominal variables are deflated to real variables by using a price index for the year 2000.

In Figure 1 we will show the total energy consumption in China from 1998 to 2012, while sort out the descriptive analysis of total input and output in Table 1.

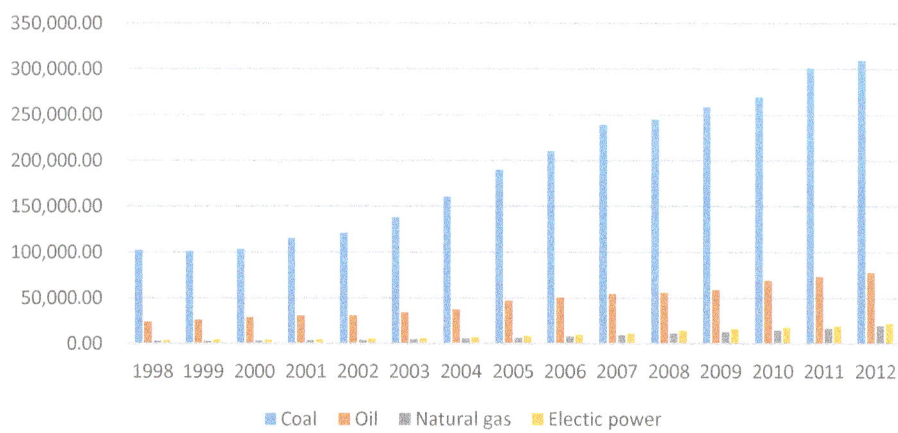

Figure 1. Total energy consumption in China.

Table 1. Descriptive analysis of total input and output.

	Count	Mean	SD	Min	Max
Coal (Tce)	450	6365.87	5337.52	121.81	27,992.19
Oil (Tce)	450	1559.95	1559.51	82.33	8532.70
Natural gas (Tce)	450	277.54	348.70	0.00	2330.96
Electic power (Tce)	450	341.49	437.86	0.00	2963.72
Labor (10,000 persons)	450	2395.31	1610.40	254.80	6554.30
Capital (100 million yuan)	450	18,670.77	18,031.69	953.21	110,485.88
GNP (100 million yuan)	450	6645.87	6667.03	223.88	42,876.30

Note: (1) Electric power only consists of hydropower and nuclear power, excluding the electricity generated from coal and gas. (2) Some regions did have little natural gas or electric power consumption in 1990s. For example, Neimenggu had little hydropower and nuclear power from 1998 to 2000, and Zhejiang had no natural gas consumption in 1998.

5. Results and Discussions

5.1. Estimated Results ANALYSIS

In Table 2, we standardize the process of input and output data of 30 areas in China from 1998 to 2012, then estimate the parameters of the directional distance function by linear programming model.

Table 2. Parameter estimation of quadratic directional distance function.

Parameter	Estimated Value	Parameter	Estimated Value	Parameter	Estimated Value
α	−0.1485	γ_{11}	−0.1281	η_{13}	−0.0087
α_1	−1.4090	γ_{12}	0.0686	η_{14}	−0.1128
α_{11}	−0.1246	γ_{13}	0.0443	η_{21}	0.0192
β_1	0.1523	γ_{14}	0.0151	η_{22}	−0.0456
β_2	0.3132	γ_{22}	−0.0386	η_{23}	0.0241
β_{11}	0.3786	γ_{23}	−0.0292	η_{24}	0.0023
β_{12}	−0.1259	γ_{24}	−0.0008	δ_{11}	−0.0022
β_{22}	−0.0388	γ_{33}	−0.0186	δ_{21}	0.1035
γ_1	0.5776	γ_{34}	0.0035	μ_{11}	−0.0658
γ_2	−0.0063	γ_{44}	−0.0178	μ_{21}	−0.0347
γ_3	0.1422	η_{11}	0.0531	μ_{31}	0.0266
γ_4	0.2864	η_{12}	0.0685	μ_{41}	0.0739

5.2. Results Analysis of Shadow Prices

Shadow prices of factors reflect the marginal output of energy; in other words, the scarcity of energy. When energy shadow price is higher than market price, firms would choose more energy input for higher profit during production. Therefore, shadow price is the key to work out reasonable energy price, regulate the supply–demand relation of energy market and promote enterprises to raise the energy using efficiency.

On the perspective of national condition, shadow price of coal, gas and electricity kept stable from 1998 to 2012. The shadow prices of the three types of energy decreased slightly annually before 2007, while had a modest rise since then. As shown in Figure 2, the shadow price of coal fluctuated dramatically, though the price maintained at a level of 0.400. Meanwhile, shadow prices of gas and electricity fluctuated around 0.150 and 0.100, respectively.

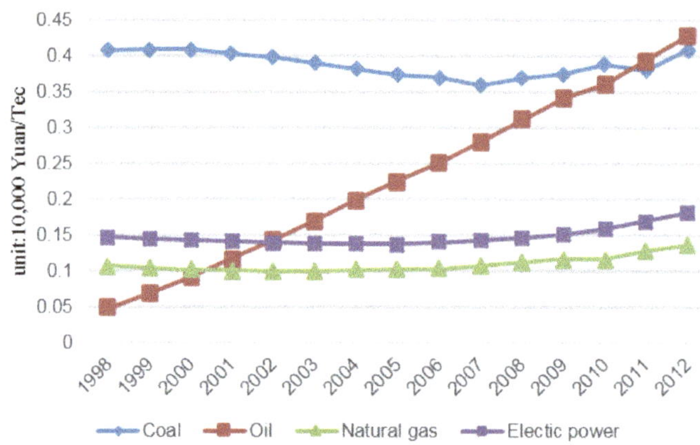

Figure 2. Variation trend of shadow price of energy in China.

Remarkably, the shadow price of oil increase sharply, from 0.047 to 0.428 during the 15 years. It means that the marginal output soared rapidly. Over the same period, the price of crude oils soared from $20 per barrel in 1998 to $90 per barrel. Aiming at the increase on price of oil products in China, we refer to the explanation of Zheng *et al.* [31] and conclude the reasons as followed. For one thing, the cost of oil processing would raise due to higher emission standard (Chinese emission standard at Phase II to Phase VI) and higher requirement for oil quality. Moreover, improving the oil quality would increase the marginal output and also the shadow price of oil. Furthermore, as the oil price kept climbing, not only have small vehicles with low emissions become popular with consumers, but they have also been supported by national policy like tax allowance (half purchase tax for household car lower than 1.6 L). On such occasion, unit oil output would grow, and accordingly the shadow price of oil would increase.

As is shown in Table 3, shadow price of the four types of energy differs from each other. On the perspective of coal, areas with high shadow prices of coal (over 5000 Yuan/Tce) are Sichuan (0.822), Beijing (0.564), Xinjiang (0.516), Chongqing (0.509) and Guangdong (0.504). For example, shadow price of coal in Sichuan rose from 5330 Yuan/Tce to 18,020 Yuan/Tce, similar to Hunan, Beijing and Guangdong.

That means the marginal output of coal increases significantly. In contrast, areas with low shadow price of coal (less than 3000 Yuan/Tce) are Hebei (0.184), Shanxi (0.192), Neimenggu (0.229), Shandong (0.261) and Jiangsu (0.293). Out of these areas, shadow price of Hebei is the lowest, whose value reached 560 Yuan/Tce, far lower than national average (3880 Yuan/Tce). Besides, shadow prices of coal in Heilongjiang, Hebei and Neimenggu had the greatest reduction, which means the marginal output of coal declined sharply in recent years.

As for oil, the areas with high shadow price of oil (over 3000 Yuan/Tce) include Hunan (0.351), Henan (0.309), Guangxi (0.309), Yunnan (0.307) and Hebei (0.305). Namely, shadow price of Hunan rose from 880 Yuan/Tce to 8450 Yuan/Tce, similar to that of Hubei, Henan, Jiangsu, Shanxi, Neimenggu, Sichuan and Chongqing, which means that the marginal output increased remarkably. On the contrary, areas with low oil shadow prices (less than 1500 Yuan/Tce) are Guangdong (0.054), Shanghai (0.097), Beijing (0.142) and Jiangsu (0.150). Specifically, the shadow price of Guangdong, at only 540 Yuan/Tce, is the lowest, far lower than the national average (2280 Yuan/Tce). Areas with growth rates that are far from the national average include Guangdong, Shandong, Jiangsu, Zhejiang, Liaoning.

On perspective of gas, areas with high shadow price of gas (over 1500 Yuan/Tce) are Hebei (0.183), Jiangsu (0.168), Henan (0.156), Shandong (0.153) and Hunan (0.150). Specifically, shadow price in Heibei is always higher than national average, increasing significantly in recent years, which is similar to Shanxi, Neimenggu, Zhejiang, Shandong and Henan. On the contrary, areas with low gas shadow prices (less than 800 Yuan/Tce) include Xinjiang (0.045), Hainan (0.056), Sichuan (0.066), Qinghai (0.067), Beijing (0.079) and Heilongjiang (0.078). For example, shadow price of Xinjiang decreased from a low value recently.

When it comes to electricity (*i.e.*, nuclear power and hydropower), areas with high shadow price of electricity (over 2000 Yuan/Tce) include Shanghai (0.227) and Beijing (0.218). Besides, other areas such as Tianjin (0.198), Neimenggu (0.186), Hainan (0.180) and Xinjiang (0.191) have high electricity output. In contrast, areas with low shadow price of electricity (less than 1.000 Yuan/Tce) include Sichuan (0.060), Henan (0.061), Hunan (0.064), Anhui (0.089), Hubei (0.094) and Guangxi (0.092).

Based on previous studies, we categorize energy into four types, finding that shadow prices of the same type of energy differ in different areas, which also shows marginal output of energy in different areas varies. For example, when shadow price of coal reached 18,020 Yuan/Tce in Sichuan, shadow price of coal in Hebei is only 550 Yuan/Tce. In a perfectly competitive market, price of the same type of energy should be the same. This situation indicates the existence of trade barrier, which means energy factor is not complete circulation in different regions.

Table 3. Shadow price of each area (Unit: 10,000 Yuan/Tce).

	Coal			Oil			Natural Gas			Electric Power		
	1998–2002	2003–2007	2008–2012	1998–2002	2003–2007	2008–2012	1998–2002	2003–2007	2008–2012	1998–2002	2003–2007	2008–2012
Beijing	0.447	0.514	0.733	0.041	0.140	0.245	0.104	0.084	0.050	0.198	0.206	0.251
Tianjin	0.405	0.384	0.398	0.057	0.191	0.324	0.092	0.085	0.083	0.192	0.191	0.211
Hebei	0.301	0.171	0.081	0.140	0.303	0.473	0.128	0.171	0.252	0.123	0.125	0.155
Shanxi	0.200	0.153	0.139	0.060	0.302	0.441	0.069	0.144	0.140	0.098	0.160	0.167
Neimenggu	0.342	0.249	0.096	0.107	0.250	0.402	0.110	0.116	0.157	0.171	0.176	0.212
Liaoning	0.393	0.310	0.323	0.073	0.203	0.250	0.102	0.115	0.118	0.159	0.164	0.204
Jilin	0.380	0.347	0.318	0.088	0.225	0.388	0.100	0.092	0.115	0.169	0.161	0.183
Heilongjiang	0.424	0.383	0.344	0.055	0.184	0.324	0.082	0.074	0.078	0.165	0.160	0.170
Shanghai	0.381	0.436	0.509	0.034	0.098	0.160	0.105	0.082	0.071	0.202	0.215	0.265
Jiangsu	0.350	0.269	0.261	0.103	0.177	0.171	0.120	0.147	0.236	0.109	0.145	0.273
Zhejiang	0.402	0.383	0.428	0.078	0.182	0.253	0.112	0.136	0.179	0.141	0.150	0.201
Anhui	0.372	0.323	0.255	0.142	0.286	0.437	0.108	0.107	0.119	0.100	0.083	0.085
Fujian	0.413	0.375	0.394	0.078	0.212	0.338	0.106	0.107	0.124	0.161	0.156	0.180
Jiangxi	0.411	0.378	0.351	0.106	0.256	0.441	0.099	0.096	0.107	0.136	0.120	0.113
Shandong	0.363	0.239	0.180	0.124	0.212	0.228	0.114	0.141	0.205	0.077	0.101	0.172
Henan	0.385	0.286	0.283	0.162	0.304	0.463	0.112	0.134	0.224	0.054	0.054	0.074
Hubei	0.407	0.402	0.412	0.128	0.268	0.431	0.111	0.112	0.145	0.101	0.084	0.097
Hunan	0.426	0.386	0.545	0.134	0.294	0.625	0.109	0.119	0.223	0.090	0.072	0.031
Guangdong	0.443	0.476	0.597	0.032	0.077	0.055	0.088	0.092	0.159	0.123	0.126	0.220
Guangxi	0.452	0.445	0.493	0.120	0.282	0.526	0.105	0.102	0.143	0.113	0.090	0.072
Hainan	0.425	0.452	0.489	0.065	0.201	0.384	0.088	0.060	0.022	0.190	0.181	0.170
Chongqing	0.459	0.499	0.569	0.080	0.222	0.423	0.091	0.080	0.081	0.157	0.155	0.167
Sichuan	0.573	0.664	1.229	0.102	0.226	0.413	0.078	0.062	0.058	0.066	0.057	0.058
Guizhou	0.397	0.325	0.260	0.123	0.286	0.463	0.106	0.103	0.100	0.136	0.121	0.121
Yunnan	0.432	0.413	0.418	0.112	0.283	0.526	0.106	0.109	0.132	0.128	0.105	0.073
Shaanxi	0.424	0.422	0.432	0.090	0.217	0.347	0.096	0.077	0.067	0.147	0.142	0.163
Gansu	0.422	0.419	0.402	0.100	0.253	0.455	0.096	0.084	0.075	0.148	0.136	0.123
Qinghai	0.423	0.457	0.516	0.073	0.220	0.454	0.092	0.068	0.042	0.189	0.179	0.162
Ningxia	0.399	0.383	0.340	0.077	0.234	0.437	0.094	0.080	0.066	0.190	0.177	0.167
Xinjiang	0.465	0.522	0.561	0.037	0.152	0.321	0.076	0.043	0.017	0.189	0.189	0.195
Eastern	0.393	0.364	0.399	0.075	0.181	0.262	0.105	0.111	0.136	0.152	0.160	0.209
Central	0.376	0.341	0.324	0.101	0.253	0.414	0.096	0.102	0.119	0.127	0.125	0.132
Western	0.435	0.436	0.483	0.093	0.239	0.433	0.095	0.084	0.085	0.149	0.139	0.138
Total	0.405	0.375	0.384	0.093	0.225	0.367	0.102	0.102	0.122	0.143	0.139	0.161

Note: The data are divided into three periods. Due to the space limitation, we did not give all data in the body part, but shadow prices in each year can be found in the Appendix.

Besides, in areas with rich energy resources, energy marginal output is low in most areas. For example, in Shanxi and Neimenggu, the major coal producing provinces, coal shadow prices are lower than national average. This result is similar to the conclusion that the richer energy resource is, the lower energy price is [32], due to the lack of an integrated energy market. This factor has seriously hindered the scale economy of regional industries and has resulted in the loss of total factor energy efficiency.

Meanwhile, in Guangdong, Shanghai and Beijing, areas with large amount of oil importation, processing and consumption, marginal output of oil is lower than national average. Obviously, there are not rich resources in these areas. However, since most state-owned large petrochemical project located in these areas due to policy support, oil efficiency is too low to save resources. Therefore, compared with the lack of energy resources, misallocation of factor market resulting from administrative interference leads to the inefficiency of energy more seriously [33].

5.3. Results Analysis of Energy Efficiency

According to the average data value of the past years, areas with high energy efficiency are Qinghai (0.972), Ningxia (0.970), Guangxi (0.942), Gansu (0.952), Sichuan (0.952) and Hainan (0.951). Energy efficiencies of all these areas are higher than 0.950. On the contrary, areas with low energy efficiency are Liaoning (0.724), Yunnan (0.783), Jiangsu (0.806), Neimenggu (0.814) and Guizhou (0.845). Energy efficiencies of all these areas are lower than 0.850. All the Energy efficiency of each province and national energy efficiency decomposition is disposed in Table 4 and Figure 3.

The sorted result is consistent with results from other research [12,34]. Besides, total factor energy efficiency of Jiangsu, Guizhou, Shandong and Beijing shows an increasing tendency in fluctuation, reaching an advanced level of the country, especially Jiangsu, whose value increased from 0.413 to 1.000 during the 15 years. In contrast, there is a significant decline of energy efficiency in Liaoning, Yunnan, Xinjiang and Zhejiang. For example, total factor energy efficiency of Xinjiang is far lower than the national average due to the decreased fluctuation from 1998.

Table 4. Energy efficiency of each province and national energy efficiency decomposition.

	1998	1999	2000	2001	2002	2003	2004	2005	2006	2007	2008	2009	2010	2011	2012
Beijing	0.797	0.851	0.848	0.925	0.929	0.950	0.975	0.949	0.966	0.976	0.988	0.990	1.000	1.000	0.994
Tianjin	0.846	0.870	0.875	0.890	0.943	0.985	0.983	0.989	0.976	0.972	0.985	1.000	0.947	0.978	0.972
Hebei	0.863	0.932	0.947	0.991	1.000	1.000	0.993	0.771	0.898	0.928	0.991	1.000	0.853	0.810	0.821
Shanxi	0.899	0.975	1.000	0.698	0.916	0.921	0.933	0.970	0.968	0.920	0.963	0.895	0.909	0.948	0.926
Neimenggu	0.894	0.924	0.946	0.969	0.984	1.000	0.886	0.841	0.770	0.706	0.632	0.579	0.640	0.687	0.756
Liaoning	0.658	0.629	0.608	0.694	0.845	0.889	1.000	0.758	0.788	0.802	0.857	0.847	0.644	0.512	0.335
Jilin	0.884	0.921	0.953	0.974	1.000	0.977	0.991	0.911	0.907	0.818	0.926	0.975	0.999	1.000	0.998
Heilongjiang	0.789	0.834	0.895	0.942	1.000	0.997	0.989	0.955	0.884	0.918	0.964	0.869	0.861	0.871	0.962
Shanghai	0.833	0.855	0.883	0.948	0.979	1.000	0.968	0.899	0.859	0.882	0.900	0.891	0.944	1.000	0.942
Jiangsu	0.413	0.520	0.598	0.684	0.761	0.810	0.828	0.887	0.985	1.000	0.919	0.952	1.000	1.000	0.740
Zhejiang	0.861	0.933	0.982	1.000	0.967	0.993	0.964	0.908	0.930	0.969	1.000	0.979	0.816	0.746	0.658
Anhui	0.714	0.747	0.756	0.777	0.800	0.788	0.839	0.845	0.858	0.879	0.891	0.921	0.982	1.000	0.941
Fujian	0.815	0.853	0.883	0.906	0.949	0.986	0.993	0.877	0.910	0.960	0.955	1.000	0.853	0.911	0.870
Jiangxi	0.794	0.826	0.859	0.870	0.892	0.903	0.887	0.864	0.875	0.878	0.908	0.937	0.920	1.000	0.974
Shandong	0.723	0.835	0.965	0.923	1.000	0.977	0.979	0.442	0.605	0.740	0.788	0.896	0.928	1.000	1.000
Henan	1.000	0.968	0.850	0.891	0.911	0.920	0.779	0.752	0.792	0.872	0.919	1.000	0.933	0.944	0.850
Hubei	0.895	0.922	0.936	1.000	1.000	0.967	0.943	0.843	0.801	0.790	0.722	0.750	0.915	0.915	0.876
Hunan	0.798	0.921	0.981	0.984	0.961	1.000	0.910	0.765	0.768	0.794	0.825	0.867	0.927	0.931	1.000
Guangdong	0.843	0.934	0.924	0.977	1.000	0.963	1.000	0.960	1.000	1.000	0.942	0.698	1.000	1.000	0.585
Guangxi	0.922	0.957	0.958	0.975	1.000	0.988	0.930	0.956	0.927	0.929	0.961	0.978	0.975	0.953	0.936
Hainan	0.865	0.926	0.905	0.932	0.920	0.914	0.913	0.967	0.959	1.000	0.982	0.994	0.990	0.995	1.000
Chongqing	0.689	0.671	0.678	0.883	0.914	0.995	1.000	0.848	0.849	0.876	0.791	0.838	0.863	0.909	0.944
Sichuan	0.786	0.884	0.968	1.000	0.987	0.908	0.904	0.989	0.978	0.994	0.966	0.929	1.000	1.000	0.981
Guizhou	0.734	0.799	0.797	0.758	0.783	0.750	0.741	0.847	0.829	0.852	0.909	0.927	0.965	0.992	1.000
Yunnan	0.821	0.867	0.885	0.893	0.887	0.844	1.000	0.689	0.725	0.704	0.693	0.700	0.695	0.680	0.668
Shaanxi	0.929	0.988	1.000	0.968	0.936	0.952	0.838	0.941	0.915	0.872	0.838	0.860	0.797	0.799	0.802
Gansu	0.924	0.936	0.955	0.998	1.000	0.982	0.952	0.940	0.931	0.934	0.935	0.928	0.971	0.948	0.948
Qinghai	0.942	0.943	0.969	1.000	0.992	0.989	0.972	0.980	0.979	0.990	0.983	1.000	0.971	0.942	0.934
Ningxia	0.980	1.000	0.982	0.964	0.946	0.929	0.973	1.000	0.981	0.926	0.996	0.993	0.984	0.950	0.949
Xinjiang	0.964	0.973	1.000	0.914	0.996	0.976	0.927	0.884	0.853	0.839	0.846	0.821	0.711	0.594	0.450
Eastern	0.774	0.831	0.856	0.897	0.936	0.952	0.963	0.855	0.898	0.930	0.937	0.932	0.907	0.905	0.811
Central	0.855	0.890	0.894	0.886	0.930	0.923	0.909	0.888	0.881	0.884	0.909	0.918	0.939	0.959	0.941
Western	0.871	0.904	0.922	0.938	0.948	0.938	0.920	0.901	0.885	0.875	0.868	0.868	0.870	0.859	0.852
A T E	0.802	0.851	0.872	0.814	0.935	0.937	0.928	0.827	0.857	0.878	0.885	0.883	0.886	0.885	0.822
A A E	0.968	0.964	0.959	0.964	0.930	0.918	0.892	0.886	0.857	0.809	0.842	0.842	0.866	0.872	0.879
R E	1.171	1.158	1.160	1.237	1.150	1.156	1.168	1.218	1.199	1.216	1.171	1.164	1.093	1.043	1.054
H	0.909	0.950	0.969	0.972	1.000	0.994	0.968	0.893	0.882	0.864	0.872	0.865	0.838	0.805	0.761

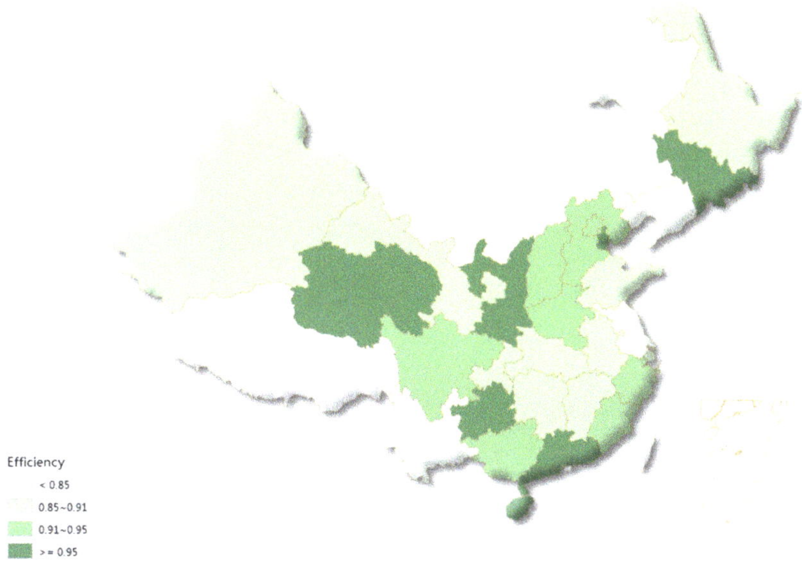

Figure 3. Total factor energy efficiency in China.

Based on the condition of the three regions, efficiency of the east area have fluctuated greatly with the decline in recent years. The efficiency drops from 0.937 in 2008 to 0.811 in 2012, illustrating the rise of extensive development mode. In the middle area, the result of significant decline is similar to the result of Li *et al* [12]. Li believes that the reason for the sharp decline is the drop of management efficiency in the middle area. On the contrary, energy efficiency performs well, even though the support policy is not as good as in the east area and energy endowment is not as good as that in the middle area. According to Lin *et al* [23], the result is due to the contribution of western development policy on energy efficiency.

Figure 4 shows the variation trend of national efficiency and its decomposition. As we see, macro-energy efficiency index (H) presents an invert U [35,36], which steadily rises from 1998 to 2003. After reaching the peak in 2002 and 2003, the energy efficiency keeps declining, even lower than 0.800 (*i.e.*, 2007: 0.761) in recent years. These results imply that extensive mode of energy consumption rises in recent decades and the total energy utilization is not in an optimistic condition.

By decomposing macro-energy efficiency index (H) into intra-provincial technical efficiency (ATE), allocation efficiency of energy input structure (AAE) and inter-provincial energy allocation efficiency (RE), we found that ATE has similar trends to macro-energy efficiency nationwide, which presents an invert U as well. Moreover, the value of ATE decreases sharply recently, which is consistent with the result that technical inefficiency tends to increase [3]. AAE drops from 96.8% to 80.9% in the ten years since 1998 and presents a V. Not until 2008 did the value start to rise again and go back to 0.879. This fluctuation indicates that the improvement of energy structure could raise the energy efficiency effectively. Therefore, it is crucial for the government to adjust the policy of energy structure, which has already achieved some results. As for RE, its value is 1.171 in 1998, which means that energy allocation efficiency could increase by 0.170 if energy could flow freely among areas. Since 1998, energy allocation efficiency among areas increased in fluctuation, so that RE decreased from 1.171 to 1.054 during the 15 years accordingly. These results imply that marketization could break the inter-provincial energy barrier. During the past decades, liberalization in energy markets has developed remarkably.

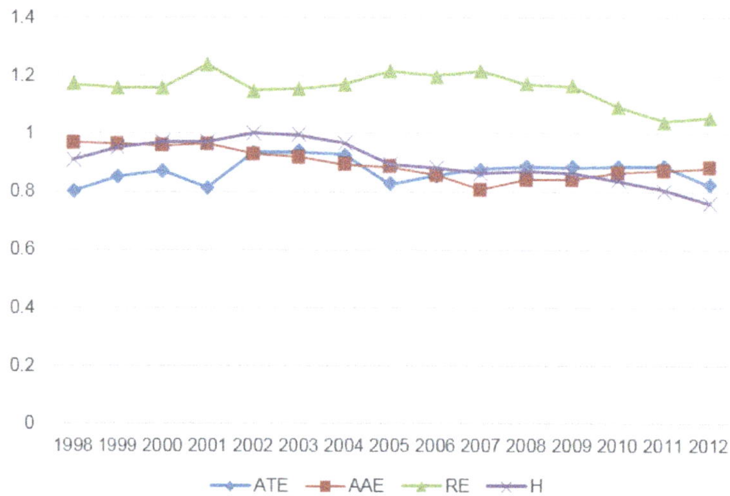

Figure 4. Variation trend of national energy efficiency and its decomposition.

6. Conclusions

In this paper, an energy-input based directional distance function was established for measuring total factor energy efficiency and shadow price of four types of energy (*i.e.*, coal, oil, gas and electricity) among 30 areas in China from 1998 to 2012. Moreover, a macro-energy efficiency index in China was estimated and divided into intra-provincial technical efficiency, allocation efficiency of energy input structure and inter-provincial energy allocation efficiency. The main conclusions are as follows:

- By extending the work of prior research, we break down the shadow price into four sources of energy, and find that the shadow prices of different kinds of energy are quite different, and the same kind of energy also has different shadow prices in different regions. The situation indicates the existence of energy trade barriers among regions and the different contribution of the several types of energy to economy.

- Among the kinds of energy mentioned, the shadow price of oil has climbed dramatically, which means the efficiency of oil consumption has greatly improved. The increasingly rigid vehicle emissions standards and increasing tax allowance on smaller vehicles, may lead to the increase of marginal oil output.
- In areas with rich energy resources, energy marginal output is low in most areas due to the lack of an integrated energy market, resulting in the reduction of total factor energy efficiency. Besides, misallocation of factor market owing to administrative interference remarkably brings about the inefficiency of energy.
- Both the structure of energy input and inter-provincial allocation of energy impact on energy efficiency. Increasing values of the two indices in recent years imply the effect of China's adjusted energy policy. Inter-provincial energy barrier will be broken for the marketization of energy factor and better negotiability of energy among areas. Specifically, it is of great necessity to realize marketing disposition of energy, energy structure adjustment and energy efficiency improvement for achieving reduction goal of carbon emission.

Thus, we call for a reform on energy factor market, in order to reduce the impact of administrative interference and make the market allocation effective. Besides, we suggest a need for greater environment regulation and greater emission standards to improve the total factor energy efficiency.

There are some limitations in this paper. First of all, the energy categories are roughly defined, which can be better classified in future research. Besides, we do not discuss carbon dioxide or other byproducts and, thus, may neglect the influence of bad outputs. Furthermore, the lack of a comparison between market prices and shadow prices needs to be extended in the future.

Acknowledgments: The authors are grateful for the financial support provided by GDUPS(2015), the National Science Foundation of China (71473105), National Social Science Foundation of China (14ZDB44), New Century Excellent Talents in University (NCET-110856), and Guangdong Project of Key Research Institute of Humanities and Social Sciences at Universities-IRESD (2012JDXM0009).

Author Contributions: Peihao Lai contributed to the data acquisition, methodological design and drafting of the article; Minzhe Du provided the statistical analysis and data interpretation; Bing Wang made substantial contributions to the concept and design of the article, helped to revise the manuscript and approved its final publication; and Ziyue Chen assisted in the literature review and conclusions.

Conflicts of Interest: The authors declare no conflict of interest.

Appendix

Appendix A1. Efficiency Decompositon Framework

According to Figure A1, a typical province could only reach point a instead of the production frontier of the province. Similarly, a country, as a whole, could only reach Point B instead of Point D in Figure A2. To reach the national optimal Point D, it is vital to improve the technical efficiency of the province, which means that $ATE = \frac{R^{TE}}{R^0}$ achieves the optimal value and Point A is replaced by Point B. Then, improving provincial allocation efficiency is of great importance as well, which means $AAE = \frac{R^{AE}}{R^{TE}}$ achieves the optimal value and Point B moves to Point C. Finally, interprovincial efficiency is to increase, which means $RE = \frac{H}{R^{AE}}$ reaches the optimal value and Point C moves to Point D.

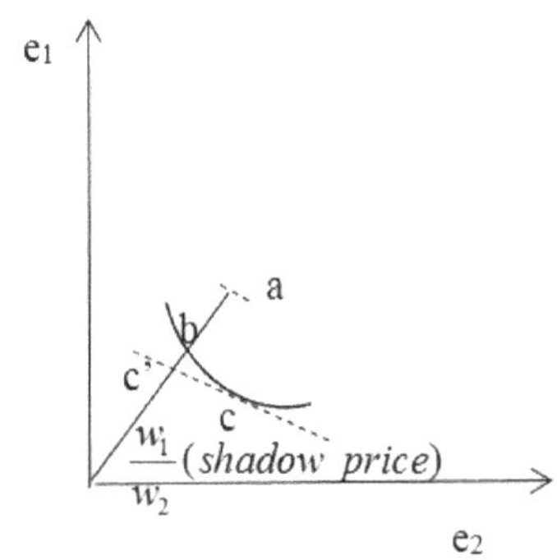

Figure A1. Production frontier of a province.

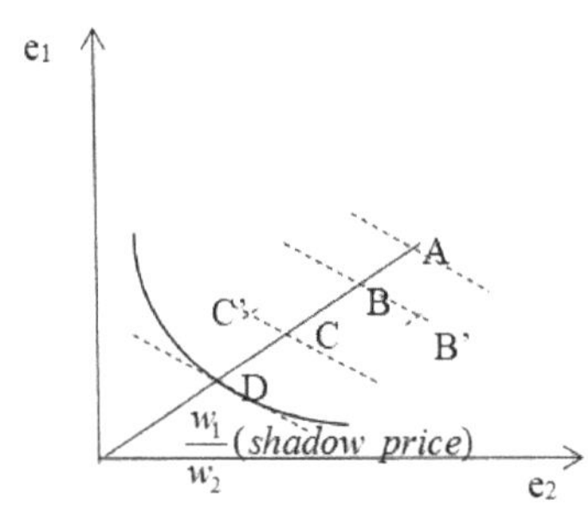

Figure A2. Production frontier of a country.

Appendix A2. Depreciation Rate of Different Areas

We introduced the value of depreciation rate according to Wu's research [37], where the value of depreciation rate differs from areas as is in Table A1.

Table A1. Depreciation rate of different areas.

Province	Beijing	Tianjin	Hebei	Shanxi	Neimenggu	liaoning	Jilin	Heilongjiang	Shanghai	Jiangsu
Depreciation rate (%)	3.4	3.7	4.3	4	4.3	5.8	5.1	6	3.4	4.2
Province	**Zhejiang**	**Anhui**	**Fujian**	**Jiangxi**	**Shandong**	**Henan**	**Hubei**	**Hunan**	**Guangdong**	**Guangxi**
Depreciation rate (%)	4	5	4.5	3.7	5	4.1	4.5	4.5	6.9	3.3
Province	**Hainan**	**Chongqing**	**Sichuan**	**Guizhou**	**Yunnan**	**Shaanxi**	**Gansu**	**Qinghai**	**Ningxia**	**Xinjiang**
Depreciation rate (%)	2.2	4.6	4.6	2.8	2.7	3.3	2.7	2.4	2.8	2.6

Appendix A3. The Shadow Price of Each Energy from 1998 to 2012 are as Follow

Table A2. The shadow price of coal.

	1998	1999	2000	2001	2002	2003	2004	2005	2006	2007	2008	2009	2010	2011	2012
Beijing	0.437	0.442	0.451	0.442	0.463	0.461	0.486	0.511	0.541	0.569	0.628	0.678	0.721	0.763	0.876
Tianjin	0.405	0.406	0.412	0.408	0.396	0.395	0.384	0.378	0.381	0.384	0.386	0.382	0.403	0.400	0.421
Hebei	0.326	0.316	0.307	0.290	0.265	0.233	0.204	0.173	0.135	0.111	0.094	0.080	0.103	0.056	0.070
Shanxi	0.321	0.324	0.313	0.209	0.250	0.226	0.196	0.182	0.161	0.000	0.147	0.149	0.166	0.126	0.108
Neimenggu	0.358	0.351	0.344	0.336	0.322	0.305	0.271	0.239	0.220	0.208	0.158	0.153	0.136	0.034	0.000
Liaoning	0.380	0.418	0.407	0.398	0.361	0.338	0.296	0.330	0.311	0.276	0.269	0.247	0.317	0.351	0.432
Jilin	0.391	0.384	0.382	0.375	0.369	0.363	0.351	0.343	0.332	0.346	0.331	0.329	0.325	0.291	0.312
Heilongjiang	0.426	0.430	0.426	0.424	0.416	0.402	0.379	0.379	0.382	0.374	0.344	0.374	0.362	0.335	0.306
Shanghai	0.383	0.376	0.384	0.379	0.384	0.386	0.410	0.432	0.471	0.481	0.478	0.496	0.505	0.507	0.559
Jiangsu	0.361	0.351	0.352	0.346	0.337	0.330	0.298	0.246	0.231	0.241	0.265	0.271	0.276	0.219	0.277
Zhejiang	0.408	0.404	0.401	0.396	0.403	0.405	0.390	0.390	0.375	0.355	0.363	0.380	0.440	0.465	0.491
Anhui	0.386	0.380	0.373	0.363	0.356	0.343	0.335	0.324	0.314	0.297	0.283	0.264	0.249	0.238	0.239
Fujian	0.422	0.416	0.413	0.414	0.402	0.388	0.381	0.378	0.371	0.354	0.360	0.349	0.432	0.394	0.436
Jiangxi	0.413	0.412	0.409	0.411	0.407	0.392	0.381	0.385	0.373	0.355	0.359	0.349	0.361	0.334	0.352
Shandong	0.381	0.370	0.388	0.354	0.322	0.302	0.270	0.246	0.201	0.179	0.162	0.172	0.181	0.183	0.200
Henan	0.392	0.393	0.388	0.383	0.369	0.354	0.325	0.277	0.252	0.222	0.229	0.237	0.268	0.258	0.423
Hubei	0.411	0.409	0.411	0.403	0.399	0.393	0.388	0.415	0.415	0.400	0.480	0.472	0.373	0.338	0.398
Hunan	0.418	0.430	0.436	0.420	0.426	0.405	0.401	0.376	0.377	0.370	0.431	0.458	0.557	0.561	0.716
Guangdong	0.450	0.443	0.438	0.440	0.445	0.449	0.449	0.469	0.488	0.527	0.563	0.732	0.560	0.514	0.615
Guangxi	0.449	0.448	0.451	0.453	0.461	0.461	0.449	0.432	0.442	0.441	0.465	0.455	0.476	0.516	0.555
Hainan	0.428	0.413	0.424	0.425	0.435	0.446	0.456	0.448	0.460	0.451	0.465	0.459	0.473	0.526	0.523
Chongqing	0.459	0.465	0.480	0.446	0.447	0.458	0.465	0.509	0.526	0.536	0.548	0.545	0.578	0.557	0.615
Sichuan	0.533	0.565	0.572	0.594	0.603	0.594	0.612	0.658	0.711	0.745	0.741	0.798	1.099	1.802	1.707
Guizhou	0.391	0.392	0.394	0.408	0.400	0.371	0.347	0.321	0.299	0.287	0.296	0.274	0.265	0.245	0.221
Yunnan	0.428	0.431	0.437	0.434	0.431	0.423	0.436	0.421	0.392	0.393	0.399	0.378	0.397	0.432	0.484
Shaanxi	0.408	0.414	0.427	0.435	0.439	0.430	0.453	0.390	0.405	0.429	0.448	0.426	0.435	0.431	0.421
Gansu	0.426	0.425	0.425	0.417	0.414	0.416	0.416	0.421	0.424	0.420	0.415	0.434	0.395	0.377	0.391
Qinghai	0.419	0.421	0.423	0.420	0.432	0.441	0.453	0.464	0.465	0.461	0.466	0.474	0.502	0.551	0.590
Ningxia	0.402	0.402	0.400	0.397	0.395	0.392	0.393	0.374	0.373	0.382	0.358	0.356	0.345	0.314	0.324
Xinjiang	0.445	0.448	0.447	0.516	0.467	0.481	0.506	0.523	0.544	0.555	0.541	0.509	0.567	0.591	0.596
eastern	0.398	0.396	0.398	0.390	0.383	0.376	0.366	0.364	0.360	0.357	0.367	0.386	0.401	0.398	0.446
central	0.396	0.393	0.391	0.374	0.375	0.365	0.351	0.344	0.336	0.306	0.330	0.329	0.322	0.306	0.333
western	0.429	0.433	0.436	0.441	0.437	0.434	0.436	0.432	0.436	0.442	0.440	0.436	0.472	0.532	0.537
Total	0.408	0.408	0.409	0.402	0.398	0.389	0.381	0.374	0.369	0.359	0.369	0.375	0.389	0.380	0.408

Table A3. The shadow price of oil.

	1998	1999	2000	2001	2002	2003	2004	2005	2006	2007	2008	2009	2010	2011	2012
Beijing	0.005	0.022	0.042	0.061	0.078	0.106	0.122	0.142	0.157	0.175	0.187	0.211	0.240	0.283	0.303
Tianjin	0.014	0.035	0.051	0.079	0.106	0.131	0.160	0.189	0.222	0.253	0.281	0.313	0.318	0.337	0.371
Hebei	0.091	0.113	0.137	0.164	0.194	0.226	0.261	0.309	0.345	0.376	0.405	0.448	0.475	0.505	0.532
Shanxi	0.075	0.096	0.123	0.188	0.195	0.229	0.263	0.292	0.325	0.403	0.377	0.406	0.431	0.476	0.515
Neimenggu	0.054	0.080	0.105	0.133	0.163	0.190	0.223	0.254	0.280	0.303	0.339	0.360	0.387	0.442	0.481
Liaoning	0.042	0.042	0.061	0.093	0.125	0.154	0.210	0.193	0.214	0.246	0.260	0.288	0.253	0.239	0.209
Jilin	0.035	0.061	0.087	0.114	0.141	0.169	0.202	0.224	0.251	0.282	0.323	0.353	0.376	0.416	0.473
Heilongjiang	0.015	0.033	0.051	0.076	0.101	0.129	0.161	0.183	0.216	0.231	0.281	0.304	0.326	0.345	0.364
Shanghai	0.000	0.019	0.034	0.051	0.066	0.082	0.090	0.102	0.101	0.113	0.135	0.149	0.160	0.172	0.184
Jiangsu	0.069	0.088	0.100	0.120	0.139	0.154	0.173	0.184	0.185	0.189	0.188	0.194	0.174	0.171	0.128
Zhejiang	0.039	0.057	0.076	0.097	0.122	0.141	0.164	0.184	0.201	0.222	0.247	0.263	0.239	0.243	0.273
Anhui	0.089	0.114	0.141	0.169	0.197	0.227	0.256	0.286	0.316	0.347	0.380	0.413	0.441	0.467	0.486
Fujian	0.030	0.053	0.077	0.103	0.127	0.153	0.177	0.213	0.245	0.271	0.313	0.334	0.323	0.336	0.383
Jiangxi	0.057	0.081	0.106	0.129	0.158	0.186	0.225	0.252	0.288	0.327	0.365	0.409	0.441	0.468	0.523
Shandong	0.084	0.105	0.115	0.143	0.173	0.186	0.202	0.211	0.223	0.238	0.259	0.251	0.221	0.218	0.190
Henan	0.108	0.133	0.164	0.188	0.217	0.245	0.277	0.305	0.332	0.360	0.394	0.437	0.469	0.489	0.526
Hubei	0.078	0.101	0.127	0.155	0.180	0.207	0.240	0.267	0.303	0.321	0.366	0.401	0.421	0.450	0.516
Hunan	0.089	0.106	0.128	0.161	0.184	0.219	0.258	0.296	0.331	0.364	0.452	0.509	0.624	0.693	0.845
Guangdong	0.000	0.013	0.035	0.049	0.062	0.077	0.085	0.082	0.081	0.060	0.067	0.000	0.068	0.068	0.071
Guangxi	0.066	0.092	0.120	0.147	0.175	0.211	0.247	0.273	0.318	0.360	0.417	0.459	0.508	0.596	0.650
Hainan	0.012	0.044	0.065	0.090	0.114	0.138	0.165	0.202	0.234	0.268	0.304	0.347	0.386	0.415	0.468
Chongqing	0.033	0.054	0.076	0.106	0.133	0.156	0.182	0.223	0.256	0.291	0.342	0.389	0.428	0.454	0.504
Sichuan	0.059	0.076	0.102	0.123	0.148	0.181	0.209	0.229	0.241	0.270	0.316	0.329	0.294	0.554	0.570
Guizhou	0.069	0.091	0.120	0.153	0.184	0.221	0.258	0.281	0.319	0.351	0.381	0.423	0.459	0.503	0.551
Yunnan	0.060	0.084	0.111	0.137	0.168	0.203	0.237	0.291	0.322	0.364	0.413	0.460	0.500	0.578	0.681
Shaanxi	0.049	0.070	0.087	0.109	0.134	0.159	0.174	0.231	0.250	0.271	0.284	0.322	0.344	0.373	0.410
Gansu	0.046	0.072	0.095	0.126	0.158	0.180	0.220	0.250	0.287	0.326	0.370	0.420	0.450	0.491	0.546
Qinghai	0.022	0.047	0.073	0.099	0.123	0.150	0.183	0.215	0.254	0.298	0.337	0.386	0.452	0.513	0.579
Ningxia	0.025	0.050	0.076	0.103	0.131	0.160	0.191	0.232	0.270	0.315	0.344	0.389	0.431	0.487	0.533
Xinjiang	0.000	0.024	0.047	0.026	0.091	0.112	0.128	0.147	0.172	0.202	0.247	0.304	0.318	0.344	0.393
eastern	0.035	0.054	0.072	0.095	0.119	0.141	0.164	0.183	0.201	0.219	0.241	0.254	0.260	0.272	0.283
middle	0.059	0.083	0.108	0.139	0.163	0.191	0.223	0.251	0.283	0.317	0.349	0.384	0.411	0.441	0.484
western	0.044	0.067	0.092	0.115	0.146	0.175	0.205	0.239	0.270	0.305	0.345	0.385	0.416	0.485	0.536
Total	0.048	0.069	0.091	0.117	0.143	0.170	0.198	0.225	0.251	0.280	0.311	0.341	0.361	0.393	0.428

Table A4. The shadow price of natural gas.

	1998	1999	2000	2001	2002	2003	2004	2005	2006	2007	2008	2009	2010	2011	2012
Beijing	0.112	0.107	0.106	0.100	0.094	0.096	0.088	0.084	0.077	0.073	0.059	0.053	0.051	0.053	0.036
Tianjin	0.100	0.096	0.089	0.089	0.088	0.084	0.084	0.084	0.085	0.085	0.084	0.088	0.078	0.079	0.087
Hebei	0.124	0.126	0.126	0.129	0.134	0.141	0.153	0.174	0.188	0.200	0.212	0.235	0.250	0.272	0.291
Shanxi	0.122	0.119	0.119	0.141	0.128	0.131	0.135	0.136	0.139	0.178	0.136	0.135	0.132	0.145	0.152
Neimenggu	0.113	0.111	0.109	0.108	0.108	0.109	0.112	0.117	0.119	0.121	0.132	0.135	0.144	0.176	0.195
Liaoning	0.117	0.099	0.092	0.098	0.104	0.107	0.132	0.104	0.108	0.121	0.128	0.142	0.118	0.110	0.091
Jilin	0.103	0.102	0.101	0.099	0.097	0.095	0.096	0.090	0.089	0.092	0.103	0.106	0.108	0.120	0.136
Heilongjiang	0.089	0.085	0.079	0.079	0.077	0.076	0.078	0.072	0.075	0.067	0.082	0.078	0.077	0.077	0.078
Shanghai	0.109	0.109	0.105	0.103	0.098	0.095	0.087	0.085	0.072	0.073	0.076	0.075	0.073	0.069	0.064
Jiangsu	0.117	0.120	0.116	0.121	0.125	0.131	0.140	0.146	0.152	0.168	0.185	0.214	0.232	0.262	0.287
Zhejiang	0.115	0.113	0.109	0.110	0.116	0.121	0.129	0.136	0.143	0.153	0.168	0.177	0.167	0.176	0.207
Anhui	0.110	0.108	0.107	0.107	0.106	0.106	0.106	0.106	0.107	0.109	0.112	0.116	0.120	0.124	0.122
Fujian	0.109	0.108	0.106	0.106	0.103	0.102	0.100	0.107	0.112	0.116	0.130	0.132	0.112	0.112	0.132
Jiangxi	0.105	0.102	0.100	0.096	0.094	0.091	0.097	0.094	0.096	0.100	0.102	0.108	0.106	0.106	0.113
Shandong	0.110	0.112	0.107	0.114	0.125	0.126	0.132	0.137	0.146	0.162	0.186	0.196	0.193	0.220	0.230
Henan	0.113	0.111	0.110	0.111	0.114	0.117	0.124	0.131	0.141	0.155	0.172	0.198	0.223	0.245	0.280
Hubei	0.112	0.110	0.111	0.111	0.109	0.108	0.111	0.110	0.117	0.113	0.121	0.129	0.140	0.154	0.181
Hunan	0.115	0.109	0.106	0.109	0.104	0.108	0.114	0.120	0.124	0.128	0.160	0.177	0.222	0.249	0.308
Guangdong	0.087	0.085	0.089	0.088	0.090	0.094	0.093	0.087	0.093	0.092	0.111	0.089	0.154	0.182	0.259
Guangxi	0.110	0.108	0.106	0.103	0.100	0.102	0.102	0.096	0.101	0.106	0.116	0.123	0.134	0.162	0.179
Hainan	0.093	0.095	0.090	0.085	0.078	0.071	0.063	0.061	0.054	0.049	0.041	0.037	0.027	0.005	0.000
Chongqing	0.098	0.094	0.091	0.088	0.087	0.082	0.078	0.083	0.080	0.077	0.083	0.087	0.083	0.079	0.075
Sichuan	0.086	0.081	0.079	0.074	0.071	0.073	0.072	0.064	0.051	0.052	0.062	0.054	0.000	0.080	0.094
Guizhou	0.111	0.107	0.106	0.105	0.103	0.104	0.105	0.102	0.104	0.102	0.096	0.099	0.098	0.101	0.107
Yunnan	0.110	0.108	0.106	0.102	0.102	0.103	0.102	0.113	0.111	0.113	0.117	0.122	0.121	0.137	0.162
Shaanxi	0.106	0.101	0.095	0.090	0.086	0.081	0.070	0.087	0.079	0.071	0.060	0.067	0.064	0.068	0.076
Gansu	0.099	0.099	0.094	0.094	0.093	0.084	0.087	0.084	0.083	0.080	0.080	0.078	0.074	0.072	0.070
Qinghai	0.100	0.097	0.094	0.089	0.083	0.077	0.073	0.066	0.063	0.061	0.054	0.049	0.048	0.036	0.025
Ningxia	0.101	0.098	0.094	0.091	0.087	0.083	0.079	0.081	0.078	0.078	0.071	0.068	0.064	0.068	0.060
Xinjiang	0.089	0.087	0.084	0.047	0.072	0.065	0.053	0.041	0.032	0.026	0.029	0.038	0.015	0.001	0.000
eastern	0.109	0.106	0.103	0.104	0.105	0.106	0.109	0.110	0.112	0.117	0.125	0.131	0.132	0.140	0.153
middle	0.106	0.104	0.102	0.104	0.100	0.099	0.101	0.100	0.102	0.108	0.108	0.113	0.117	0.122	0.133
western	0.102	0.099	0.096	0.090	0.090	0.088	0.085	0.085	0.082	0.081	0.082	0.084	0.077	0.089	0.095
Total	0.106	0.104	0.101	0.100	0.100	0.099	0.101	0.101	0.103	0.107	0.112	0.116	0.117	0.127	0.136

Table A5. The shadow price of electric power.

	1998	1999	2000	2001	2002	2003	2004	2005	2006	2007	2008	2009	2010	2011	2012
Beijing	0.195	0.197	0.198	0.201	0.200	0.201	0.199	0.203	0.209	0.216	0.226	0.238	0.250	0.258	0.285
Tianjin	0.194	0.192	0.193	0.192	0.190	0.190	0.190	0.190	0.191	0.192	0.195	0.201	0.208	0.218	0.234
Hebei	0.124	0.125	0.123	0.122	0.122	0.123	0.124	0.121	0.123	0.131	0.137	0.143	0.150	0.165	0.181
Shanxi	0.167	0.165	0.163	0.164	0.161	0.159	0.159	0.159	0.158	0.164	0.157	0.159	0.167	0.172	0.178
Neimenggu	0.175	0.173	0.171	0.169	0.167	0.170	0.171	0.173	0.178	0.186	0.192	0.204	0.212	0.219	0.231
Liaoning	0.165	0.158	0.156	0.155	0.160	0.162	0.163	0.159	0.164	0.171	0.181	0.189	0.196	0.216	0.238
Jilin	0.174	0.172	0.169	0.167	0.165	0.162	0.161	0.160	0.161	0.160	0.168	0.174	0.184	0.190	0.199
Heilongjiang	0.167	0.165	0.166	0.165	0.163	0.163	0.161	0.160	0.158	0.161	0.164	0.163	0.166	0.173	0.181
Shanghai	0.199	0.201	0.201	0.204	0.205	0.206	0.206	0.214	0.219	0.228	0.236	0.246	0.263	0.280	0.299
Jiangsu	0.107	0.108	0.109	0.110	0.112	0.116	0.125	0.143	0.162	0.180	0.204	0.228	0.261	0.308	0.363
Zhejiang	0.144	0.144	0.141	0.140	0.138	0.139	0.143	0.145	0.157	0.165	0.171	0.178	0.196	0.218	0.242
Anhui	0.107	0.103	0.099	0.096	0.093	0.089	0.086	0.083	0.080	0.080	0.079	0.080	0.084	0.090	0.094
Fujian	0.164	0.163	0.161	0.159	0.158	0.158	0.157	0.154	0.154	0.157	0.159	0.168	0.181	0.191	0.199
Jiangxi	0.141	0.138	0.137	0.135	0.129	0.127	0.123	0.118	0.116	0.114	0.112	0.111	0.109	0.115	0.115
Shandong	0.079	0.077	0.073	0.076	0.079	0.084	0.092	0.097	0.111	0.120	0.132	0.148	0.168	0.190	0.220
Henan	0.070	0.061	0.048	0.047	0.046	0.045	0.046	0.053	0.059	0.065	0.066	0.065	0.068	0.083	0.089
Hubei	0.107	0.104	0.100	0.097	0.094	0.091	0.087	0.082	0.078	0.083	0.073	0.078	0.101	0.116	0.118
Hunan	0.097	0.094	0.091	0.088	0.082	0.080	0.073	0.071	0.068	0.067	0.049	0.044	0.022	0.029	0.012
Guangdong	0.124	0.126	0.122	0.121	0.122	0.119	0.119	0.116	0.127	0.151	0.163	0.201	0.204	0.243	0.289
Guangxi	0.123	0.119	0.113	0.109	0.103	0.098	0.094	0.093	0.085	0.082	0.073	0.073	0.071	0.065	0.078
Hainan	0.195	0.191	0.189	0.187	0.186	0.185	0.184	0.180	0.178	0.175	0.173	0.169	0.167	0.173	0.169
Chongqing	0.158	0.157	0.156	0.156	0.157	0.157	0.157	0.153	0.154	0.155	0.156	0.158	0.164	0.174	0.183
Sichuan	0.076	0.069	0.064	0.060	0.059	0.058	0.056	0.054	0.058	0.059	0.058	0.071	0.093	0.015	0.054
Guizhou	0.147	0.145	0.139	0.127	0.122	0.119	0.116	0.125	0.123	0.124	0.121	0.120	0.122	0.121	0.121
Yunnan	0.137	0.134	0.128	0.124	0.120	0.116	0.110	0.101	0.101	0.095	0.087	0.084	0.078	0.066	0.053
Shaanxi	0.152	0.150	0.147	0.146	0.142	0.141	0.143	0.139	0.141	0.144	0.150	0.153	0.161	0.170	0.183
Gansu	0.153	0.150	0.148	0.145	0.142	0.141	0.136	0.138	0.135	0.132	0.127	0.119	0.123	0.123	0.122
Qinghai	0.194	0.191	0.188	0.187	0.186	0.184	0.182	0.180	0.177	0.172	0.169	0.166	0.159	0.157	0.156
Ningxia	0.194	0.191	0.189	0.188	0.186	0.183	0.181	0.177	0.174	0.170	0.170	0.167	0.166	0.166	0.164
Xinjiang	0.191	0.190	0.189	0.184	0.189	0.188	0.191	0.187	0.189	0.190	0.190	0.190	0.192	0.198	0.204
eastern	0.154	0.153	0.151	0.152	0.152	0.153	0.155	0.157	0.163	0.171	0.180	0.192	0.204	0.224	0.247
middle	0.141	0.137	0.134	0.132	0.130	0.128	0.126	0.124	0.123	0.125	0.124	0.125	0.131	0.139	0.143
western	0.155	0.152	0.149	0.145	0.143	0.141	0.140	0.138	0.138	0.137	0.136	0.137	0.140	0.134	0.141
Total	0.147	0.145	0.142	0.140	0.139	0.138	0.137	0.137	0.139	0.143	0.146	0.151	0.159	0.169	0.181

References

1. Hu, J.L.; Kao, C.H. Efficient energy-saving targets for APEC economies. *Energy Policy* **2007**, *35*, 373–382.
2. Zhang, X.P.; Cheng, X.M.; Yuan, J.H.; Gao, X.S. Total-factor energy efficiency in developing countries. *Energy Policy* **2011**, *39*, 644–650.
3. Sheng, P.F. The explanation for the low energy efficiency of China: Allocation inefficiency or technology inefficiency. *Ind. Econ. Res.* **2015**, *1*, 9–20.
4. Wang, F.; Feng, G.F. Contribution of improving energy mix to carbon intensity target in China: Potential assessment. *China Ind. Econ.* **2011**, *4*, 127–137.
5. Shi, D. Regional Differences in China's Energy Efficiency and Conservation Potentials. *China Ind. Econ.* **2006**, *10*, 49–58.
6. Fan, Y.; Liao, H.; Wei, Y.M. Can market oriented economic reforms contribute to energy efficiency improvement? Evidence from China. *Energy Policy* **2007**, *35*, 2287–2295.
7. Hang, L.; Tu, M. The impacts of energy prices on energy intensity: Evidence from China. *Energy Policy* **2007**, *35*, 2978–2988.
8. Zhang, Z.X. Why did the energy intensity fall in China's industrial sector in the 1990s? The relative importance of structural change and intensity change. *Energy Econ.* **2003**, *25*, 625–638.
9. Hu, J.L.; Wang, S.C. Total-factor energy efficiency of regions in China. *Energy Policy* **2006**, *34*, 3206–3217.
10. Lin, B.Q.; Du, K. Technology gap and China's regional energy efficiency: A parametric metafrontier approach. *Energy Econ.* **2013**, *40*, 529–536.
11. Fujii, H.; Kaneko, S.; Managi, S. Changes in environmentally sensitive productivity and technological modernization in China's iron and steel industry in the 1990s. *Environ. Dev. Econ.* **2010**, *15*, 485–504.
12. Li, L.B.; Hu, J.L. Ecological total-factor energy efficiency of regions in China. *Energy Policy* **2012**, *46*, 216–224.
13. Li, K.; Lin, B. Metafroniter energy efficiency with CO_2 emissions and its convergence analysis for China. *Energy Econ.* **2015**, *48*, 230–241.
14. Ouyang, X.L.; Sun, C. Energy savings potential in China's industrial sector: From the perspectives of factor price distortion and allocative inefficiency. *Energy Econ.* **2015**, *48*, 117–126.
15. Sheng, P.F.; Yang, J. The Heterogeneity and convergence of energy's shadow price in China—The estimation of nonparametric input distance function. *Ind. Econ. Res.* **2014**, *1*, 70–80.
16. Sheng, P.F.; Yang, J.; Shackman, J.D. Energy's Shadow Price and Energy Efficiency in China: A Non-Parametric Input Distance Function Analysis. *Energies* **2015**, *8*, 1975–1989.
17. Kumar, S.; Fujii, H.; Managi, S. Substitute or complement? Assessing renewable and nonrenewable energy in OECD countries. *Appl. Econ.* **2015**, *47*, 1438–1459.
18. Kaneko, S.; Fujii, H.; Sawazu, N.; Fujikura, R. Financial allocation strategy for the regional pollution abatement cost of reducing sulfur dioxide emissions in the thermal power sector in China. *Energy Policy* **2010**, *38*, 2131–2141.

19. Ishinabe, N.; Fujii, H.; Managi, S. The true cost of greenhouse gas emissions: Analysis of 1000 global companies. *PLoS ONE* **2013**, *8*, e78703. PubMed]
20. Yagi, M.; Fujii, H.; Hoang, V.; Hoeng, V.; Managi, S. Environmental efficiency of energy, materials, and emissions. *J. Environ. Manag.* **2015**, *161*, 206–218. PubMed]
21. Molinos-Senante, M.; Hanley, N.; Sala-Garrido, R. Measuring the CO_2 shadow price for wastewater treatment: A directional distance function approach. *Appl. Energy* **2015**, *144*, 241–249.
22. Zhou, P.; Ang, B.W.; Zhou, D.Q. Measuring economy-wide energy efficiency performance: A parametric frontier approach. *Appl. Energy* **2012**, *90*, 196–200.
23. Lin, B.Q.; Du, K.R. The energy effect of factor market distortion in China. *Econ. Res. J.* **2013**, *9*, 125–136.
24. Yuan, P.; Cheng, S. Estimating shadow pricing of industrial pollutions in China. *Stat. Res.* **2011**, *28*, 66–73.
25. Wang, B.; Huang, R.J. Regional green development efficiency and green total productivity growth in China: From 2000 to 2010—Base on parametric metafrontier analysis. *Ind. Econ. Rev.* **2014**, *5*, 16–35.
26. Li, S.; Ng, Y.C. Measuring the productive efficiency of a group of firms. *Int. Adv. Econ. Res.* **1995**, *1*, 377–390.
27. National Bureau of Statistics. *China Statistical Yearbook*; China Statistics Press: Beijing, China, 1998–2012.
28. Department of Energy Statistics, National Bureau of Statistics. *China Energy Statistical Yearbook*; China Statistics Press: Beijing, China, 1998–2012.
29. Wang, B.; Liu, G.T. Energy conservation and emission reduction and China's green economic growth—Based on a total factor productivity perspective. *China Ind. Econ.* **2015**, *5*, 57–69.
30. Zhang, J.; Wu, G.Y.; Zhang, J.P. The Estimation of China's provincial capital stock: 1952–2000. *Econ. Res. J.* **2004**, *10*, 35–44.
31. Zheng, K.; Yao, G. The Main Reason for Surge of Domestic Oil price—A VARA Approach. *South China J. Econ.* **2006**, *5*, 83–94.
32. Wang, J.; Zhong, W.Z. The Research of Regional Energy Intensity Difference in China—From the Factor Endowment Perspectcive. *Ind. Econ. Res.* **2009**, *6*, 44–51.
33. Shi, B.; Shen, K.R. The Government Intervention, the Economic Agglomeration and the Energy Efficiency. *Manag. World* **2008**, *10*, 6–18.
34. Wang, W.G.; Fan, D. Influential factor and convergence of total factor energy efficiency in China based on the malmquist-luenber index. *Resour. Sci.* **2012**, *34*, 1816–1824.
35. Tang, L.; Yang, Z.L. Energy efficiency and industrial economic transition. *J. Quant. Tech. Econ.* **2009**, *10*, 34–48.
36. Wang, K.L.; Yang, L.; Yang, B.C.; Cheng, Y.H. Energy economic efficiency, the energy environmental performance and regional economic growth. *J. Manag.* **2013**, *26*, 86–99.
37. Wu, Y.R. The role of productivity in China's growth: New estimates. *China Econ. Q.* **2008**, *2*, 827–842.

Estimating the Contribution of Industry Structure Adjustment to the Carbon Intensity Target: A Case of Guangdong

Ping Wang and Bangzhu Zhu

Abstract: Industry structure adjustment is an effective measure to achieve the carbon intensity target of Guangdong Province. Accurately evaluating the contribution of industry structure adjustment to the carbon intensity target is helpful for the government to implement more flexible and effective policies and measures for CO_2 emissions reduction. In this paper, we attempt to evaluate the contribution of industry structure adjustment to the carbon intensity target. Firstly, we predict the gross domestic product (GDP) with scenario forecasting, industry structure with the Markov chain model, CO_2 emissions with a novel correlation mode based on least squares support vector machine, and then we assess the contribution of industry structure adjustment to the carbon intensity target of Guangdong during the period of 2011–2015 under nine scenarios. The obtained results show, in the ideal scenario, that the economy will grow at a high speed and the industry structure will be significantly adjusted, and thus the carbon intensity in 2015 will decrease by 25.53% compared to that in 2010, which will make a 130.94% contribution to the carbon intensity target. Meanwhile, in the conservative scenario, the economy will grow at a low speed and the industry structure will be slightly adjusted, and thus the carbon intensity in 2015 will decrease by 23.89% compared to that in 2010, which will make a 122.50% contribution to the carbon intensity target.

Reprinted from *Sustainability*. Cite as: Wang, P.; Zhu, B. Estimating the Contribution of Industry Structure Adjustment to the Carbon Intensity Target: A Case of Guangdong. *Sustainability* **2016**, *8*, 355.

1. Introduction

Climate change is one of the most serious challenges for global sustainable development in the 21st century. To address the global climate change, the Chinese government announced that carbon intensity, defined by the comparison between CO_2 emissions and gross domestic product (GDP), should be reduced by 40%–45% from 2005 to 2020. For the purpose of achieving this target, China's state council issued the greenhouse gas emissions control scheme during the 12th five-year plan (2011–2015) (China made the first "five-year plan (1953–1957)" in 1953; so far, China has published 12 five-year plans, and the 13th five-year plan (2016–2020) will be developed in 2016) to allocate the CO_2 emissions reduction tasks for every province in 2012, in which Guangdong's carbon intensity should decrease 19.5% from 2010 to 2015. Industry structure adjustment, which is trying to coordinate the proportions of primary, secondary and tertiary industries, naturally becomes an effective way for achieving the carbon intensity target [1], because it can effectively reduce the carbon intensity. Through industry structure adjustment, realizing the carbon intensity target requires solving two questions: how much will industry structure adjustment reduce the carbon intensity, and how much will industry structure adjustment realize the carbon intensity target?

Located in southern China, Guangdong Province is adjacent to the special administrative regions of Hong Kong and Macao. In 2014, Guangdong's GDP reached 6.78 billion Yuan at the current price, accounting for over 10% of that of the whole country. At that time, it had ranked first in terms of economic performance for over 20 successive years. The permanent population of Guangdong was 107.24 million in 2014, ranking first in China. During 1990–2014, Guangdong's energy consumption increased rapidly, from 36.90 to 256.36 million tons standard coal, implying a mean annual increase of 8.41%. Furthermore, such a large increase in energy consumption will certainly cause a large increase in CO_2 emissions. Thus, there will be tremendous pressure on Guangdong to realize this goal during the period of the 12th five-year plan (2011–2015).

During the past few years, some scholars have developed related research on the realization of the carbon intensity target from different angles using various methods [2–15]. These studies show that industry structure adjustment is one of the effective measures for decreasing carbon intensity, which can provide some important references for making out energy conservation and emission reduction policies. However, the existing research mainly focused on the national level, and ignored the larger differences among resource endowments and industry foundations of various provinces [16]. Their obtained conclusions and relevant policies cannot have a good credibility and feasibility; thus, it is difficult for various provinces to guide their practices. Moreover, the existing studies mainly used the input-output method and structure decomposition method. The former needs to use the input-output table data, which usually applies to the national level rather than provincial level for the lack of enough data, while the latter is usually used to explore the inner mechanism of carbon intensity changes, and is seldom used to estimate the contribution of industry structure adjustment to the carbon intensity target. So far, rare literature has been found in terms of evaluating the contribution of industry structure adjustment to the carbon intensity target of Guangdong.

In order to fill this gap, this study mainly contributes to and aims at evaluating the contribution of industry structure adjustment to the carbon intensity target by an integrated use of multidisciplinary approaches. Firstly, we predict GDP with scenario forecasting, industry structure with the Markov chain model, CO_2 emissions with a novel correlation model based on least squares support vector machines (LSSVM), and then we assess the contribution of industry structure adjustment to the carbon intensity target of Guangdong in 2015 under nine scenarios. The obtained results show, in the ideal scenario, that the economy will grow at a high speed and the industry structure will be significantly adjusted, and thus the carbon intensity will decrease by 25.53% from 2010 to 2015, which will make a 130.94% contribution to the carbon intensity target. Meanwhile, in the conservative scenario, the economy will grow at a low speed and the industry structure will be slightly adjusted, and thus the carbon intensity will decrease by 23.89% from 2010 to 2015, which will make a 122.50% contribution to the carbon intensity target. This study can help answer the above two questions, and providing theoretical support for decision-making for the related governments.

2. Methods and Data

As a relative index, carbon intensity is defined as: carbon intensity = CO_2 emission/GDP. Thus, we will forecast CO_2 emissions and GDP respectively. Firstly, we will set three scenarios to predict the GDP. Secondly, we will employ the Markov chain model to predict the industry structure. Thirdly, we will calculate and forecast CO_2 emissions with a novel correlation model based on LSSVM under various industry structure scenarios. Finally, we will evaluate the contribution of industry structure adjustment to the carbon intensity target.

2.1. GDP Forecasting

Using scenario prediction, GDP is forecasted by:

$$GDP_{2014+i} = GDP_{2014} \times (1+r)^i, i = 1, 2, \cdots$$

where r is the average annual growth rate of the GDP.

2.2. Industry Structure Prediction with the Markov Chain Model

In this study, we use the Markov chain model [17] to predict the industry structure by 2015. The state vector of the industry structure at year n is defined as $S(n)$:$S(n) = \{s_a(n), s_b(n), s_c(n)\}$, where $s_a(n), s_b(n), s_c(n)$ are the respective proportions of the primary, secondary and tertiary industries.

The one-step transition probability matrix of industry structure from n to $n+1$ is defined as:

$$\begin{bmatrix} M_{a\to a}(n) & M_{a\to b}(n) & M_{a\to c}(n) \\ M_{b\to a}(n) & M_{b\to b}(n) & M_{b\to c}(n) \\ M_{c\to a}(n) & M_{c\to b}(n) & M_{c\to c}(n) \end{bmatrix}$$

where the elements in the leading diagonals of the transition probability matrix are, respectively, the probabilities of three industries for maintaining their original structures. Except the leading elements, other elements are the transition probabilities; the row elements show the probability of one industry transferring to the other industries. The column element represents the probability of specific gravity of other industries transferring to one certain industry; the sum of the transition probability in each row is 1.

For convenience, the elements in the leading diagonals of the transition probability matrix are called reserved probability elements. Except the reserved probability elements, other elements in the rows are called the transition probability elements. Elements in the column, except the reserved probability elements, are named after absorptive elements.

We use the Markov chain model to predicate the industry structure as follows:

(1) Calculate the reserved probability element. For the industry structure, if the proportion of one industry increases from the nth year to the n+1th year, the reserved probability of this industry in the transition probability matrix is 1. Otherwise, the reserved probability is the ratio of the proportion of the n+1th year to that of the nth year. Taking coal as an example:

$$S_a(n+1) \geqslant S_a(n) \Rightarrow M_{a\to a}(n) = 1$$

$$S_a(n+1) < S_a(n) \Rightarrow M_{a\to a}(n) = S_a(n+1)/S_a(n)$$

(2) Compute the transition probability elements in the row in which an element with the reserved probability is 1. The value of the reserved probability elements in a certain row is 1, which shows that the proportion of one certain energy represented by this row is unchanged or increases, and thus the certain energy cannot transfer to the other energies. Besides, the sum of the transition probability in each row is 1. Therefore, each transition probability element in this row is 0. Taking primary industry as an example:

$$M_{a\to a}(n) = 1 \Rightarrow \{M_{a\to b}(n) = 0; M_{a\to c}(n) = 0\}$$

(3) Calculate the absorptive probability elements in the column where the reserved probability is less than 1. The reserved probability element in a certain column is less than 1, which implies that the proportion of the certain industry represented by this column decreases, and other industries cannot transfer to this industry. Therefore, each absorptive probability element in this column is 0. Taking primary industry as an instance:

$$M_{a\to a}(n) < 1 \Rightarrow \{M_{b\to a}(n) = 0; M_{c\to a}(n) = 0\}$$

(4) Compute the elements of non-zero transition probability in the row where the reserved probability is less than 1. As for primary industry, taking the scenario where the reserved probability is less than 1 as an example:

$$\left.\begin{array}{l} M_{a\to a}(n) < 1 \\ M_{a\to b}(n) \neq 0 \\ M_{a\to c}(n) \neq 0 \end{array}\right\} \Rightarrow \left\{\begin{array}{l} M_{a\to b}(n) = \frac{[1-M_{a\to a}(n)]\times[S_b(n+1)-S_b(n)]}{[S_b(n+1)-S_b(n)]+[S_c(n+1)-S_c(n)]+[S_d(n+1)-S_d(n)]} \\ M_{a\to c}(n) = \frac{[1-M_{a\to a}(n)]\times[S_c(n+1)-S_c(n)]}{[S_b(n+1)-S_b(n)]+[S_c(n+1)-S_c(n)]+[S_d(n+1)-S_d(n)]} \end{array}\right.$$

(5) Repeat the above steps (1)–(4); the transition probability matrixes of the industry structure in each year are, respectively, $M(1), M(2), \cdots, M(n)$, from the initial year to the nth year. The average matrix of the transition probability is achieved as:

$$M = \left[\prod_{i=1}^{n} M(i)\right]^{1/n}$$

(6) Forecast the industry structure in the n+lth year:

$$S(n+l) = S(n) \cdot M^l$$

2.3. *Estimating CO_2 Emissions*

Energy consumption is the main source of CO_2 emissions, accounting for more than 90% of the total CO_2 emissions in China. Inspired by the reference method for estimating CO_2 emissions of the International Panel on Climate Change [18], we calculate the CO_2 emissions of primary energy consumption as:

$$CE = \sum_{i=1}^{4} E_i \times Coe_i \times \frac{44}{12}$$

where CE is the total CO_2 emissions, E1, E2, E3 and E4 are consumptions of the coal, petroleum, natural gas and non-fossil energies, respectively; Coe1, Coe2, Coe3 and Coe4, respectively, are carbon emission coefficients of the coal, petroleum, natural gas and non-fossil energies.

2.4. *Forecasting CO_2 Emissions*

Inspired by the method proposed by [19], we set up a new correlation model incorporating industry structure and LSSVM to predict the CO_2 emissions under various industry structure scenarios, defined by:

$$CE = f(x),\ x = (x_1, x_2, x_3)$$

where x_1, x_2 and x_3 are the proportions of the primary, secondary and tertiary industries, respectively, and $f(\cdot)$ is a nonlinear function determined by the LSSVM model.

As a novel machine learning approach for small-sample data, for a given training set $\{x_i, y_i\}, i = 1, 2, \cdots, l$ with input data x_i, and output data y_i, LSSVM defines the regression function as [20]:

$$\begin{aligned} & min\, Q(\omega, b, e) = \tfrac{1}{2}||\omega||^2 + \tfrac{C}{2}\sum_{i=1}^{l} e_i^2 \\ & s.t. \\ & \quad y_i = \omega^T \varphi(x_i) + b + e_i,\ i = 1, 2, \cdots, l \end{aligned}$$

where w is the weight vector, C is the penalty parameter, e_i is the approximation error, $\varphi(\cdot)$ is the nonlinear mapping function and b is the bias term. The corresponding Lagrange function can be obtained:

$$L(w, e, \alpha, b) = J(w, e) - \sum_{i=1}^{l} \alpha_i \left\{ w^T \varphi(x_i) + b + e_i - y_i \right\}$$

where α_i is the Lagrange multiplier. Using the Karush–Kuhn–Tucker (KKT) conditions, the solutions can be obtained by partially differentiating with respect to w, b, e_i and α_i:

$$\begin{cases} \frac{\partial L}{\partial \omega} = 0 \Rightarrow \omega = \sum_{i=1}^{l} a_i \varphi(x_i) \\ \frac{\partial L}{\partial b} = 0 \Rightarrow \sum_{i=1}^{l} a_i = 0 \\ \frac{\partial L}{\partial e_i} = 0 \Rightarrow a_i = Ce_i \\ \frac{\partial L}{\partial a_i} = 0 \Rightarrow \omega^T \varphi(x_i) + b + e_i - y_i = 0 \end{cases}$$

By eliminating w and e_i, the equations can be changed into

$$\begin{bmatrix} 0 & 1_v^T \\ 1_v & \Omega + C^{-1} I \end{bmatrix} \begin{bmatrix} b \\ a \end{bmatrix} = \begin{bmatrix} 0 \\ y \end{bmatrix}$$

where $y = [y_1, y_2, \cdots, y_l]^T$, $I_v = [1, 1, \cdots, 1]^T$, $\alpha = [\alpha_1, \alpha_2, \cdots, \alpha_l]^T$, and the Mercer condition has been applied to matrix Ω with $\Omega_{km} = \varphi(x_k)^T \varphi(x_k), k, m = 1, 2, \cdots, l$. Therefore, the LSSVM for regression can be obtained:

$$y(x) = \sum_{i=1}^{l} \alpha_i K(x, x_i) + b$$

where $K(x, x_i)$ is the kernel function.

The establishment of a LSSVM model requires two key issues to be addressed [21]. These are, firstly, how does one make an appropriate choice of the kernel function? Secondly, how does one accurately determine the model parameters? In this study, the first issue is solved by using the radial basis kernel function (RBF), $K(x,y) = \exp\left(-||x-y||^2/2\sigma^2\right)$, to build the LSSVM model. This is because this function generally yields good predictions [22]. For the second issue, particle swarm optimization (PSO), proposed by Kennedy and Eberhart [23], is adopted to find out the best values of C and σ [24].

PSO represents each candidate solution by the state of a particle in a search space. M particles constitute a swarm in a two-dimensional space. Vectors $x_i = (x_{i1}, x_{i2})$ and $v_i = (v_{i1}, v_{i2})$, respectively, express the position and velocity of particle i. Other vectors $p_{best} = (p_{i1}, p_{i2})$ and $g_{best} = (g_{i1}, g_{i2})$, respectively, express the optimum positions of particle i and all particles at the present stage. The position and velocity of every particle are renewed according to the following formulas:

$$x_{id}(t+1) = \begin{cases} -p_{\max}, & x_{id} < -p_{\max} \\ x_{id}(t) + v_{id}(t+1), & -p_{\max} \leqslant x_{id} \leqslant p_{\max} \\ p_{\max}, & x_{id} > p_{\max} \end{cases} \tag{1}$$

$$v_{id}(t+1) = \begin{cases} -v_{\max}, & v_{id} < -v_{\max} \\ w(t)\cdot v_{id}(t) + c_1 \cdot r_1 \cdot [p_{id}(t) - x_{id}(t)] + c_2 \cdot r_2 \cdot [p_{gd}(t) - x_{id}(t)], & -v_{\max} \leqslant v_{id} \leqslant v_{\max} \\ v_{\max}, & v_{id} > v_{\max} \end{cases} \tag{2}$$

where $1 \leqslant i \leqslant m$, $1 \leqslant d \leqslant M$, $v_{id}(t)$ denotes the velocity of particle i at iteration t, and $x_{id}(t)$ signifies the position of particle i at iteration t. Also, p_{id} is the previous optimum position of particle i, p_{gd} is all particles' greatest position, and w is the inertia weight. The latter is generally set as a function that decreases linearly with time as:

$$w(t) = w_{\max} - \frac{w_{\max} - w_{\min}}{t_{\max}} \times t \tag{3}$$

where $w_{\max}$ and $w_{\min}$ are, respectively, the biggest and smallest inertia weights.

In this study, PSO is introduced to find the optimal LSSVM model parameters (C and σ) to enhance search efficiency and prediction accuracy. Figure 1 illustrates the process of LSSVM parameter searching using PSO. In greater detail, the steps unfold as follows:

Step 1: Establish training and testing sets. The data gathered during a certain period are, respectively, separated into a training set employed in building the model, and a test set used for checking the model's predictive performance.

Step 2: PSO initialization. First, m particles can be randomly produced by coding C and σ with real values. Then, PSO parameters are set as: $t_{\max}$ (maximal iterations), $p_{\max}$ (maximal position), $v_{\max}$ (maximal velocity), $w_{\max}$ (maximal inertia weight), $w_{\min}$ (minimum inertia weight), c_1 and c_2 (acceleration coefficients),

C_{min} (minimum C) and C_{max} (maximal C), σ_{min} (minimum σ) and σ_{max} (maximal σ). Training begins when t = 0.

Step 3: Choose the root mean square error (RMSE) as each particle's fitness function:

$$RMSE = \sqrt{\frac{1}{n}\sum_{i=1}^{n}(x_i - \hat{x}_i)^2} \tag{4}$$

In this formula, n is the amount of training samples, x_i is the real value, and $\hat{x}_i$ is the predictive value.

Step 4: Appraise fitness. Compute the fitness values of entire particles using Equation (4). Then, find out p_{best} and g_{best} at the current generation.

Step 5: Renew the positions and velocities of the whole particles using Equations (1)–(3).

Step 6: Check terminal conditions. The search process is terminated when the termination criteria, such as the maximum number of iterations, *etc.*, are satisfied. If fulfilled, output the optimal values. Otherwise, repeat Step 7 iteratively.

Step 7: Let $t = t + 1$, and recur to Step 4.

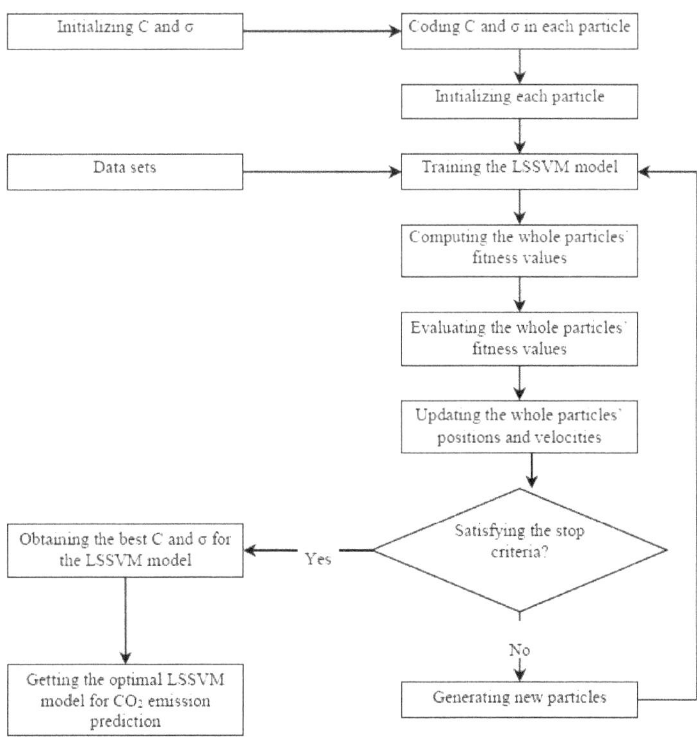

Figure 1. The searching process for LSSVM model parameters by PSO.

2.5. Evaluating the Contribution of the Carbon Intensity Target

The contribution of the carbon intensity target is defined as:

$$d = \frac{CIR}{CIT} = \frac{(CI_{2015} - CI_{2010})/CI_{2010}}{CIT} \times 100\%$$

where d is the contribution of the carbon intensity target; CIR is the carbon intensity reduction amplitude; CIT is the carbon intensity reduction target; CI_{2015} and CI_{2010} are the carbon intensities in 2015 and in 2010, respectively.

By combining the scenarios of different economic growth and industry structure, various combined scenarios can be formed in total. The value of d in a certain scenario is equal to or greater than 100%, which indicates that the carbon intensity target can be accomplished under this scenario. If not, the value of d is less than 100%, which shows that Guangdong cannot achieve the target.

2.6. Data

Selecting the data from 1990–2014 as the study samples, except for carbon emission coefficients and CO_2 emissions, we have obtained all the data from the Guangdong Statistical Yearbook (2015) [25]. GDP data (unit: 100 million Yuan) are adjusted as the constant price in 1978. E is converted into standard coal, with the unit of 10,000 tons of standard coal.

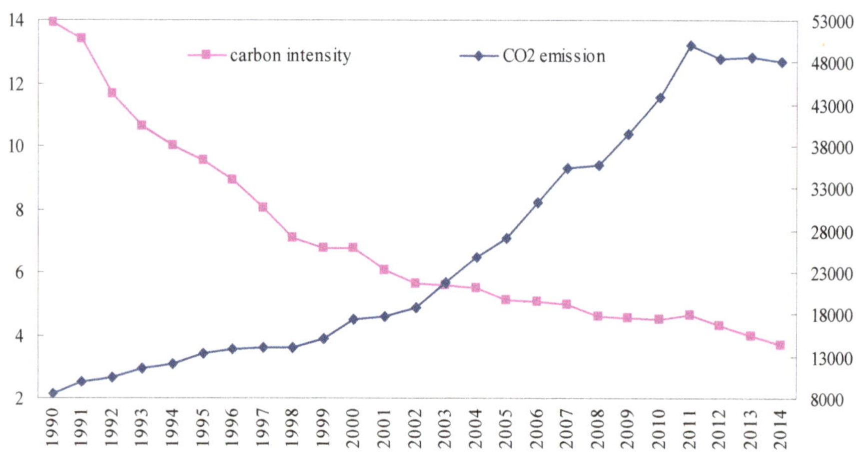

Figure 2. CO_2 emissions and carbon intensity in Guangdong from 1990–2014.

The carbon emission coefficients are the values of those recommended by the State Development and Reform Commission in China, namely Coe1 = 0.7476, Coe2 = 0.5825, Coe3 = 0.4435 and Coe4 = 0, with the unit of t(C)/t. CO_2 emissions from 1990 to 2014 in Guangdong estimated by this study are illustrated in Figure 2.

It can be observed that CO_2 emissions in Guangdong have increased sharply from 84.98 million tons in 1990 to 499.52 million tons in 2011, down to 483.98 million tons in 2012, up to 485.72 million tons in 2013, and down to 481.09 million tons in 2014. At the same time, the carbon intensity in Guangdong has decreased significantly from 13.89 ton/10^4 RMB Yuan in 1990 to 3.69 ton/10^4 RMB Yuan in 2014 at the fixed price of 1978.

3. Results and Discussion

3.1. Prediction on GDP of Guangdong in 2015

Although Guangdong clearly stated that the expected average annual growth rate of GDP should be above 8% during 2010–2015, the real one reached 7.58% at the fixed price of 1978 during the period of 2010–2014. The main reasons may be two-fold: one is that the global economic downturn has hit Chinese economic growth, and the other is that China is committed to the industrial structure transformation, which has brought a slower economic growth than ever. Economic growth is expected to fall further by 2015. Thus, we set up three scenarios for economic growth: high-speed with an average annual growth rate of 7.7%, medium-speed with an average annual growth rate of 7.5%, and low-speed with an average annual growth rate of 7.3%. On this basis, the GDP of Guangdong in 2015 will reach 14,057.98 (high-speed scenario), 14,031.87 (medium-speed scenario) and 14,005.77 billion RMB Yuan (low-speed scenario) at the fixed price of 1978.

3.2. Prediction on Industry Structure of Guangdong in 2015

The industry structure data in Guangdong from 2010 to 2014 are used to calculate the transition probability matrix of the Markov chain model. In this way, we can obtain the transition probability matrixes of the energy structure in each step during 2010 to 2014, which can be used to calculate the average transition probability matrix as follows:

$$M = \begin{bmatrix} 0.9816 & 0 & 0.0184 \\ 0 & 0.9792 & 0.0208 \\ 0 & 0 & 1 \end{bmatrix}$$

Then, we can forecast the industry structure in 2015 under the natural evolution: $S(2015) = S(2014+1) = S(2014) \cdot M$. Thus, we can obtain the industry structure in 2015: primary industry (4.58%), secondary industry (45.38%) and tertiary industry (50.04%).

The future trend of the industry structure is merely forecasted based on the evolution law of the historical industry structure instead of the adjustment of the national and provincial policies. Here, we define this as the industry structure "minor adjustment" scenario. Thus, we can obtain the industry structure under the "minor adjustment" scenario: primary industry (4.58%), secondary industry (45.38%) and tertiary industry (50.04%).

Taking this as an important strategic task, Guangdong has implemented an innovation-driven development strategy to promote economic structural adjustment and industrial transformation and upgrading since 2013. It is supposed that on the basis of minor adjustment, the proportion of tertiary industry will increase 0.5% in 2015, which will obtained by reducing the proportion of secondary industry. In this study, this scenario is called "medium adjustment". Under "medium adjustment", Guangdong's industry structure will be: primary industry (4.58%), secondary industry (44.88%) and tertiary industry (50.54%) by 2015.

Guangdong has not set up a target for primary industry by 2015. However, the proportion of primary industry in Guangdong has steadily decreased during recent years, which is in accordance with the industry structure evolution law of developed countries, namely economic development will inevitably result in the decrease of the proportion of primary industry. In fact, the proportion of tertiary industry rises at the cost of the synergistic decreases of the proportions of primary industry and secondary industry, rather than secondary industry alone. The continued increase in tertiary industry can alleviate CO_2 emissions to some extent. To achieve the CO_2 emissions reduction target of Guangdong, it appears necessary to increase the proportion of tertiary industry as soon as possible. Suppose that the proportion of primary industry will decrease from 5.03% in 2010 to 4.5% in 2015, and that the proportional decrease will be achieved by increasing the proportion of tertiary industry. Thus, the industry structure with planning restrictions and primary industry target restriction is predicted. In this study, this scenario is called "substantial adjustment". Accordingly, Guangdong's industry structure under "substantial adjustment" by 2015 is: primary industry (4.50%), secondary industry (44.88%) and tertiary industry (50.62%).

3.3. *Prediction on CO_2 Emissions of Guangdong in 2015*

For the data from 1990 to 2014 in Guangdong, we use the five-fold cross-validation method [22] to train the LSSVM model, and we obtain the best values of the model parameters: $C = 359.8661, \sigma = 4.1248$. Once the LSSVM model has been trained, the future industry structure is input into the trained LSSVM model so as to obtain the CO_2 emissions in the corresponding year. Thus, we input the industry structures under the "minor adjustment", "medium adjustment" and "substantial adjustment" scenarios for 2015, and obtain successively the forecasted CO_2 emissions of 478.82, 473.86 and 470.21 million tons for 2015.

3.4. Prediction on Carbon Intensity of Guangdong in 2015

Guangdong's CO_2 emissions were 437.69 million tons and the constant-price GDP was 974.44 billion RMB Yuan in 2010. Thus, the carbon intensity was 4.4917 ton/10^4 RMB Yuan in 2010. Table 1 lists the predicted values of Guangdong's carbon intensity under different scenarios in 2015 with the unit of ton/10^4 RMB Yuan. Taking the high-speed economic growth as an example, under the "minor adjustment" scenario, the carbon intensity will reduce from 4.4917 ton/10^4 RMB Yuan in 2010 to 3.4061 ton/10^4 RMB Yuan, decreased by 24.17%. Under the "medium adjustment" scenario, the carbon intensity will reduce from 4.4917 ton/10^4 RMB Yuan in 2010 to 3.3708 ton/10^4 RMB Yuan, decreased by 24.95%. Under the "substantial adjustment" scenario, the carbon intensity will reduce from 4.4917 ton/10^4 RMB Yuan in 2010 to 3.3448 ton/10^4 RMB Yuan, decreased by 25.53%.

Table 1. The predicted values of Guangdong's carbon intensity under different scenarios in 2015 (Unit: ton/104 Yuan).

	High-Speed	Medium-Speed	Low-Speed
Minor adjustment	3.4061	3.4124	3.4188
Medium adjustment	3.3708	3.3771	3.3834
Substantial adjustment	3.3448	3.3510	3.3573

3.5. Prediction on CO_2 Emissions of Guangdong in 2015

According to "China's 12th five-year control scheme for greenhouse gas emissions", in 2015, Guangdong's carbon intensity should be 19.5% lower than that in 2010. By combining the above scenarios of economic growth with industry structure adjustment, nine combined scenarios are formed in total. We calculate the amplitude of the carbon intensity reduction, as well as the potential of the industry structure adjustment to reach the carbon intensity target under each combined scenario compared with 2010. The specific results are shown in Table 2 (in %).

It can be found that, for a given level of economic growth, the larger the industry structure adjustment, the larger its "reduction amplitude" and "contribution" to achieving the carbon intensity target. For a given industry structure adjustment, the higher the economic growth, the larger its "reduction amplitude" and "contribution" to realizing the carbon intensity target. Under all scenarios, each of the contributions to the carbon intensity target will exceed 100%, which means a 19.5% decline in carbon intensity can be realized by industry structure adjustment. We can cautiously conclude that it appears to be feasible for Guangdong's government to reach its carbon intensity goal by 2015.

Table 2. The reduction amplitudes and contribution of industry structure adjustment to the carbon intensity target (Unit: %).

	High-Speed	Medium-Speed	Low-Speed
Minor adjustment	−24.17 (123.95)	−24.03 (123.22)	−23.89 (122.50)
Medium adjustment	−24.95 (127.97)	−24.82 (127.26)	−24.68 (126.54)
Substantial adjustment	−25.53 (130.94)	−25.39 (130.23)	−25.26 (129.51)

Note: The data in the table are denoted by a (b), wherein a represents a reduction amplitude of carbon intensity; b refers to the contribution to the carbon intensity target.

4. Conclusions

Using the data from 1990–2014 in Guangdong as well as the integrated use of multidisciplinary approaches, including the Markov chain model, least squares support vector machine and scenario prediction, we calculate Guangdong's GDP, industry structure, CO_2 emissions, and carbon intensity by 2015. Moreover, we evaluate the contribution of industry structure adjustment to the "12th five-year" carbon intensity target under nine combined scenarios. From this, we make the following conclusions:

(i) Industry structure adjustment is an effective measure for driving the reduction of carbon intensity. For a given level of economic growth, the larger the industry structure adjustment, the larger the "reduction amplitude" of the carbon intensity. For a given industry structure adjustment, the higher the economic growth, the larger the "reduction amplitude" of the carbon intensity.

(ii) Under the ideal scenario (*i.e.*, "high-speed economic growth" and "substantial industry structure adjustment"), industry structure adjustment contributes most to the realization of the carbon intensity goal, with a contribution of 130.94%. Carbon intensity would be reduced by 25.53% in 2015 as compared to 2010. Under the conservative scenario (*i.e.*, "low-speed economic growth" and "minor industry structure adjustment"), the contribution of industry structure adjustment to meeting the carbon intensity goal will reach 122.50%. At the same time, the carbon intensity in 2015 will decrease by 23.89% as compared to 2010.

(iii) The reduction by 19.5% of the carbon intensity goal can be achieved under all the combined scenarios through industry structure adjustment. Thus, it can be concluded that the set target appears scientific and reasonable for Guangdong's government to reach its carbon intensity goal by 2015.

(iv) Although the obtained results show that the goal of reducing 19.5% of Guangdong's carbon intensity can be achieved, there are some limitations and uncertainties. In this paper, we have not taken into account the adoption and use of novel low carbon policies, which will change over time. How to capture them is the focus of one of our next works.

Acknowledgments: We thank two anonymous referees for their valuable comments. Our heartfelt thanks should also be given to the National Natural Science Foundation of China (NSFC) (71201010, 71303174 and 71473180), National Philosophy and Social Science Foundation of China (14AZD068, 15ZDA054), Natural Science Foundation for Distinguished Young Talents of Guangdong (2014A030306031), Soft Science Foundation of Guangdong (2014A070703062), Social Science Foundation of Guangdong (GD14XYJ21), Distinguished Young Teachers of Guangdong ([2014]145), High-level Personnel Project of Guangdong ([2013]246), Guangdong key base of humanities and social science: Enterprise Development Research Institute and Institute of Resource, Environment and Sustainable Development Research, and Guangzhou key base of humanities and social science: Centre for Low Carbon Economic Research for funding supports.

Author Contributions: This work was designed, analyzed, and written by the first author. The data were gathered and disposed by both the authors. The LSSVM modeling was built by the second author.

Conflicts of Interest: The authors declare no conflict of interest.

References

1. Xiang, N.; Xu, F.; Sha, J.H. Simulation Analysis of China's Energy and Industrial Structure Adjustment Potential to Achieve a Low-carbon Economy by 2020. *Sustainability* **2013**, *5*, 5081–5099.
2. Uwasu, M.; Jiang, Y.; Saijo, T. On the Chinese Carbon Reduction Target. *Sustainability* **2010**, *2*, 1553–1557.
3. Yi, W.J.; Zou, L.L.; Guo, J.; Wang, K.; Wei, Y.M. How can China reach its CO_2 intensity reduction targets by 2020? A regional allocation based on equity and development. *Energy Policy* **2011**, *39*, 2407–2415.
4. Li, H.Q.; Wang, L.M.; Shen, L.; Chen, F.N. Study of the potential of low carbon energy development and its contribution to realize the reduction target of carbon intensity in China. *Energy Policy* **2012**, *41*, 393–401.
5. Yuan, J.H.; Hou, Y.; Xu, M. China's 2020 carbon intensity target: Consisteney, implementations, and policy imp1ications. *Renew. Sustain. Energy Rev.* **2012**, *16*, 4970–4981.
6. Liu, L.W.; Zong, H.J.; Zhao, E.D.; Chen, C.X.; Wang, J.Z. Can China realize its carbon emission reduction goal in 2020: From the perspective of thermal power development. *Appl. Energy* **2014**, *124*, 199–212.
7. Jiao, J.L.; Qi, Y.Y.; Qun, C.; Liu, L.C.; Liang, Q.M. China's targets for reducing the intensity of CO_2 emissions by 2020. *Energy Strateg. Rev.* **2013**, *2*, 176–181.
8. Zhou, L.; Zhang, X.L.; Qi, T.Y.; He, J.K.; Luo, X.H. Regional disaggregation of China's national carbon intensity reduction target by reduction pathway analysis. *Energy Sustain. Dev.* **2014**, *23*, 25–31.

9. Cui, L.B.; Fan, Y.; Zhu, L.; Bi, Q.H. How will the emissions trading scheme save cost for achieving China's 2020 carbon intensity reduction target? *Appl. Energy* **2014**, *136*, 1043–1052.
10. Wang, X.W.; Cai, Y.P.; Xu, Y.; Zhao, H.Z.; Chen, J.J. Optimal strategies for carbon reduction at dual levels in China based on a hybrid nonlinear grey-prediction and quota-allocation model. *J. Clean. Prod.* **2014**, *83*, 185–193.
11. Yu, H.; Hua, L.; Wei, Y.M. Is China's carbon reduction target allocation reasonable? An analysis based on carbon intensity convergence. *Appl. Energy* **2015**, *142*, 229–239.
12. Zhu, B.Z.; Wang, K.F.; Chevallier, J.; Wang, P.; Wei, Y.M. Can China achieve its carbon intensity target by 2020 while sustaining economic growth? *Ecol. Econ.* **2015**, *119*, 209–216.
13. Jiusto, S. The differences that methods make: Cross-border power flows and accounting for carbon emissions from electricity use. *Energy Policy* **2006**, *34*, 2915–2928.
14. Yi, H. Clean Energy Policies and Electricity Sector Carbon Emissions in the U.S. States. *Util. Policy* **2015**, *34*, 19–29.
15. Cai, W.; Wang, C.; Chen, J.; Wang, S. Green economy and green jobs: Myth or reality? The case of China's power generation sector. *Energy* **2011**, *36*, 5994–6003.
16. Yi, H.; Liu, Y. Green economy in China: Policy drivers and regional variations. *Glob. Environ. Change* **2015**, *31*, 11–19.
17. Huang, Y.F.; Wang, C.N.; Dang, H.S.; Lai, S.T. Predicting the Trend of Taiwan's Electronic Paper Industry by an Effective Combined Grey Model. *Sustainability* **2015**, *7*, 10664–10683.
18. IPCC. *Greenhouse Gas Inventory: IPCC Guidelines for National Greenhouse Gas Inventories*; United Kingdom Meteorological Office: Bracknell, UK, 2006.
19. Zhang, L.; Li, Y.M.; Huan, Y.X.; Wuan, Y.M. Analysis on character and potential of energy saving and carbon reducing by structure evolution in China. *Chin. Soft Sci.* **2011**, *2*, 42–51.
20. Suykenns, J.A.K.; Vandewalle, J. Least squares support vector machine. *Neural Process. Lett.* **1999**, *9*, 293–300.
21. Smola, A.J. Learing with Kernels. Ph.D. Thesis, Department of Computer Science, Technical University, Berlin, Germany, 1998.
22. Silva, D.A.; Silva, J.P.; Neto, A.R.R. Novel approaches using evolutionary computation for sparse least square support vector machines. *Neurocomputing* **2015**, *168*, 908–916.
23. Kennedy, J.; Eberhart, R.C. Particle swarm optimization. In Proceedings of the IEEE Conference on Neural Network, Perth, Australia, 27 November–1 December 1995; Volume 4, pp. 1942–1948.
24. Zhu, B.Z.; Wei, Y.M. Carbon price prediction with a hybrid ARIMA and least squares support vector machines methodology. *Omega* **2013**, *41*, 517–524.
25. Statistics Bureau of Guangdong Province. *Guangdong Statistical Yearbook (2015)*; China Statistics Press: Beijing, China, 2015.

Measuring the Total-Factor Carbon Emission Performance of Industrial Land Use in China Based on the Global Directional Distance Function and Non-Radial Luenberger Productivity Index

Wei Wang, Hualin Xie, Tong Jiang, Daobei Zhang and Xue Xie

Abstract: Industry is a major contributor to carbon emissions in China, and industrial land is an important input to industrial production. Therefore, a detailed analysis of the carbon emission performance of industrial land use is necessary for making reasonable carbon reduction policies that promote the sustainable use of industrial land. This paper aims to analyze the dynamic changes in the total-factor carbon emission performance of industrial land use (TCPIL) in China by applying a global directional distance function (DDF) and non-radial Luenberger productivity index. The empirical results show that the eastern region enjoys better TCPIL than the central and western regions, but the regional gaps in TCPIL are narrowing. The growth in NLCPILs (non-radial Luenberger carbon emission performance of industrial land use) in the eastern and central regions is mainly driven by technological progress, whereas efficiency improvements contribute more to the growth of NLCPIL in the western region. The provinces in the eastern region have the most innovative and environmentally-friendly production technologies. The results of the analysis of the influencing factors show implications for improving the NLCPIL, including more investment in industrial research and development (R&D), the implementation of carbon emission reduction policies, reduction in the use of fossil energy, especially coal, in the process of industrial production, actively learning about foreign advanced technology, properly solving the problem of surplus labor in industry and the expansion of industrial development.

Reprinted from *Sustainability*. Cite as: Wang, W.; Xie, H.; Jiang, T.; Zhang, D.; Xie, X. Measuring the Total-Factor Carbon Emission Performance of Industrial Land Use in China Based on the Global Directional Distance Function and Non-Radial Luenberger Productivity Index. *Sustainability* **2016**, *8*, 336.

1. Introduction

China's industrial economy has achieved remarkable progress since its reform and opening up in the late 1970s. According to the *China Statistical Yearbook 2013*, in the past 30 years, the average annual growth rate of the industrial gross domestic product (GDP) was approximately 16%, which is an amazing number compared to the industrial GDPs of most other countries [1]. However, the rapid industrial economic growth has been accompanied by an increased consumption of fossil energy. The amount of fossil energy consumption for production was as high as 3.62 billion tons of standard coal in 2012, and a large part of the coal consumption was used for industrial production [2]. In addition, coal consumption accounts for the main part of the fossil energy consumption in China and releases more harmful gases (e.g., carbon dioxide and sulfur dioxide) compared to some types of clean energy (e.g., nuclear and wind power) for the same given amount [3]. In fact, China became the largest energy consumer and carbon emitter in the world prior to 2010 [4], which indicate that China must face considerable pressure to promote energy-saving and low carbon industrial production. Although the famous *Kyoto Protocol* does not give China a mandatory carbon emissions reduction task, as a responsible country that has been trying to achieve sustainable development for a long time, China should pay more attention to optimizing its energy structure and effectively reducing its carbon emissions [5].

Fortunately, the central government of China has already started to pay attention to this problem and has tried to solve it. Many policies aimed at addressing excessive carbon emissions have been launched in recent years, such as the *Special Action of Response to Climate Change Science and Technology* issued in 2007, which sought to speed up the development of cleaner production technologies and reduce carbon emissions in China [6]. In addition, the *National Program on Addressing Climate Change* was released in the same year and sought to limit the total amount of greenhouse gas emissions and punish enterprises and local governments that did not comply with those policies [7]. Two years later, the central government of China promised a reduction of 45% in carbon emission intensity by 2020 compared to that in 2005, which was included in the long-term planning of national economic and social development in China [8]. This promise has undoubtedly been conducive to the sustainable development of China's industries. However, whether those policies have achieved remarkable effects remains unknown.

According to the *China Urban Construction Statistical Yearbook*, industrial land refers to land for industrial production in built-up urban areas including factories and workshops [5–9]. As an important input in the industrial production process, the amount of industrial land consumption has shown an obvious increasing trend in recent years. In 2012, industrial land accounted for approximately 20% of the total area of urban built-up zone in China [10]. However, the proportion of industrial land in most developed countries is less than 10% (e.g., France and Japan). This implies excessive industrial land use in China. In fact, the vigorous land expansion of recent years across China has led to a seriously inefficient use of industrial land, and a large amount of industrial lands have not been used for many years in some cities. The poor economic performance of industrial land is common in China [11]. In addition, industrial land is the main carrier of industrial production activities and suffers most from industrial pollutants. That is to say, industrial land is constantly used to create industrial products, but it also suffers serious problems of ecological environmental pollution. As the famous *Outline of the 12th Five-Year Plan for Economic and Social Development* noted, improving land use efficiency was a key problem of urban development [12]. Therefore, how to quantify and improve the economic and environmental performance of industrial land use in China is a key issue to realize the sustainable development of Chinese industry.

Regarding the research method, many previous studies of land efficiency preferred using single factor indexes such as the economic output per km^2 of land [13]. However, as noted in some previous studies, land cannot produce products unless accompanied by other inputs such as labor and capital, and land efficiency computed under the total factor framework is more reasonable [14]. Therefore, the data envelopment analysis (DEA) model is widely adopted by current studies because the input and output variables are incorporated into the model. Chen *et al.* [15] have analyzed industrial land use efficiency in China using a DEA model at the provincial level and found that China's eastern region enjoys a higher industrial land use efficiency than in the central and western regions. Xiong *et al.* [16] have reached a similar conclusion and suggested that excessive inputs of industrial land were the main reasons for the inefficient use of industrial land. However, these studies consider only the economic efficiency of industrial land use, and negative outputs were not included in the models. Therefore, they can be considered as partial analyses because they ignored the negative impacts on the environment caused by industrial production. Guo *et al.* [17] have incorporated three main negative outputs (*i.e.*, the amounts of industrial sulfur dioxide, industrial wastewater, and industrial dust emissions) into a DEA model to measure the combined efficiency of economy and environment for 33 typical cities in China. They found that the values of efficiency computed from models that incorporated negative outputs were always lower than those from models that ignored negative outputs. Xie *et al.* [18] have applied a similar model to explore industrial land use efficiency at the city level in China, and they found that cities located in economically-developed regions performed much better than those in economically undeveloped regions. However, these results were based on contemporaneous production technology, and they were in fact static analyses, which can be referred to as cross-sectional rather than time series analysis. Therefore, the efficiencies in different years could not be compared. Zhang *et al.* [19] have applied an advanced directional distance function (DDF) approach based on global benchmark technology. Global technology enveloped all of the contemporaneous technologies, which indicated that the model actually provides a dynamic analysis, and the results for different periods therefore could be compared to one another. Moreover, Zhang *et al.* [20] noted that the traditional radial DDF model had a limitation that reduced inputs and expanded outputs at the same rate. This is clearly not consistent with reality, and they proposed a non-radial DDF to reduce inputs and expand outputs at different rates. Therefore, their model was capable of providing a more reasonable assessment.

To obtain more insights into the dynamic changes in land use efficiency, two popular approaches, the Malmquist and Luenberger indices, have been widely used in recent related studies. Both indices can decompose productivity changes into efficiency and technological changes to explore the main contributors to changes in productivity. Chung *et al.* [21] have proposed a relatively advanced approach, the Malmquist-Luenberger (ML) index, to incorporate negative outputs. Many later studies adopted this method to model environmentally sensitive productivity growth at the national level [22], regional level [23] and industry level [24]. However, as Boussemart *et al.* [25] have noted, the Malmquist index tends to overestimate productivity changes because it calculates the growth in productivity as a ratio. The results measured by the Luenberger index seem more reasonable because they are calculated in an additive way and are only half of those computed using the Malmquist index approach. The difference in the two methods arises because the former adopts the geometric means of the distance functions, whereas the latter uses the arithmetic means of the distances in two periods. Many recent studies have noted that the results based on the Luenberger index are more robust than those based on the Malmquist index [26], and Chang *et al.* [27] have made a further improvement by applying a non-radial Luenberger index. This approach was widely applied in later studies to model environmentally sensitive productivity [28].

Unfortunately, there are no studies on the dynamic changes in the total-factor carbon emission performance of industrial land use (TCPIL) in China. You *et al.* [29] have measured the carbon emission efficiencies in China's 30 regions by employing a traditional DEA model. However, they treated the amount of carbon emissions as an input in the production process, which was inconsistent with the actual production. Cui *et al.* [30] have modeled the carbon emission performance of urban non-agricultural land for China using a Malmquist index. However, the study was actually a static analysis because it was based on contemporaneous production technology, and it could not depict dynamic changes in the carbon emission performance of land use. In addition, many previous studies ignored carbon emission performance when exploring the negative impacts of pollutants on the environment in the process of industrial production. However, considering the serious threat to human health and socioeconomic development caused by greenhouse gases, which are mainly composed of carbon dioxide, it is obviously meaningful to consider carbon emissions.

This paper aims to apply a global DDF and non-radial Luenberger productivity index to analyze the dynamic changes in TCPIL for China. This total factor index can be referred to as the non-radial Luenberger carbon emission performance of industrial land use (NLCPIL). We then explore the main contributors to the growth in NLCPIL by decomposing the NLCPIL into two indices, *i.e.*, efficiency change (EC) and technological change (TC), and we further find which provinces have made innovations in carbon performance. Lastly, we explore the impacts on the NLCPIL of energy utilization, production technology and environmental policy factors in Chinese industry to present some policy implications.

Therefore, this paper makes three main contributions to the relevant studies. Firstly, we compute the TCPIL for each province in China under a global environmental technology framework. Secondly, we compute the NLCPIL to measure the dynamic changes in the TCPIL and determine which NLCPIL component index, *i.e.*, EC and TC, is the main contributor to the growth of NLCPIL. Lastly, we learn which provinces have provided innovation and leadership in environmentally friendly production technologies as examples of provinces that have poor TCPILs.

The remainder of this paper is organized as follows: Section 2 introduces the methods and data, Section 3 shows the results of the empirical analysis, and Section 4 concludes the paper with some policy implications.

2. Methods and Data

2.1. Non-Radial Directional Distance Function (NDDF)

We assume that there are N provinces in our study and that each province has M inputs (x) to produce J desirable outputs (y) and K undesirable outputs (b), and the production possibility set $T(x)$ can be expressed as

$$T\left(x\right)=\left\{\left(x,y,b\right)x\ can\ produce\ \left(y,b\right),\ x\geqslant X\lambda,\ y\leqslant Y\lambda, b=B\lambda, \lambda\geqslant 0\right\} \quad (1)$$

where the production possibility set $T(x)$ is assumed to satisfy the production function theory [31]. This theory states that reducing undesirable outputs during the production process is costly, and industrial production will inevitably bring about carbon dioxide emissions [32]. In addition, the traditional radial DDF approach always assumes that the solution of the linear programming implies an assumption that we should reduce the inputs (or undesirable outputs) and expand the outputs at the same rate β, as expressed in Equation (2), and are $g=\left(-g_x,g_y,-g_b\right)$ the direction vectors [33]. However, this is almost impossible in real production. To address this shortcoming, a NDDF approach has been developed and is widely used in studies of resource efficiency evaluations [34]. In Equation (3), $w^T=\left(x,y,b\right)^T$ is the standard weight matrix of inputs and outputs. $\beta=\left(-\beta_x,\beta_y,-\beta_b\right)$ refers to the adjustment ratios of all inputs, outputs and undesirable outputs, and they are nonnegative numbers. *diag* is the diagonal matrix. Using the NDDF, the adjustment ratios of the inputs and outputs can be different, which is consistent with actual production. Equation (4) represents the linear programming functions for the NDDF model.

$$\vec{D}\,(x,y,b;\,g) = \sup\{\beta : ((x,y,b) + g \times \beta) \in T\} \tag{2}$$

$$\vec{D}\,(x,y,b;\,g) = \sup\left\{w^T\beta : ((x,y,b) + g \times diag\,(\beta)) \in T\right\} \tag{3}$$

$$\vec{D}\,(\mathrm{x,y,b;\,g}) = \max\left(w_{LD}\beta_{LD} + w_Y\beta_Y + w_{CO2}\beta_{CO2}\right)$$
$$s.t. \begin{cases} \sum\limits_{n=1}^{N} \lambda_n LD_n \leqslant (1-\beta_{LD})LD_0\,,\ \sum\limits_{n=1}^{N} \lambda_n L_n \leqslant L_0\,,\ \sum\limits_{n=1}^{N} \lambda_n K_n \leqslant K_0\,,\ \sum\limits_{n=1}^{N} \lambda_n E_n \leqslant E_0, \\ \sum\limits_{n=1}^{N} \lambda_n Y_n \geqslant (1+\beta_Y)Y_0\,,\ \sum\limits_{n=1}^{N} \lambda_n CO2_n = (1-\beta_{CO2})CO2_0 \\ \beta_{LD} \geqslant 0, \beta_Y \geqslant 0, \beta_{CO2} \geqslant 0, \\ n = 1,2,...,N; t = 1,2,...,T;\ \lambda_n \geqslant 0,\ \sum\limits_{n=1}^{N} \lambda_n = 1 \end{cases} \tag{4}$$

In Equation (4), the superscripts *LD*, *Y* and *CO2* represent the industrial land used for industrial production, the industrial GDP, and the amount of carbon dioxide emissions during the industrial production process, respectively. *L* and *K* refer to the industrial labor and industrial capital. It is worth noting that we set the weight vector to (0, 0, 1/3, 1/3, 1/3) to remove the diluting effects of capital and labor from the constraints. The superscript 0 refers to the province under estimation. The symbols β_{LD}, β_{CO2} and β_Y are the reduction ratios of the industrial land and carbon dioxide emissions and the expand ratio of the industrial GDP, respectively. λ is a non-negative vector, and we impose a constraint of $\sum\limits_{n=1}^{N} \lambda = 1$ according to the assumption of variable returns to scale (VRS). The superscript *n* refers to the number of provinces in the sample. Thus, the TCPIL can be expressed as Equation (5)

$$\begin{aligned} TCPIL &= \frac{1 - 0.5 \times \left(\beta^*_{LD} + \beta^*_{CO2}\right)}{1 + \beta^*_Y} \\ &= \frac{0.5 \times \left(1 - \beta^*_{LD}\right)}{1 + \beta^*_Y} + \frac{0.5 \times \left(1 - \beta^*_{CO2}\right)}{1 + \beta^*_Y} = ECPIL + ENPIL \end{aligned} \tag{5}$$

where β^*_{LD}, β^*_{CO2} and β^*_Y are the optimal solutions of inputs and outputs for the province under estimation, and the province would be located along the production technology frontier in the *g* direction if its β^*_{LD}, β^*_{CO2} and β^*_Y have zero values. That is to say, the $\vec{D}^*$ $(LD,\ L,\ K,\ Y,\ CO2;\ g) = 0$. In addition, we decompose the TCPIL into two parts, the economic performance of industrial land use (ECPIL) and the environmental performance of industrial land use (ENPIL), in order to find out whether the ECPIL or the ENPIL is the main contributor to the imperfect TCPIL. The value for TCPIL ranges from 0 to unity, while those for ECPIL and ENPIL range from 0–0.5.

2.2. Non-Radial Luenberger Productivity Index

Considering that Equation (5) only presents a static analysis for the TCPIL, we should employ the non-radial Luenberger productivity index approach to perform a dynamic analysis. Because the traditional Malmquist-Luenberger (ML) index has the problematic potential to provide no solutions when dealing with extreme data, Oh [35] has combined the concept of productivity and the DDF and constructed a global Malmquist-Luenberger (GML) index instead of the traditional ML index. Therefore, the NLCPIL can be decomposed into two indices based on contemporaneous and global environmental production technologies, which can be denoted as T_n^C and T_n^G. T_n^C refers to the environmental production technology of a given group R_n at time t, n =1, 2, ..., N, and T_n^G envelopes the technologies for all groups for the entire study period. Therefore, we can define the global NDDF approach as in Equation (6):

$$\vec{D}^G(x,y,b;\ g) = \sup\left\{w^T\beta^G : \left((x,y,b) + g \times diag(\beta^G)\right) \in T_n^G\right\} \tag{6}$$

To compute and decompose the NLCPIL, four different NDDFs should be solved: $\vec{D}^C(x^s,y^s,b^s;\ g^s)$ and $\vec{D}^G(x^s,y^s,b^s;\ g^s)$, $s = t, t+1$. Therefore, we can solve the four NDDFs based on Equation (7):

$$\vec{D}^d(x^s,y^s,b^s;\ g) = \max\left(w_{LD}\beta^d{}_{LD} + w_Y\beta^d{}_Y + w_{CO2}\beta^d{}_{CO2}\right)$$
$$s.t.\begin{cases}\sum\limits_{con}\lambda_n^s LD_n^s \leqslant (1-\beta^s{}_{LD})LD_0,\ \sum\limits_{con}\lambda_n^s L_n^s \leqslant L_0,\ \sum\limits_{con}\lambda_n^s K_n^s \leqslant K_0,\ \sum\limits_{con}\lambda_n^s E_n^s \leqslant E_0,\\ \sum\limits_{con}\lambda_n^s Y_n^s \geqslant (1+\beta^s{}_Y)Y_0,\ \sum\limits_{con}\lambda_n^s CO2_n^s = (1-\beta^s{}_{CO2})CO2_0,\\ \beta^s{}_{WA} \geqslant 0, \beta^s{}_Y \geqslant 0, \beta^s{}_{CO2} \geqslant 0,\\ n = 1,2,\ldots,N;\ t = 1,2,\ldots,T;\ \lambda_n \geqslant 0,\ \sum\limits_{n=1}^{N}\lambda_n = 1\end{cases} \tag{7}$$

where the superscript d on $\vec{D}^d(x^s,y^s,b^s;\ g)$ refers to the type of NDDF, *i.e.*, contemporaneous and global. In addition, the symbol *con* under $\sum$ refers to the conditions for constructing the two environmental production technologies. Thus, we can construct the contemporaneous NDDF by having $d \equiv C$ and $con = \{n \in R_n\}$ and the global NDDF by having $d \equiv G$, and $con = \{n \in R_1 \cup R_2 \cup ... \cup R_N, s \in [1,2,...,t,...T]\}$. We can then solve the four NDDFs using Equation (8)

$$TCPIL^d\left(x^s, y^s, b^s;\ g\right) = \left[\frac{1 - 0.5 \times \left(\beta_{LD}^* + \beta_{CO2}^*\right)}{1 + \beta_Y^*}\right]^s \tag{8}$$

where $d \equiv (C, G)$ and $s = t, t + 1$. Then, we can define the NLCPIL as in Equation 9, which measures the dynamic changes of TCPIL. In addition, values of NLCPIL greater than, equal to or less than 1 indicate that the province under estimation is moving toward the global environmental production technology, is not changing, or is moving far away from the global environmental production technology, respectively. The two decomposition indices of NLCPIL are efficiency change (EC) and technological change (TC). During the period t and $t + 1$ if the EC is greater than, equal to or less than 0, this indicates that the technical efficiency has gained, has no change or has lost; and if the TC is greater than 0, this indicates a technological progress, and vice versa. In addition, the method solves the problem of infeasibility in linear programming whereby the results become circular.

$$\begin{aligned} & NLCPIL^d\left(x^s, y^s, b^s;\ g\right) = TCPIL^G\left(x^{t+1},\ y^{t+1},\ b^{t+1}\right) - TCPIL^G\left(x^t,\ y^t,\ b^t\right) \\ & = \left[TCPIL^{t+1}\left(.^{t+1}\right) - TCPIL^t\left(.^t\right)\right] \\ & + \left\{\left[TCPIL^G\left(.^{t+1}\right) - TCPIL^{t+1}\left(.^{t+1}\right)\right] - \left[TCPIL^G\left(.^t\right) - TCPIL^t\left(.^t\right)\right]\right\} \\ & = EC + TC \end{aligned} \tag{9}$$

In addition, by combining Equation (5) with Equation (9), we can explore whether the changes of ECPIL or ENPIL as Equations (10) and (11).

$$TCPIL^d\left(x^s, y^s, b^s;\ g\right) = ECPIL^d\left(x^s, y^s, b^s;\ g\right) + ENPIL^d\left(x^s, y^s, b^s;\ g\right) \tag{10}$$

therefore,

$$\begin{aligned} & NLCPIL^d\left(x^s, y^s, b^s; g\right) = NLCECPIL^d\left(x^s, y^s, b^s; g\right) + NLCENPIL^d\left(x^s, y^s, b^s; g\right) \\ & = ECPIL^G\left(x^{t+1},\ y^{t+1},\ b^{t+1}\right) - ECPIL^G\left(x^t,\ y^t,\ b^t\right) + ENPIL^G\left(x^{t+1},\ y^{t+1},\ b^{t+1}\right) - ENPIL^G\left(x^t,\ y^t,\ b^t\right) \end{aligned} \tag{11}$$

where NLCECPIL and NLCENPIL refer to the non-radial Luenberger economic performance of industrial land use, and the non-radial Luenberger environmental performance of industrial land use, respectively.

2.3. Data

According to geographical closeness and industrial development, we divided the provinces across China into three regions: eastern (E), central (C) and western (W). The eastern region is composed of three municipalities (Beijing, Tianjin, and

Shanghai) and nine coastal provinces (Hebei, Liaoning, Jiangsu, Zhejiang, Fujian, Shandong, Guangdong, Taiwan, and Hainan). This region enjoys the highest level of industrial development in China, with advanced industrial production technology and most of the foreign industrial enterprises in the whole country. The industrial GDP in this region accounted for more than 55% of the national industrial GDP in 2012 [36]. The central region consists of eight inland provinces (Shanxi, Jilin, Heilongjiang, Anhui, Jiangxi, Henan, Hubei, and Hunan). This region is the main base of heavy industry in China and is famous for its high resource consumption and large pollutant emissions. The western region consists of one municipality (Chongqing) and eleven inland provinces and autonomous regions (Inner Mongolia, Guangxi, Sichuan, Guizhou, Yunnan, Shaanxi, Gansu, Qinghai, Ningxia, Xinjiang, and Tibet). This area covers more territory than the other two areas, but the industry sectors in this area are less developed compared to the other two areas. Some regions in this area have faced serious water shortages. Because complete data over the study period could not be obtained, Tibet and Taiwan were not included in our sample.

The TCPIL is affected by economic and environmental factors, and we should therefore account for environmental outputs as well as economic outputs. We constructed an indicator system to evaluate the TCPIL using the following input and output indicators.

(1) The inputs consist of industrial land, industrial capital and industrial labor according to production function theory [32]. Industrial land refers to the area of industrial land, and industrial labor refers to the number of workers in industrial sectors. The industrial capital data cannot be obtained from the *China Statistical Yearbook* [36], and we can compute the data using the perpetual inventory method according to Zhang *et al.* [37]

$$K_t = (1 - \delta) K_{t-1} + I_t \tag{12}$$

where K_t, K_{t-1}, I_t and δ refer to industrial capital stock at times t and $t - 1$, I_t represents the investment in industrial fixed assets, and δ refers to the rate of depreciation at time t. We used the industrial capital stock in 2003 as the initial industrial capital stock, because 2003 was the first year in our study. The investments in industrial fixed assets and depreciation rates can be obtained from the *China Statistical Yearbook 2004–2013* [36]. In addition, the rates of depreciation and capital stock numbers in this paper are procured from Wu [38], and we estimate the data after 2006 using the perpetual inventory method since they were not available.

(2) Regarding output indicators, we selected the industrial GDP as the desirable output, and the amount of carbon emissions as the undesirable output. Because data on carbon emissions are unavailable, we compute the carbon emissions for each

province based on regional energy balance tables and the IPCC guidelines [39,40]. The formula for computing carbon emissions is shown in Equation (13)

$$CO_2 = \sum_{i=1}^{n} A \times CCF_i \times HE_i \times COF_i \times \left(\frac{44}{12}\right) \quad (13)$$

where A refers to the amount of carbonaceous fuel combusted during the process of industrial production, which can be obtained from the *China Energy Statistics Yearbook* [36]. CCF, HE and COF represent the carbon content factor, heat equivalent and carbon oxidation factor of the carbonaceous fuel, which can be found in the IPCC guidelines. The number (44/12) refers to the ratio of weight of carbon dioxide (44 atomic mass units) to the molecular weight of carbon (12 atomic mass units). We selected representative carbonaceous fuels such as coal, petrol, kerosene, diesel, fuel oil and nature gas. By doing so, the carbon emissions can be calculated. Moreover, to exclude the impact of price, the GDP and capital variables were converted into year 2003 constant pieces with their deflators.

Additionally, to put forward effective advice for improving the NLCPIL, we selected influencing indicators from the aspects of economy, society and policy. The econometric model is

$$Y_{it} = \alpha_{it} + \beta_1 EI_{it} + \beta_2 ES_{it} + \beta_3 RI_{it} + \beta_4 FI_{it} + \beta_5 LS_{it} + \beta_6 IS + \beta_7 POL + \varepsilon_{it} \quad (14)$$

where i and t (t = 2003, ..., 2013) represent the i-th province and year t in each zone, respectively. The term ε_{it} is the random error term. Y is the NLCPIL. Data on the influencing indicators were obtained from the *China City Statistical Yearbook 2004–2013* and *China Energy Statistics Yearbook 2004–2013* [41,42], and are as follows.

(1) We selected two indicators for energy utilization, energy intensity (EI) and energy structure (ES), which refer to the share of fossil energy used for industrial production in the total fossil energy consumption in the country and the share of coal consumption in the total fossil energy consumption for industrial production, respectively. It is generally accepted that less energy consumption used for the same GDP is more suitable for sustainable development [23]. Therefore, to produce the same amount of industrial product, smaller EIs always lead to better performance in TCPIL, and we can assume that the coefficient is negative. In addition, coal is the most widely used and possibly the most polluting resource. According to the *China Energy Statistics Yearbook* [42], the share of coal consumption in all fossil energy resource consumption has always exceeded 70% in recent years, which should be responsible for the worsening air condition across China. Therefore, the coefficient is expected to be negative.

(2) According to previous studies, technology plays an important role in driving industrial economic development, and more investment in research and development

(R&D) could lead to obvious positive impacts on industrial sustainable development by saving energy and reducing greenhouse gas emissions (e.g., carbon dioxide) [43]. Therefore, we selected two indices, R&D intensity (R&D) and foreign funded industrial enterprises introduction (FI), which refer to the share of industrial R&D investment in the industrial GDP and the share of the GDP produced by foreign funded industrial enterprises in the national industrial GDP, respectively. The coefficients of the two indices are expected to be positive.

(3) Regarding industrial structure, we selected two indices, industrial labor share (LS) and industrial GDP share (IS), which refer to the share of workers in the industrial sectors to the total number of workers nationwide and the share of GDP in industrial sectors to the total GDP, respectively. According to previous studies, China has faced serious problems of labor surpluses, and the industrial sector has also experienced these problems [44]. Thanks to inexpensive labor, most industrial enterprises in China are labor-intensive and are engaged in low-tech production activities. However, this inevitably impedes the large-scale use of new technologies and upgrading of China's industries and negatively impacts NLCPIL. Thus, we assumed that there is a negative relationship between LS and NLCPIL. In addition, many studies have suggested that China has entered a middle and late stage of industrialization, which implies that China's industrial development will shift from a simple emphasis on economic output to improvements in the quality of industrial development. This might lead to a reduction of the share of GDP in industrial sectors in the total GDP. Thus, we can assume that an increase in IS has a negative impact on the improvement of NLCPIL.

(4) Regarding environmental policies, the central government of China promised an ambitious plan to reduce its carbon emissions per unit GDP by 40%–45% based on 2005 levels in late 2009 [36]. Subsequently, a series of policies aimed at saving energy and reducing carbon emissions was introduced, and the central government of China also strengthened the supervision of local governments to implement those policies [45]. Therefore, the environmental policy variable was assigned a value of 1 from 2009 onward and a value of 0 before 2009.

Table 1 shows the definitions, descriptions and expected impacts of the influencing indicators on the NLCPIL.

Table 1. Descriptions of the influential factors.

Variable	Definition	Description	Expected Effect
EI	Energy intensity	The share of fossil energy used for industrial production in the total consumption in the country	Negative
ES	Energy structure	The share of coal consumption in the total fossil energy consumption for industrial production	Negative
RI	Research and development intensity	The share of investment in industrial research and development in the industrial GDP	Positive
FI	Foreign funded industrial enterprises introduction	The share of GDP produced by foreign funded industrial enterprises in the national industrial GDP	Positive
LS	Industrial labor share	The share of workers in industrial sectors in the total number of workers nationwide	Negative
IS	Industrial GDP share	The share of GDP in industrial sectors in the total GDP	Negative
POL	Carbon emission reduction policy	The ambitious policies to reduce carbon emissions since 2009	Positive

3. Empirical Results

In this section, we first show the status of the TCPIL during the study period. We then calculate the NLCPIL and its decomposition indices to measure the dynamic changes in TCPIL at the national and regional levels. Lastly, we learn which provinces offer innovative examples for improving environmentally-friendly production technologies.

3.1. TCPIL

Using Equations (4) and (5), we computed the TCPIL for the three regions across China. From Table 2, we note that the average TCPIL was 0.4939. Thus, China has a large potential to improve its TCPIL. As shown in Figure 1, the TCPIL showed an upward trend over the study period in China, except for 2005 and 2009. This may be due to the rapid development of heavy industry in western China approximately in 2005 and the economic stimulus plans introduced by the central government of China that were aimed at reducing the negative impact on China's industrial economy caused by the international financial crisis in 2008 [46]. These two policies quickly achieved their expected effects, but the amount of energy consumption and carbon emissions increased significantly. In addition, according to Equation (5) the average values of ECPIL and ENPIL are 0.278 and 0.216, respectively. The value of ENPIL for China is much lower than that of ECPIL in every year of the study period. Thus, solving the environmental problems of excessive carbon dioxide emissions in industrial production would be very helpful to improve the TCPIL. The results also prove that environmental policies, especially those on the control of carbon dioxide emissions, are necessary.

At the regional level, all three regions shared trends similar to those at the national level. The eastern region enjoyed the highest TCPIL over the study period, with an average value of 0.6143. It was followed by the central and western regions, which had average values of 0.4371 and 0.4149, respectively. We found that the eastern region had a much better TCPIL than the other two regions, which may be due to the eastern region's relatively developed industrial production technologies and effective environmental protections. This is consistent with the observations of Xiong *et al.* [16]. At the provincial level, Hainan had the highest average TCPIL, 0.9485, followed by Zhejiang, Guangdong and Fujian, which had average TCPILs of 0.7483, 0.7359 and 0.7339, respectively. The four provinces are located in the central region, but Ningxia, which suffered the poorest TCPIL, 0.2587, is in the western region. Ningxia's poor TCPIL may be the result of its relatively underdeveloped industry development, which is in turn due to its lack of natural resources and relatively small industrial sector. According to the *China Statistical Yearbook 2013* [36], the industrial GDP in Ningxia only accounted for 0.35% of the national industrial GDP. Thus, provinces with relatively underdeveloped industrial economies (e.g., Ningxia and Gansu) and a preference for heavy industry (e.g., Liaoning, Jilin and Heilongjiang) always suffered relatively poor TCPILs, whereas those with developed industrial economies (e.g., Zhejiang, Guangdong and Jiangsu) and better environments (e.g., Hainan and Fujian) always enjoyed better TCPILs. To analyze the dynamic changes in TCPIL in detail, the NLCPIL for China and its three regions is discussed in the next subsection.

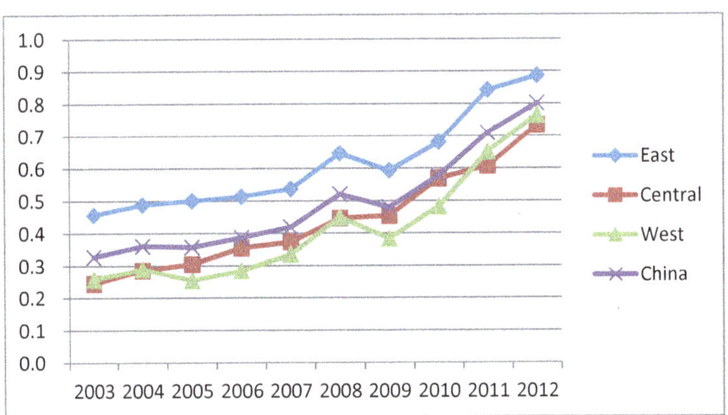

Figure 1. Trends in total-factor carbon emission performance of industrial land use (TCPIL) for China and its three regions, 2003–2012.

Table 2. TCPIL at the provincial level for all three regions, East (E), Central (C) and West (W) China 2003–2012.

Province	Region	2003	2004	2005	2006	2007	2008	2009	2010	2011	2012	Mean
Beijing	E	0.2375	0.2977	0.3540	0.3288	0.3351	0.3523	0.3715	0.3861	0.8488	1.0000	0.4512
Tianjin	E	0.3304	0.3692	0.4679	0.5151	0.5588	0.6391	0.6252	0.7452	0.9101	1.0000	0.6161
Hebei	E	0.3073	0.3581	0.3366	0.3878	0.4301	0.4763	0.4631	0.5453	0.7900	1.0000	0.5095
Liaoning	E	0.1875	0.2213	0.2634	0.2845	0.3519	0.5135	0.4135	0.4637	0.4915	0.5143	0.3705
Shanghai	E	0.3483	0.3707	0.3654	0.3870	0.4048	0.4437	0.3991	0.4114	0.4031	0.3864	0.3920
Jiangsu	E	0.5231	0.5041	0.4888	0.5245	0.5536	0.5939	0.6449	0.7138	1.0000	1.0000	0.6547
Zhejiang	E	0.6141	0.5800	0.5879	0.6369	0.6854	0.7553	0.7469	0.8762	1.0000	1.0000	0.7483
Fujian	E	0.5982	0.6607	0.5679	0.6494	0.6320	0.7354	0.7245	0.7974	1.0000	0.9730	0.7339
Shandong	E	0.4153	0.4488	0.4458	0.4873	0.5401	0.5989	0.6309	0.7008	0.8207	0.8775	0.5966
Guangdong	E	0.4588	0.5564	0.6241	0.6814	0.6880	1.0000	0.5039	0.8463	1.0000	1.0000	0.7359
Hainan	E	1.0000	1.0000	1.0000	0.7651	0.7203	1.0000	1.0000	1.0000	1.0000	1.0000	0.9485
Shanxi	C	0.1771	0.2325	0.3212	0.3230	0.3829	0.4618	0.4119	0.5212	0.6352	0.5856	0.4052
Jilin	C	0.1762	0.2136	0.2225	0.2431	0.3049	0.3719	0.4037	0.4622	0.5313	0.6473	0.3577
Heilongjiang	C	0.3026	0.3560	0.3111	0.3275	0.3193	0.3701	0.3336	0.4356	0.5379	0.5465	0.3840
Anhui	C	0.2102	0.2574	0.2477	0.3283	0.3223	0.3787	0.4140	0.5077	0.4579	0.7501	0.3874
Jiangxi	C	0.2588	0.3063	0.4082	0.5491	0.4559	0.5052	0.5356	0.6749	0.6810	0.7908	0.5166
Henan	C	0.3156	0.3568	0.4097	0.4146	0.4917	0.6114	0.6066	0.8383	0.9934	1.0000	0.6038
Hubei	C	0.2536	0.2733	0.2378	0.3190	0.3221	0.3628	0.4228	0.4351	0.4265	0.5471	0.3600
Hunan	C	0.2628	0.2855	0.2794	0.3382	0.3839	0.5039	0.5009	0.6723	0.5973	1.0000	0.4824
Inner Mongolia	W	0.1685	0.2226	0.2742	0.3109	0.3836	0.4766	0.4895	0.7098	1.0000	1.0000	0.5036
Guangxi	W	0.2103	0.2557	0.2772	0.3131	0.3824	0.6516	0.4884	0.6788	0.8272	0.8775	0.4962
Chongqing	W	0.1810	0.2193	0.1839	0.2195	0.2558	0.3081	0.3311	0.3727	0.4308	0.5373	0.3040
Sichuan	W	0.1508	0.1801	0.2001	0.2660	0.3164	0.3351	0.3516	0.3745	0.4927	0.6646	0.3332
Guizhou	W	0.1647	0.1746	0.2114	0.2384	0.2840	0.3540	0.3427	0.3987	0.3663	0.5419	0.3077
Yunnan	W	0.3251	0.4284	0.3547	0.3830	0.3562	0.3984	0.3820	0.4672	0.5172	0.9554	0.4568
Shaanxi	W	0.2476	0.2214	0.3429	0.4050	0.4049	1.0000	0.5325	0.6141	0.6956	1.0000	0.5464
Gansu	W	0.1502	0.1841	0.1782	0.2064	0.2443	0.2810	0.2746	0.3620	0.3997	0.4253	0.2706
Qinghai	W	1.0000	1.0000	0.4277	0.4128	0.5240	0.5242	0.5125	0.6008	1.0000	1.0000	0.7002
Ningxia	W	0.0959	0.1445	0.1357	0.1576	0.3069	0.2305	0.2730	0.3990	0.4322	0.4120	0.2587
Xinjiang	W	0.1410	0.1507	0.2041	0.2069	0.2211	0.3713	0.2427	0.3238	1.0000	1.0000	0.3862
East		0.4564	0.4879	0.5002	0.5134	0.5364	0.6462	0.5930	0.6806	0.8422	0.8865	0.6143
Central		0.2446	0.2852	0.3047	0.3554	0.3729	0.4457	0.4536	0.5684	0.6076	0.7334	0.4371
West		0.2577	0.2892	0.2536	0.2836	0.3345	0.4483	0.3837	0.4819	0.6511	0.7649	0.4149
China		0.3271	0.3610	0.3577	0.3870	0.4188	0.5202	0.4791	0.5778	0.7095	0.8011	0.4939

3.2. NLCPIL and Its Decompositions

As shown in Figure 2 and Table 3, the values of NLCPIL were above zero in most years in the study period for China, and the average value was 0.0527, which indicated that the NLCPIL increased by approximately 5.27% per year over the study period. This implies an obvious improvement in TCPIL. The NLCPIL was less than zero only for the periods 2004–2005 and 2008–2009, and had values of −0.0033 and −0.0411, respectively. This indicates that the TCPIL decreased by approximately 0.33% and 4.11% in the two periods, respectively. With regard to the decomposition indices of the NLCPIL, Table 3 shows that the average EC and TC were 0.0258 and 0.0268 under the NLCPIL framework. This indicates that the environmental efficiency for provinces in China increased by approximately 2.58% per year during the study period, and the contemporaneous technology frontier moved toward the global technology frontier at an annual average rate of approximately 2.68% per year for the same period. It is worth noting that the NLCPIL and TC shared trends, that of EC, especially after 2009. In addition, the EC values were greater than zero prior to the period 2008–2009 and were higher than those of TC. However, the TC values noticeably increased and remained above zero after 2008–2009, whereas the EC values sharply decreased in the period 2010–2011. Therefore, we have found that the growth in NLCPIL was mainly driven by EC before 2009 and by TC afterwards. This may have been due to increased production costs caused by increasing investments in R&D as a result of environmental regulations and the use of expensive clean energy (e.g., electricity), but it would promote technological progress and improve the enterprise competitiveness in the long run. Thus, this result demonstrates the effectiveness of carbon emission reduction policies since 2009, and it is consistent with the findings of Zheng *et al.* [47]. The results also provide evidence for the Porter hypothesis, which states that strict environmental regulation is conducive to the improvement of enterprise competitiveness and resource use efficiency [48]. Additionally, according to Equations (10) and (11), we have found that during the study period, the average values of NLCECPIL and NLCENPIL for China are 0.0273 and 0.0254, respectively. Thus, the ECPIL makes faster progress than the ENPIL, which implies more effective policies and regulations on industrial carbon emission reduction need to be issued urgently.

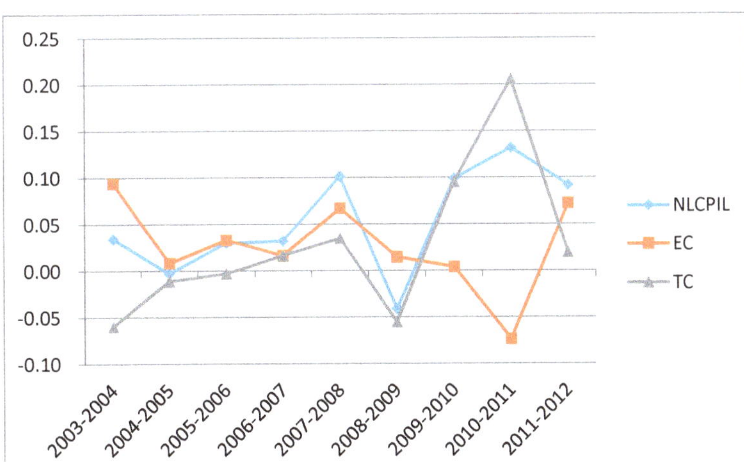

Figure 2. Trends in non-radial Luenberger carbon emission performance of industrial land use (NLCPIL) and its decomposition indices for China, 2003–2013, including efficiency change (EC) and technological change (TC).

Table 3. NLCPIL and its decomposition indices for the three regions.

	NLCPIL	EC	TC
E	0.0478	0.0122	0.0356
C	0.0543	0.0187	0.0357
W	0.0563	0.0446	0.0117
China	0.0527	0.0258	0.0268

Tables 3–6 provide detailed information. As Table 3 shows, at the regional level, the NLCPIL and its two decomposition indices were greater than zero in the three regions, which indicates that the three regions enjoyed continuous progress in NLCPIL, EC and TC as a whole during the study period. Specifically, the western region enjoyed the highest NLCPIL with a value of 0.0563, and it was followed by the central and eastern regions, which had values of 0.0543 and 0.0478, respectively. This indicates that the annual average growth rates of NLCPIL for the three regions were 5.63%, 5.43% and 4.78%, respectively. This is in contrast with the TCPIL's results; because the eastern region enjoyed the best TCPIL and the western region suffered the poorest TCPIL. This may because NLCPIL measures dynamic changes in TCPIL, and the central and western regions were trying their best to catch up with the eastern region. Therefore, the regional gaps in TCPIL seem to be narrowing, which is helpful for realizing the balanced and sustainable utilization of industrial lands in China. This is consistent with the findings of Xie *et al.*, which reached a similar conclusion at both the provincial and city levels [9].

Specifically, as shown in Table 4, the average NLCPIL for the provinces were greater than zero, with the exception of Hainan and Qinghai, whose values were equal to zero. This implies that the TCPIL remained unchanged for the two provinces over the study period as a whole. In addition, Xinjiang enjoyed the best performance of NLCPIL, with an average value of 0.0954; it was followed by Inner Mongolia (0.0924) and Beijing (0.0847). This may be partly because the industrial land input in those places showed obvious reductions during the last few years of the study period, which implies that the local governments there successfully blocked the blind expansion of industrial lands and improved land utilization efficiency. In addition, Beijing, which is a large emitter of industrial carbon emissions, has best endeavored to reduce carbon emissions, and it has achieved notable success in carbon emission reductions since the Olympic Games by limiting the number of motor vehicles and by moving heavily polluting enterprises out of the city. Moreover, Xinjiang and Inner Mongolia have been committed to the development of industries with high economic output and low carbon emissions in recent years, such as wind power technology. This is consistent with the findings of Zhang *et al.* [28].

With regard to the decomposition indices of NLCPIL, Table 3 shows that the western region ranks first in average EC with the value of 0.0446, followed by the eastern and central regions, which have values of 0.0187 and 0.0122, respectively. This indicates that the western region enjoyed the best "catch-up" effect in terms of changes in the efficiency of carbon emissions for industrial land use. In addition, as shown in Table 5, Xinjiang had the highest EC, 0.072, and it was followed by Inner Mongolia (0.0701) and Guizhou (0.0687). Heilongjiang, Shanghai and Shandong suffered relatively poor ECs of −0.0264, −0.0186 and −0.0123, respectively, which indicates that efficiency receded in those places. This may be due to the promotion of environment-friendly industrial production technologies that have brought additional costs to industry. Regarding TC, the central region shows the highest average TC of 0.0357, followed by the eastern (0.0356) and western regions (0.0117). Because TC measures technological progress over time, the results therefore imply that the production frontiers of the central, eastern and western regions are moving toward global low carbon production technology by approximately 3.57%, 3.56% and 1.17% per year, respectively. Henan had the highest TC, 0.076, following by Yunnan and Jiangxi, which had values of 0.07 and 0.0677, respectively. Ningxia suffered the poorest performance in TC, −0.0289, which implies a noticeable regression in environmental production technology.

Interestingly, the western region had relatively high ECs, whereas the eastern and central regions had relatively high TCs. Therefore, the growth in NLCPIL in the western region was mainly driven by improved efficiencies, and the eastern and central regions were more dependent on technological progress.

To establish good examples of environmentally friendly production technology across China, we try to find provinces that have made outstanding progress in the carbon emission performance of industrial land use and have pushed the technology frontier outward. We used three conditions to distinguish the innovators [35,49]:

$$\begin{aligned} & TC^{t,t+1} > 0, \\ & \vec{D}^{t}\left(x^{t+1}, y^{t+1}, b^{t+1}\right) < 0, \\ & \vec{D}^{t+1}\left(x^{t+1}, y^{t+1}, b^{t+1}\right) = 0. \end{aligned} \tag{15}$$

Equation (15) suggests that to become an innovator, the contemporaneous environmental technology frontier should move toward the global environmental technology frontier. In addition, the industrial production activity of an innovative region in period $t + 1$ should be outside of the contemporaneous environmental technology frontier in period t, and the region under estimation should be located on the contemporaneous environmental technology frontier in period $t + 1$. We have listed the innovators in the eastern, central and western regions in Table 6.

Table 4. Changes in the NLCPIL at the provincial level, 2003–2012. Positive values are marked grey.

Province	Region	03–04	04–05	05–06	06–07	07–08	08–09	09–10	10–11	11–12	Mean
Beijing	E	0.0601	0.0563	−0.025	0.0063	0.0172	0.0191	0.0147	0.4627	0.1512	0.0847
Tianjin	E	0.0388	0.0987	0.0472	0.0437	0.0803	−0.013	0.1201	0.1648	0.0899	0.0744
Hebei	E	0.0508	−0.021	0.0512	0.0423	0.0462	−0.013	0.0822	0.2447	0.2100	0.0770
Liaoning	E	0.0338	0.0421	0.0211	0.0674	0.1616	−0.100	0.0503	0.0277	0.0229	0.0363
Shanghai	E	0.0224	−0.005	0.0216	0.0177	0.0390	−0.044	0.0123	−0.0083	−0.016	0.0042
Jiangsu	E	−0.0190	−0.015	0.0357	0.0290	0.0403	0.0510	0.0689	0.2862	0.0000	0.0530
Zhejiang	E	−0.034	0.0079	0.0490	0.0485	0.0700	−0.008	0.1292	0.1238	0.0000	0.0429
Fujian	E	0.0624	−0.092	0.0814	−0.017	0.1034	−0.010	0.0729	0.2026	−0.027	0.0416
Shandong	E	0.0335	−0.002	0.0415	0.0528	0.0588	0.0320	0.0699	0.1199	0.0568	0.0514
Guangdong	E	0.0976	0.0677	0.0573	0.0066	0.3120	−0.496	0.3424	0.1537	0.0000	0.0601
Hainan	E	0.0000	0.0000	−0.2349	−0.044	0.2797	0.0000	0.0000	0.0000	0.0000	0.0000
Shanxi	C	0.0553	0.0887	0.0018	0.0599	0.0789	−0.049	0.1094	0.1140	−0.049	0.0454
Jilin	C	0.0374	0.0089	0.0206	0.0618	0.0670	0.0318	0.0585	0.0691	0.1161	0.0524
Heilongjiang	C	0.0534	−0.045	0.0165	−0.008	0.0508	−0.036	0.1020	0.1023	0.0087	0.0271
Anhui	C	0.0472	−0.009	0.0806	−0.006	0.0565	0.0353	0.0937	−0.049	0.2922	0.0600
Jiangxi	C	0.0475	0.1019	0.1408	−0.093	0.0492	0.0304	0.1393	0.0061	0.1099	0.0591
Henan	C	0.0412	0.0529	0.0049	0.0770	0.1197	−0.004	0.2317	0.1551	0.0066	0.0760
Hubei	C	0.0197	−0.035	0.0812	0.0032	0.0407	0.0600	0.0123	−0.008	0.1206	0.0326
Hunan	C	0.0227	−0.006	0.0589	0.0457	0.1199	−0.003	0.1714	−0.075	0.4027	0.0819
Inner Mongolia	W	0.0541	0.0516	0.0367	0.0727	0.0931	0.0129	0.2204	0.2902	0.0000	0.0924
Guangxi	W	0.0454	0.0215	0.0359	0.0693	0.2692	−0.163	0.1903	0.1484	0.0503	0.0741
Chongqing	W	0.0382	−0.035	0.0356	0.0363	0.0523	0.0230	0.0415	0.0581	0.1065	0.0396
Sichuan	W	0.0293	0.0200	0.0660	0.0504	0.0187	0.0165	0.0229	0.1182	0.1719	0.0571
Guizhou	W	0.0099	0.0368	0.0270	0.0457	0.0700	−0.011	0.0560	−0.032	0.1756	0.0419
Yunnan	W	0.1033	−0.073	0.0282	−0.026	0.0421	−0.016	0.0851	0.0501	0.4382	0.0700
Shaanxi	W	−0.026	0.1215	0.0621	0.0000	0.5951	−0.467	0.0816	0.0815	0.3044	0.0836
Gansu	W	0.0339	−0.005	0.0282	0.0379	0.0366	−0.006	0.0874	0.0377	0.0256	0.0306
Qinghai	W	0.0000	−0.5723	−0.014	0.1112	0.0002	−0.011	0.0883	0.3992	0.0000	0.0000
Ningxia	W	0.0486	−0.008	0.0219	0.1492	−0.076	0.0425	0.1260	0.0333	−0.020	0.0351
Xinjiang	W	0.0097	0.0534	0.0027	0.0142	0.1502	−0.128	0.0811	0.6762	0.0000	0.0954
China		0.0339	−0.003	0.0294	0.0317	0.1014	−0.041	0.0987	0.1317	0.0916	0.0527

As shown in Table 7, it was found that provinces from the eastern region were determined to be innovators 50 times during the study period, whereas those in the western and central regions appeared 24 and 18 times, respectively. Specifically, Henan ranked first and was registered as an innovator 8 times. Beijing, Guangdong, Jiangxi and Shandong each appeared 7 times. Jiangsu, Xinjiang and Zhejiang each appeared 6 times. Fujian and Hebei appeared 5 times. Inner Mongolia and Tianjin appeared 4 times. Guangxi, Qinghai, Shaanxi and Yunnan appeared 3 times. Heilongjiang, Ningxia and Shanghai appeared 2 times, and Hainan and Hunan each appeared only 1 time. The result is consistent with that of Zhang *et al.* [28] and implies that the eastern region should be referenced as a benchmark for the low carbon utilization of industrial land. Moreover, provinces that are not listed in Table 6 should benchmark the innovating provinces and strive to improve their clean production technologies by learning from the innovators. In addition, the central government of China should actively create opportunities for provinces to communicate with each other to narrow regional gaps in low carbon industrial production technology and realize the balanced utilization of industrial land in the country.

Table 5. EC component of the NLCPIL at the provincial level, 2003–2012. Positive values are marked grey.

Province	Region	03–04	04–05	05–06	06–07	07–08	08–09	09–10	10–11	11–12	Mean
Beijing	E	0.5551	0.0000	0.0000	0.0000	0.0000	0.0000	0.0000	0.0000	0.0000	0.0617
Tianjin	E	0.0510	0.2648	−0.065	0.0650	0.0000	0.0000	0.0000	0.0000	0.0000	0.0351
Hebei	E	0.4358	0.0000	0.0000	0.0000	0.0000	−0.174	0.1740	0.0000	0.0000	0.0484
Liaoning	E	0.0540	0.0712	−0.024	0.0879	0.1145	−0.051	0.0157	−0.0547	−0.030	0.0202
Shanghai	E	0.3374	−0.371	0.3715	0.0000	0.0000	0.0000	−0.2716	−0.318	0.0849	−0.0186
Jiangsu	E	0.0000	0.0000	0.0000	0.0000	0.0000	0.0000	0.0000	0.0000	0.0000	0.0000
Zhejiang	E	0.0000	0.0000	0.0000	0.0000	0.0000	0.0000	0.0000	0.0000	0.0000	0.0000
Fujian	E	0.0000	0.0000	0.0000	0.0000	0.0000	0.0000	0.0000	0.0000	0.0000	0.0000
Shandong	E	0.0000	0.0000	0.0000	0.0000	0.0000	0.0000	0.0000	0.0000	−0.1107	−0.0123
Guangdong	E	0.0000	0.0000	0.0000	0.0000	0.0000	0.0000	0.0000	0.0000	0.0000	0.0000
Hainan	E	0.0000	0.0000	0.0000	0.0000	0.0000	0.0000	0.0000	0.0000	0.0000	0.0000
Shanxi	C	0.0485	0.2033	−0.045	0.1234	0.0357	−0.059	0.0758	−0.0042	−0.095	0.0314
Jilin	C	0.0916	0.0117	−0.000	0.0794	0.0726	0.0613	0.0103	−0.098	0.1426	0.0411
Heilongjiang	C	0.0000	0.0000	0.0000	−0.455	0.2494	0.0903	0.1157	−0.413	0.1757	−0.0264
Anhui	C	0.1039	0.0045	0.0937	−0.043	0.0879	0.0718	0.0004	−0.186	0.2657	0.0442
Jiangxi	C	0.0000	0.0000	0.0000	0.0000	0.0000	0.0000	0.0000	0.0000	−0.0775	−0.0086
Henan	C	0.0000	0.0000	0.0000	0.0000	0.0000	0.0000	0.0000	0.0000	0.0000	0.0000
Hubei	C	0.0183	−0.031	0.0774	−0.006	0.0070	0.0817	−0.0252	−0.097	0.0920	0.0129
Hunan	C	0.0578	−0.011	0.2200	0.0403	0.0091	0.0568	0.0506	0.0686	0.0000	0.0546
Inner Mongolia	W	0.1012	0.0827	0.0296	0.4176	0.0000	0.0000	0.0000	0.0000	0.0000	0.0701
Guangxi	W	0.4709	−0.377	0.0224	0.0650	0.2904	0.0000	0.0000	0.0000	0.0000	0.0523
Chongqing	W	−0.0714	−0.045	0.0296	0.0382	0.0398	0.1453	−0.0147	−0.105	0.1447	0.0178
Sichuan	W	0.0346	0.0453	0.0600	0.0922	0.0201	0.0352	−0.072	0.5161	−0.299	0.0479
Guizhou	W	0.0395	0.0481	0.0905	−0.019	0.2370	0.2228	−0.1448	−0.462	0.6076	0.0687
Yunnan	W	0.0000	0.0000	0.0000	−0.347	0.1191	−0.053	0.0953	−0.242	0.4294	0.0000
Shaanxi	W	−0.067	0.2546	0.0561	0.0040	0.3034	0.0000	0.0000	−0.182	0.1822	0.0611
Gansu	W	0.0366	0.0471	0.0146	0.0517	0.0815	−0.003	0.0951	−0.173	0.1836	0.0370
Qinghai	W	0.0000	0.0000	0.0000	0.0000	0.0000	0.0000	0.0000	0.0000	0.0000	0.0000
Ningxia	W	−0.135	0.0473	0.0475	0.2812	0.3359	0.0000	0.0000	−0.463	0.4636	0.0641
Xinjiang	W	0.6477	0.0000	0.0000	0.0000	0.0000	0.0000	0.0000	0.0000	0.0000	0.0720
China		0.0936	0.0081	0.0326	0.0158	0.0668	0.0141	0.0035	−0.073	0.0720	0.0258

3.3. *Determinants of the NLCPIL*

Based on Equation (12), we explored how the influencing factors impacted the NLCPIL using a regression analysis. We introduce the influencing factors from each aspect (e.g., energy utilization, technology, industrial structure and environmental policy) successively to enhance the robustness of the regression results. The results of a Hausman test showed that the fixed effect model was better than the random effect model and that the fixed effect model was able to remove the effects of regional disparity, and we therefore adopted the fixed effect model. The estimation results are shown in Table 8.

Table 6. TC component of the NLCPIL at the provincial level, 2003–2012. Positive values are marked grey.

Province	Region	TC	TC	TC	TC	TC	TC	TC	TC	TC	Mean
Beijing	E	−0.494	0.0563	−0.025	0.0063	0.0172	0.0192	0.0146	0.4627	0.1512	0.0230
Tianjin	E	−0.0122	−0.166	0.1122	−0.021	0.0803	−0.013	0.1200	0.1649	0.0899	0.0393
Hebei	E	−0.3850	−0.021	0.0512	0.0423	0.0462	0.1608	−0.091	0.2447	0.2100	0.0285
Liaoning	E	−0.0202	−0.029	0.0458	−0.020	0.0471	−0.048	0.0345	0.0825	0.0532	0.0161
Shanghai	E	−0.315	0.3662	−0.349	0.0178	0.0389	−0.044	0.2839	0.3102	−0.101	0.0229
Jiangsu	E	−0.0190	−0.015	0.0357	0.0291	0.0403	0.0510	0.0689	0.2862	0.0000	0.0530
Zhejiang	E	−0.034	0.0079	0.0490	0.0485	0.0699	−0.008	0.1293	0.1238	0.0000	0.0429
Fujian	E	0.0625	−0.092	0.0815	−0.017	0.1034	−0.010	0.0729	0.2026	−0.027	0.0416
Shandong	E	0.0335	−0.003	0.0415	0.0528	0.0588	0.0320	0.0699	0.1199	0.1675	0.0637
Guangdong	E	0.0976	0.0677	0.0573	0.0066	0.3120	−0.496	0.3424	0.1537	0.0000	0.0601
Hainan	E	0.0000	0.0000	−0.2349	−0.044	0.2797	0.0000	0.0000	0.0000	0.0000	0.0000
Shanxi	C	0.0069	−0.114	0.0468	−0.063	0.0432	0.0097	0.0335	0.1182	0.0454	0.0140
Jilin	C	−0.0542	−0.002	0.0214	−0.0176	−0.0056	−0.029	0.0482	0.1680	−0.026	0.0113
Heilongjiang	C	0.0534	−0.044	0.0164	0.4472	−0.1986	−0.1268	−0.013	0.5153	−0.167	0.0535
Anhui	C	−0.0567	−0.0142	−0.013	0.0377	−0.0315	−0.036	0.0933	0.1362	0.0265	0.0157
Jiangxi	C	0.0475	0.1019	0.1409	−0.093	0.0493	0.0304	0.1393	0.0061	0.1873	0.0677
Henan	C	0.0412	0.0529	0.0049	0.0771	0.1197	−0.004	0.2317	0.1551	0.0066	0.0760
Hubei	C	0.0014	−0.003	0.0038	0.0096	0.0337	−0.021	0.0375	0.0888	0.0286	0.0198
Hunan	C	−0.035	0.0058	−0.161	0.0054	0.1109	−0.059	0.1208	−0.143	0.4027	0.0273
Inner Mongolia	W	−0.0471	−0.031	0.0071	−0.344	0.0930	0.0129	0.2203	0.2902	0.0000	0.0223
Guangxi	W	−0.425	0.3993	0.0135	0.0043	−0.0212	−0.163	0.1904	0.1484	0.0503	0.0218
Chongqing	W	0.1097	0.0105	0.0060	−0.001	0.0125	−0.122	0.0563	0.1633	−0.038	0.0218
Sichuan	W	−0.0053	−0.025	0.0059	−0.0418	−0.0014	−0.018	0.0957	−0.397	0.4712	0.0092
Guizhou	W	−0.0296	−0.0113	−0.063	0.0652	−0.1670	−0.234	0.2008	0.4304	−0.4320	−0.0268
Yunnan	W	0.1033	−0.073	0.0283	0.3208	−0.076	0.0372	−0.010	0.2926	0.0088	0.0700
Shaanxi	W	0.0416	−0.133	0.0060	−0.004	0.2917	−0.467	0.0816	0.2637	0.1222	0.0225
Gansu	W	−0.0027	−0.053	0.0136	−0.0138	−0.0448	−0.0030	−0.007	0.2116	−0.1580	−0.0064
Qinghai	W	0.0000	−0.5723	−0.014	0.1112	0.0002	−0.011	0.0883	0.3992	0.0000	0.0000
Ningxia	W	0.1839	−0.0561	−0.0256	−0.1319	−0.412	0.0425	0.1260	0.4968	−0.4838	−0.0289
Xinjiang	W	−0.638	0.0534	0.0028	0.0142	0.1502	−0.128	0.0811	0.6762	0.0000	0.0235
China		−0.0597	−0.0114	−0.003	0.0160	0.0346	−0.055	0.0953	0.2057	0.0196	0.0268

The results show that the coefficients of the variables were statistically significant at the national level, which implies that the determinant analysis is quite robust and that the results can well explain the model. With regard to the coefficients of EI and ES, they were significantly negative for models 1–4, shown in Table 8, which is consistent with our assumptions in Section 2.3. As shown in model 4, the NLCPIL decreased by approximately 1.44% and 0.19% with 1% increases in EI and ES. This indicates that energy intensity had an obvious negative impact on NLCPIL, which corresponds to many previous studies [50]. This may be because higher EIs mean that more fossil energy is used in industrial production, which would lead to an obvious increase in carbon emissions. In addition, the development oriented towards heavy industry and heavily dependent on coal, has exerted a reverse impact on NLCPIL in China, which may be because coal releases more carbon dioxide than other forms of clean energy (e.g., natural gas) for the same given amount and could easily cause increases in temperature and pose threats to human health [51].

Table 7. Innovators in the eastern, central and western regions, 2003–2012.

Period	East	Central	West
2003–2004	Shandong, Fujian, Guangdong	Henan, Jiangxi, Heilongjiang	Yunnan
2004–2005	Zhejiang, Beijing, Guangdong	Henan, Jiangxi	Xinjiang
2005–2006	Jiangsu, Shandong, Zhejiang, Hebei, Guangdong, Fujian	Henan, Heilongjiang, Jiangxi	Xinjiang, Yunnan
2006–2007	Beijing, Guangdong, Shanghai, Jiangsu, Hebei, Zhejiang, Shandong	Henan	Xinjiang, Qinghai
2007–2008	Beijing, Shanghai, Jiangsu, Hebei, Shandong, Zhejiang, Tianjin, Fujian, Hainan, Guangdong	Jiangxi, Henan	Inner Mongolia, Xinjiang, Shaanxi
2008–2009	Beijing, Shandong, Jiangsu	Jiangxi	Inner Mongolia, Ningxia
2009–2010	Beijing, Jiangsu, Shandong, Fujian, Tianjin, Zhejiang, Guangdong	Jiangxi, Henan	Xinjiang, Shaanxi, Qinghai, Ningxia, Guangxi, Inner Mongolia
2010–2011	Shandong, Zhejiang, Guangdong, Tianjin, Fujian, Hebei, Jiangsu, Beijing	Jiangxi, Henan	Guangxi, Inner Mongolia, Qinghai, Xinjiang
2011–2012	Tianjin, Beijing, Hebei	Henan, Hunan	Guangxi, Shaanxi, Yunnan

Table 8. Regression results. *** $p < 0.001$, ** $p < 0.05$, * $p < 0.01$.

	Model 1	Model 2	Model 3	Model 4
EI	−0.0164 *** (−2.5946)	−0.0173 *** (−2.6929)	−0.0227 *** (−2.9775)	−0.0144 ** (−1.9460)
ES	−0.0013 *** (−3.4321)	−0.0012 *** (−3.2012)	−0.0019 *** (−4.7769)	−0.0019 *** (−4.5950)
RI		0.0034 *** 3.7902)	0.0134 *** (3.0684)	0.0122 *** (2.6954)
FI		−0.0001 ** (−2.2041)	−0.0001 ** (−2.0582)	−0.0001 ** (−2.0997)
LS			−0.0024 *** (−7.3504)	−0.0025 *** (−7.2698)
IS			0.0014 ** (2.3019)	0.0015 ** (2.3484)
POL				0.0266 *** (3.1863)
Constant	0.0696 *** (6.5035)	0.0656 *** (4.3790)	0.0935 *** (2.7554)	0.0813 ** (2.3186)
Adjusted R^2	0.0674	0.0487	0.2843	0.2329
F-statistic	17.3950	15.4415	7.1233	5.6660
Prob.	0.0000	0.0000	0.0000	0.0000

RI and FI, which were introduced in models 2–4, represent industrial production technology and showed different relationships with the NLCPIL. The RI coefficient was significantly positive, whereas the FI coefficient was significantly negative. This implies that a greater investment in industrial R&D was indeed helpful for saving energy, reducing carbon emissions and realizing environmentally friendly industrial development. However, the impact of FI is inconsistent with our assumptions in Section 2.3, which states that both of RI and FI have positive impacts on NLCPIL in China. As model 4 shows, the NLCPIL increased by 1.22% with a 1% increase in RI, and it decreased by 0.01% with a 1% increase in FI. A possible explanation may be that the introduction of foreign industrial enterprises has not played the expected role because of the lack of scientific planning. In fact, most Chinese local governments prefer short–term economic benefits rather than learning about and developing advanced clean production technologies. Some previous studies have even noted that autonomous product development is not only a necessary condition for industrial innovation but is also the best way to learn foreign technology, and the blind introduction of foreign industrial enterprises can be counterproductive [52].

In addition, LS and IS were introduced in models 3–4. The coefficient of LS is significantly negative, which indicates that an increase in LS could give rise to a decrease in NLCPIL. This is consistent with our expectations and may be due to the industrial labor surplus across the country caused by the continued migration of rural surplus labor forces into the cities. The industrial enterprises prefer low-tech production activities because of inexpensive labor, and they do not have enough enthusiasm to improve production technology, which is not helpful for improving NLCPIL [53]. In contrast, the IS showed a significantly positive impact on NLCPIL, which defied our expectations. A possible reason may be that China has not yet fully entered the period of post industrialization, and increases in the scale of industrial development are beneficial for improving NLCPIL.

Lastly, we introduced (carbon emission reduction policy (POL) in model 4, and its coefficient was significantly positive, which indicates that policies on carbon emission reductions since 2009 have obviously contributed to the growth in NLCPIL. Therefore, strengthening environmental regulations on local governments and enterprises, developing low carbon industries and promoting the use of clean energy would be helpful to improve the NLCPIL. This is consistent with the findings of Xie *et al.* [9].

4. Conclusions

China has become the world's largest energy consumer and carbon emitter, and industrial production is the primary contributor to carbon emissions. Industrial lands bear most of the industrial production activities and industrial pollutants, and the serious problems of environmental pollution in areas surrounding industrial land caused by industrial production therefore deserve more attention. Fortunately, the central government of China has already accepted the importance of improving the environmental and economic performance of industrial land use. However, the above polices cannot achieve the desired effects unless accompanied by an overall understanding of the actual situation. Therefore, modeling the dynamic changes in carbon emission performance of industrial land use in recent years and forwarding constructive policy implications are urgently needed.

In this study, we employed a global DDF approach to compute the TCPIL, and we then used a NLCPIL index to model the dynamic changes in the TCPIL. The results are as follows:

Firstly, the TCPILs for China and its three regions showed rising trends over the study period, and the eastern region performed much better in TCPIL than the central and western regions. However, all three regions have a large potential for improving their TCPILs. Most of the provinces in the eastern region (e.g., Hainan, Zhejiang and Guangdong) enjoyed better TCPILs, whereas those from the central and western regions (e.g., Gansu and Ningxia) suffered worse TCPILs.

Secondly, the NLCPILs for China were greater than zero in most years of the study period, and their growth was mainly driven by ECs before 2009 and by TC subsequently. The eastern and central regions showed higher TCs, whereas the western region had a better EC performance. Many provinces with poor TCPILs had higher NLCPILs, which indicates that the regional gaps in TCPIL clearly showed a narrowing trend. In addition, most of the provinces that were identified as innovators of environmentally friendly industrial production technologies were in the eastern region (e.g., Beijing, Guangdong and Shandong).

Lastly, the results of the influencing factor analysis showed that the carbon emission reduction policies since 2009 have performed as expected, and they are helpful to improve the NLCPIL, which recommends the environmental protection policies. The EI, ES and LS indicators had expected significantly negative impacts on the NLCPIL, which means that the NLCPIL could be improved by saving more fossil energy, optimizing the energy structure by reducing the use of coal and properly solving the problem of surplus labor in the industrial sectors. The RI had an expected significantly positive impact on the NLCPIL, which implies that more investment in industrial R&D is needed. However, the FI and IS indices influenced the NLCPIL opposite the expected impacts, which had negative and positive signs, respectively. This may due to incomplete learning of foreign advanced production technologies and the fact that China has not yet fully entered into the stage of middle and late industrialization.

Based on the empirical analysis, we put forward some policy implications. Firstly, the central government of China should issue more policies on low-carbon and energy-saving industrial production to effectively protect the environment given rapid industrial economic development. In addition, the central government should strictly regulate local governments to fully implement those policies. Therefore, severe punishments for local government officials such as removing administrative duties or cutting their powers are necessary. Secondly, industrial enterprises should further optimize energy structures by reducing the use of coal and increasing the use of clean energy such as nuclear and wind power. The government should spend more money on the R&D of clean energy, subsidize enterprises that use clean energy, and promote cooperation between enterprises and research institutes. Third, we should carefully study advanced industrial production and management technologies by introducing foreign investment and industrial enterprises and by trying to develop our own technologies, especially for the regions with underdeveloped environmentally friendly industrial production technology. Lastly, the regional gaps in industrial production technology and environmental protection deserve more attention, and the central government of China should create more opportunities for underdeveloped provinces to communicate with developed provinces and introduce necessary technologies and talent. This study also has some limitations. Firstly, we only adopted a ten-year sample period because of the unavailability of data. We will try to obtain more data to extend the study period to produce more convincing and meaningful results. Secondly, some factors that play important roles in determining the efficiency of industrial land use were not considered in this paper for the same reason, such as the price of industrial land, carbon emissions trading costs and human capital. We will make these improvements in future studies.

Acknowledgments: This study was supported by the National Natural Science Foundation of China, No. 41561040; the Key projects of the National Social Science Fund of China, No. 15AZD075; the Key projects of the social science fund of Jiangxi Province in China "Study on the green utilization of land resources in China"; the Natural Science Foundation of Jiangxi Province No. 20143ACB21023; the Technology Foundation of Jiangxi Education Department of China, No. KJLD14033; the Fok Ying-Tung Fund, No. 141084; the graduate innovation fund project of Jiangxi university of finance and economics of China, No.XS295, No.XS308; the graduate innovation special funds project of Jiangxi province of China, No.YC2015-S217.

Author Contributions: Wei Wang and Hualin Xie had the original idea for the study. Wei Wang was responsible for data collecting. Wei Wang, Tong Jiang, Xue Xie, Daobei Zhang and Hualin Xie carried out the analyses. All the authors drafted the manuscript, and approved the final one.

Conflicts of Interest: The authors declare no conflict of interest.

References

1. Jiang, F. The dynamic mechanism transform of China's industrial economic growth. *China Ind. Econ.* **2014**, *5*, 5–17. (In Chinese)
2. Che, L.; Han, X.; Zhao, L.; Wu, X. Coal use efficiency evaluation and decoupling analysis between coal use efficiency and economic growth in China. *China Popul. Resour. Environ.* **2015**, *3*, 104–110. (In Chinese)
3. Tollefson, J. Gas to displace coal on road to clean energy. *Nature* **2010**, *466*, 19.
4. Choi, Y.; Zhang, N.; Zhou, P. Efficiency and abatement costs of energy-related CO_2 emissions in China: A slacks-based efficiency measure. *Appl. Energy* **2012**, *98*, 198–208.
5. Liu, L.; Zong, H.; Zhao, E.; Chen, C.; Wang, J. Can China realize its carbon emission reduction goal in 2020: From the perspective of thermal power development. *Appl. Energy* **2014**, *124*, 199–212.

6. Ministry of Environmental Protection of the People's Republic of China (MEPPRC). Available online: http://www.zhb.gov.cn/gkml/hbb/gwy/200910/t20091030_180716.htm (accessed on 30 October 2009). (In Chinese)
7. Li, Y. Analysis of the mechanism and process of social conflicts in China's urbanization. *China Popul. Resour. Environ.* **2015**, *2*, 57–65. (In Chinese)
8. Cui, L.B.; Fan, Y.; Zhu, L.; Bi, Q.H. How Will the Emissions Trading Scheme Save Cost for Achieving China's 2020 Carbon Intensity Reduction Target? *Appl. Energy* **2014**, *136*, 1043–1052.
9. Xie, H.; Wang, W. Spatiotemporal differences and convergence of urban industrial land use efficiency for China's major economic zones. *J. Geogr. Sci.* **2015**, *25*, 1183–1198.
10. Lu, C.; Wen, F.; Yang, Q.; Zhang, P. Characteristics and driving factors of urban land use structure of cities at provincial level and above. *Scientia Geographica Sinica* **2011**, *5*, 599–607.
11. Zhao, X.; Huang, X.; Ma, W.; Zhang, X. Identification method and treatment suggestion for idle land. *China Land Sci.* **2011**, *9*, 3–7. (In Chinese)
12. Li, Y.; Shu, B.; Wu, Q. Urban Land Use Efficiency in China: Spatial and Temporal Characteristics, Regional Difference and Influence Factors. *Econ. Geogr.* **2014**, *34*, 133–139. (In Chinese)
13. Huang, D.; Hong, L.; Liang, J. Analysis and evaluation of industrial land efficiency and intensive use in Fujian Province. *Acta Geogr. Sin.* **2009**, *64*, 479–486. (In Chinese)
14. Hu, J.; Wang, S.; Yeh, F. Total-factor water efficiency of regions in China. *Resour. Policy* **2006**, *31*, 217–230.
15. Chen, W.; Peng, J.; Wu, Q. Spatial and temporal differences in industrial land use efficiency in China. *Resour. Sci.* **2014**, *36*, 2046–2056. (In Chinese)
16. Xiong, Q.; Guo, G. Study on the efficiency difference of city industrial land production across provinces in China. *Resour. Sci.* **2013**, *35*, 910–917. (In Chinese)
17. Guo, G.; Wen, Q. Industrial land productivity research under the environmental restriction based on unexpected outputs of 33 typical cities in China. *China Popul. Resour. Environ.* **2014**, *24*, 121–127. (In Chinese)
18. Xie, H.; Wang, W. Exploring the spatial-temporal disparities of urban land use economic efficiency in China and its influencing factors under environmental constraints based on a sequential slacks-based model. *Sustainability* **2015**, *7*, 10171–10190.
19. Zhang, N.; Kong, F.; Choi, Y. Measuring sustainability performance for China: A sequential generalized directional distance function approach. *Econ. Model.* **2014**, *41*, 392–397.
20. Zhang, N.; Kong, F.; Choi, Y.; Zhou, P. The effect of size-control policy on unified energy and carbon efficiency for Chinese fossil fuel power plants. *Energy Policy* **2014**, *70*, 193–200.
21. Chung, Y.H.; Färe, R.; Grosskopf, S. Productivity and undesirable outputs: A directional distance function approach. *J. Environ. Manag.* **1997**, *51*, 229–240.
22. Kumar, S. Environmentally sensitive productivity growth: A global analysis using Malmquist-Luenberger index. *Ecol. Econo.* **2006**, *56*, 280–293.

23. Fan, M.; Shao, S.; Yang, L. Combining global Malmquist-Luenberger index and generalized method of moments to investigate industrial total factor CO_2 emission performance: A case of Shanghai (China). *Energy Policy* **2015**, *79*, 189–201.
24. Arabi, B.; Munisamy, S.; Emrouznejad, A.; Shadman, F. Power industry restructuring and eco-efficiency changes: A new slacks-based model in Malmquist-Luenberger Index measurement. *Energy Policy* **2014**, *68*, 132–145.
25. Boussemart, J.P.; Briec, W.; Kerstens, K.; Poutineau, J.C. Luenberger and Malmquist productivity indices: Theoretical comparisons and empirical illustration. *Bull. Econ. Res.* **2003**, *55*, 391–405.
26. Chang, T.P.; Hu, J.L. Total-factor energy productivity growth, technical progress, and efficiency change: An empirical study of China. *Appl. Energy* **2010**, *87*, 3262–3270.
27. Chang, Y.; Zhang, N.; Danao, D.; Zhang, N. Environmental efficiency analysis of transportation system in China: A non-radial DEA approach. *Energy Policy* **2013**, *58*, 277–283.
28. Zhang, N.; Wei, X. Dynamic total-factor carbon emissions performance changes in the Chinese transportation industry. *Appl. Energy* **2015**, *146*, 409–420.
29. You, H.; Wu, C. Carbon emission efficiency and low carbon optimization of land use: Based on the perspective of energy consumption. *J. Nat. Sources* **2010**, *25*, 1875–1886. (In Chinese)
30. Cui, Y.; Miao, J.; Yang, J. Empirical study on the eco-efficiency of urban non-agricultural land considering carbon emission. *Syst. Eng.* **2012**, *30*, 10–18. (In Chinese)
31. Färe, R.; Grosskopf, S. *New Directions: Efficiency and Productivity*; Springer: New York, NY, USA, 2005.
32. Mei, G.; Gan, J.; Zhang, N. Metafrontier environmental efficiency for China's regions: A slack-based efficiency measure. *Sustainability* **2015**, *7*, 4004–4021.
33. Chambers, R.G.; Chung, Y.; Färe, R. Benefit and distance functions. *J. Econ. Theory* **1996**, *70*, 407–419.
34. Zhou, P.; Ang, B.W.; Wang, H. Energy and CO_2 emission performance in electricity generation: A non-radial directional distance function approach. *Eur. J. Oper. Res.* **2012**, *221*, 625–635.
35. Oh, D. A metafrontier approach for measuring an environmentally sensitive productivity growth index. *Energy Econ.* **2010**, *32*, 146–157.
36. National Bureau of Statistics of China (NBSC). Available online: http://www.stats.gov.cn/tjsj/ndsj/ (accessed on 12 March 2015). (In Chinese)
37. Zhang, N.; Kong, F.; Yu, Y. Measuring ecological total-factor energy efficiency incorporating regional heterogeneities in China. *Ecol. Indic.* **2015**, *51*, 165–172.
38. Wu, Y. *China's Capital Stock Series by Region and Sector*; The University of Western Australia Discussion Paper; University of Western Australia: Perth, Australia, 2009.
39. Zhang, N.; Choi, Y. Environmental energy efficiency of China's regional economies: A non-oriented slacks-based measure analysis. *Soc. Sci. J.* **2013**, *50*, 225–234.
40. Du, L.; Wei, C.; Cai, S. Economic development and carbon dioxide emissions in China: Provincial panel data analysis. *China Econ. Rev.* **2012**, *23*, 371–384.

41. National Bureau of Statistics of China (NBSC). Available online: http://tongji.cnki.net/kns55/Navi/ HomePage.aspx?id=N2010042092&name=YZGCA&floor=1 (accessed on 1 April 2015). (In Chinese)
42. National Bureau of Statistics of China (NBSC). Available online: http://tongji.cnki.net/kns55/Navi/ HomePage.aspx?id=N2010080088&name=YCXME&floor=1 (accessed on 1 November 2015). (In Chinese)
43. Liu, N.; Ma, Z.; Kang, J. Changes in carbon intensity in China's industrial sector: Decomposition and attribution analysis. *Energy Policy* **2015**, *87*, 28–38.
44. Chu, N.Y.; Li, S.K.; Ki, T.S. The Incidence of Surplus Labor in Rural China: A Nonparametric Estimation. *J. Comp. Econ.* **2000**, *28*, 565–580.
45. Qin, H.; Su, Q.; Khu, S.; Tang, N. Water Quality changes during rapid urbanization in the Shenzhen river catchment: an integrated view of socio-economic and infrastructure development. *Sustainability* **2014**, *6*, 7433–7451.
46. Ouyang, M.; Peng, Y. The treatment-effect estimation: A case study of the 2008 economic stimulus package of China. *J. Econ.* **2015**, *188*, 545–557.
47. Zheng, S.; Yi, H.; Li, H. The impacts of provincial energy and environmental policies on air pollution control in China. *Renew. Sustain. Energy Rev.* **2015**, *49*, 386–394.
48. Porter, M.; van der Linde, C. Toward a new conception of the environment: Competitiveness relationship. *J. Econ. Perspect.* **1995**, *9*, 120–134.
49. Färe, R.; Grosskopf, S.; Norris, M.; Zhang, Z. Productivity growth, technical progress, and efficiency change in industrialized countries. *Am. Econ. Rev.* **1994**, *84*, 66–83.
50. Balsalobre, D.; Álvarez, A.; Cantos, J.M. Public budgets for energy RD&D and the effects on energy intensity and pollution levels. *Environ. Sci. Pollut. Res.* **2014**, *22*, 4881–4892.
51. Tang, L.; Li, R.; Tokimatsu, K.; Itsubo, N. Development of human health damage factors related to CO_2 emissions by considering future socioeconomic scenarios. *Int. J. Life Cycle Assess.* **2015**, *9*, 1–12.
52. Lu, F.; Feng, K. Why is autonomous product development the best way to learn foreign technology?—Lessons from the historical experiences of Japan's and South Korea's auto industries. *Chin. Soft Sci.* **2004**, *4*, 6–11. (In Chinese)
53. Zhang, Y.; Xu, Q.; Wu, J. A Successful Anti-poverty War: Experience from China. *Econ. Res. J.* **2012**, *11*, 76–87. (In Chinese)

Analysis of the Threshold Effect of Financial Development on China's Carbon Intensity

Xiongfeng Pan, Yaobo Yan, Xiaoxue Peng and Qing Liu

Abstract: Using panel data on 30 provinces in China from 2005 to 2012, this paper conducts an empirical test on the threshold effect of the relationship between financial development and carbon emission intensity from the perspectives of financial scale and financial efficiency. The results show that at a low level of per capita GDP, the expansion of the financial scale and the enhancement of financial efficiency will increase carbon intensity. When the per capita GDP is greater than the threshold value (RMB 37,410), the expansion of the financial scale will also increase carbon intensity, but the potency of this effect will be weaker. At the same time, the improvement of financial efficiency will help reduce carbon intensity. Most provinces with per capita GDP greater than the threshold value (RMB 37,410) are located in the eastern coastal areas of China, whereas most provinces with per capita GDP less than the threshold value are located in the central and western areas of China. Both raising the level of openness and improving the industrial structure will have significantly positive effects on carbon intensity.

Reprinted from *Sustainability*. Cite as: Pan, X.; Yan, Y.; Peng, X.; Liu, Q. Analysis of the Threshold Effect of Financial Development on China's Carbon Intensity. *Sustainability* **2016**, *8*, 271.

1. Introduction

For quite some time, China's energy consumption structure has given priority to coal. Consequently, all kinds of problems relating to carbon dioxide emissions have become increasingly serious there, in line with increased levels of consumption. Environmental degradation caused by the emission of greenhouse gases, such as carbon dioxide, is an important challenge especially with regards to both sustainable economic development and climate change control [1]. The optimization of energy production systems is a relevant issue that must be considered if China is to adhere to reduction policies for fossil fuel consumption and regulations on carbon dioxide emissions [2]. Not only does this phenomenon relate to China's economic and social development, it has also become a weak point in the international interest game [3]. As such, the rapid increase in carbon dioxide emissions has captured the attention of many researchers.

Financial development, as a key economic variable, can improve the economic efficiency of a country's financial system and affect both its economic behavior and energy demand. Financial development can also play an intermediary role in integrating social capital and in pushing enterprises to introduce advanced energy-saving technology and relevant management experience. Thus, financial development is likely to have a certain impact on carbon emissions. Scholars have conducted considerable research on the relationship between financial development and carbon emissions. Tamazian *et al.* [4] found that financial development reduced carbon dioxide emissions in Brazil, Russia, India, and China. Tamazian and Rao [5] analyzed the relationship between financial development and carbon emissions using panel data of 24 transition economies from 1993 to 2004, They found that financial development had an adverse effect on carbon emissions. Sadorsky [6] examined the effects of financial development on energy consumption in a sample of nine central and eastern European frontier economies, and the results indicated a positive and statistically significant relationship between financial development and carbon dioxide emissions when financial development was measured using banking variables, such as deposit money bank assets to GDP, financial system deposits to GDP, or liquid liabilities to GDP. Shahbaz *et al.* [7] analyzed the period of 1975Q1–2011Q4 in the case of Indonesia and found that financial development can reduce carbon dioxide emissions in all income group countries. In the same year, Shahbaz *et al.* [8] dealt with the question of whether financial development reduces carbon dioxide emissions in case of Malaysia, they revealed that financial development reduced carbon dioxide emissions for the Malaysian economy in the long term using the ARDL bound testing approach. Omri [9] estimated the three-way linkages among carbon dioxide emissions, energy consumption and economic growth using financial development as an instrumental variable and confirmed that financial development increased carbon dioxide emissions. Boutabba [10] examined the long-term equilibrium and the existence and direction of a causal relationship among carbon emissions, financial development, economic growth, energy consumption, and trade openness in India, and his results indicated that financial development had a long-term positive impact on carbon emissions, thereby implying that financial development improved environmental degradation. Ziaei [11] investigated the effects of financial indicator shocks, including credit and stock market shocks, on energy consumption and carbon dioxide using panel vector auto-regression models, and he founded that the associations between financial development variables and carbon dioxide emissions were driven by feedback and growth hypothesis. Meanwhile, Al-mulali *et al.* [12] also investigated the effects of energy consumption and carbon dioxide emissions on GDP growth and financial development in 30 Sub Saharan African countries, energy consumption and carbon dioxide emissions were found to have a positive causal relationship on GDP growth and financial development. Zhang [13] used cointegration theory, the Granger causality test and variance decomposition to explore the influence of financial development on carbon emissions, and his results indicated that China's financial development acts as an important driver for carbon emissions increase. Salahuddin *et al.* [14] investigated the relationship between carbon dioxide emissions, economic growth, electricity consumption and financial development in the Gulf Cooperation Council (GCC) countries using panel data for the period of 1980–2012, and they revealed that there is no causal link between financial development and carbon emissions. Abbasi and Riaz [15] explored the influence of economic and financial development on carbon emissions in a small emerging economy using the augmented VAR approach, and CO_2 emissions were found to cointegrate with financial development indicators.

From the aforementioned literature, we see that scholars have achieved a certain level of progress in the field of the effect of financial development on carbon emission, but some questions still require further exploration. First, previous studies have ignored the nonlinear relationship between financial development and carbon emissions because of the differences among China's provinces in terms of economic development level and industrial structure, leading to differential effects between financial development and carbon emissions. Second, previous research has considered financial scale as a single measure of financial development to study the effects of financial development on energy efficiency. In fact, financial development is a function of the financial system, which should be considered as an entire process of dynamic optimization, including the expansion of financial scale and the enhancement of financial efficiency. Thus, the combination of these two aspects constitutes the entire point of financial development. Hence, we will divide financial development into two components, namely financial scale and efficiency, and use a panel threshold model to test the relationship between financial development and carbon emission, Doing so will help provide convincing evidence that finance supports low-carbon development. This paper contributes to the literature in two ways. First, we use the threshold regression model proposed by Hansen to test whether there is a nonlinear relationship between financial development and carbon intensity, and find that there is not a simple linear relationship between financial development and carbon intensity, but rather a subsectional threshold relationship divided by per capita GDP. Second, we examine the nonlinear relationship between financial development and carbon emissions from the perspectives of financial scale and financial efficiency based on the dynamic optimization of the overall function of the financial system, and find that at a low level of per capita GDP, the expansion of financial scale and the enhancement of financial efficiency will both increase carbon intensity. When per capita GDP is greater than the threshold value (RMB 37,410), the expansion of the financial scale will also increase carbon intensity, but the potency of this effect will be comparatively weaker. Our research may help to better understand how to effectively use the role of finance in reducing the carbon intensity and provide a realistic basis for formulating financial policies.

This paper is organized as follows: Section 2 introduces model specification, variable selection and data sources. Section 3 studies the threshold effect of financial development on China's carbon intensity. Finally, conclusions and suggestions are proposed in Section 4.

2. Model Specification, Variables Selection and Data Sources

2.1. Model Specification

This study focuses on the impact of financial scale and financial efficiency on carbon intensity. Considering that carbon emission can also be influenced by a series of non-financial factors, we introduce the level of openness and industrial structure as control variables. Thus, we define the basic linear model as follows:

$$\ln Q_{it} = \mu_i + \alpha_1 \ln FD_{it} + \alpha_2 \ln FE_{it} + \beta_1 \ln OPEN_{it} + \beta_2 \ln IND_{it} + \varepsilon_{it} \quad (1)$$

where Q_{it} is the carbon intensity, the core explanatory variables, FD_{it} and FE_{it} represent financial scale and financial efficiency, respectively, $OPEN_{it}$ is the level of openness, and IND_{it} is industrial structure. Moreover, the subscript i and t represent the region and time, respectively.

To examine the nonlinear threshold effect between financial development and carbon emissions, we adopt the threshold regression model developed by Hansen [16]. This model can avoid the disadvantages inherent in the traditional threshold model and has the following advantage: first, it does not need to set the nonlinear equations; second, the number of the threshold is totally determined endogenously by the sample data; third, it will calculate the confidence interval of parameters according to the asymptotic distribution theorem; four, it will estimate the statistical significance using the bootstrap method. Considering that the influence of financial development on carbon emission intensity is constrained by the level of economic development, the basic model is extended to the multiple threshold model with the level of economic development acting as the threshold. Thus, the multiple threshold model is described as follows:

$$\begin{aligned} \ln \mathrm{Q}_{it} = {} & \mu_i + \alpha_{11} \ln FD_{it}(\ln PGDP_{it} \leqslant \gamma_1) + \alpha_{12} \ln FD_{it}(\gamma_1 < \ln PGDP_{it} \leqslant \gamma_2) \\ & + \cdots + \alpha_{1,n-1} \ln FD_{it}(\gamma_{n-1} < \ln PGDP_{it} \leqslant \gamma_n) + \alpha_{1,n} \ln FD_{it}(\gamma_n < \ln PGDP_{it}) \\ & + \alpha_{21} \ln FE_{it}(\ln PGDP_{it} \leqslant \gamma_1) + \alpha_{22} \ln FE_{it} \cdot I(\gamma_1 < \ln PGDP_{it} \leqslant \gamma_2) + \cdots \\ & + \alpha_{2,n-1} \ln FE_{it}(\gamma_{n-1} < \ln PGDP_{it} \leqslant \gamma_n) + \alpha_{2,n} \ln FE_{it}(\gamma_n < \ln PGDP_{it}) \\ & + \beta_1 \ln OPEN_{it} + \beta_2 \ln IND_{it} + \varepsilon_{it} \end{aligned} \quad (2)$$

where p^{GDP} is the threshold variable, which represents the level of regional economic development, γ_1, γ_2, $\gamma_{n-1}...\gamma_n$ are the values of the threshold variables, and α_{11}, α_{12}, $\alpha_{2,n-1}...\alpha_{2,n}$ are the estimated coefficients of the threshold variables. Moreover, we adopt the fixed effects model to assume that each sample individual, μ_i is independent identically distributed, such that every individual has its own fixed and unique intercept to express the unique qualities of every observation, in so doing, we reduce the covariation coefficient and enhance the validity of the estimated results. The coefficient matrix and variable matrix, respectively, are β and χ:

$$\beta = \begin{vmatrix} \alpha_{11} \\ \alpha_{12} \\ \vdots \\ \alpha_{1,n} \\ \alpha_{21} \\ \alpha_{22} \\ \vdots \\ \alpha_{2,n} \\ \beta_1 \\ \beta_2 \end{vmatrix}, \chi = \begin{vmatrix} \text{ln}FD_{it}(\text{ln}PGDP_{it} \leqslant \gamma_1) \\ \text{ln}FD_{it}(\gamma_1 < \text{ln}PGDP_{it} \leqslant \gamma_2) \\ \vdots \\ \text{ln}FD_{it}(\gamma_n < \text{ln}PGDP_{it}) \\ \text{ln}FE_{it}(\text{ln}PGDP_{it} \leqslant \gamma_1) \\ \text{ln}FE_{it}(\gamma_1 < \text{ln}PGDP_{it} \leqslant \gamma_2) \\ \vdots \\ \text{ln}FE_{it}(\gamma_n < \text{ln}PGDP_{it}) \\ \text{ln}OPEN_{it} \\ \text{ln}IND_{it} \end{vmatrix} \quad (3)$$

Therefore, Equation (2) can be transformed into following function:

$$\text{ln}Q_{it} = \beta^T \chi_{it}(\gamma) + \mu_i + \varepsilon_{it} \quad (4)$$

The above function in matrix form is:

$$Y^* = \chi(\gamma)^{*T} \beta + e* \quad (5)$$

The coefficient for the limited variable is β:

$$\beta_\gamma = (\chi^*(\gamma)^T \chi^*(\gamma))^{-1} \chi^*(\gamma)^T Y^* \quad (6)$$

The residual for the regression function is:

$$e^*(\gamma) = Y^* - \chi^*(\gamma)\beta_\gamma^T \quad (7)$$

The residual square sum for the regression function is:

$$SSE(\gamma) = e^*(\gamma)^T e^*(\gamma) = Y^{*T}[1 - \chi^*(\gamma)^T(\chi^*(\gamma)^T \chi^*(\gamma))^{-1} \chi^*(\gamma)^T]Y^* \quad (8)$$

where the threshold point can be estimated as follows:

$$\hat{\gamma} = \underset{\gamma}{\text{argmin}} SSE_1(\gamma) \quad (9)$$

2.2. Variables and Data

2.2.1. Explained Variable

Carbon intensity is a measure of unit GDP carbon emission. A two-step calculation is performed as the data of carbon emissions cannot be directly obtained from the statistical yearbook. First, we collect the energy consumption values of various areas within China and the conversion coefficients of various energy types into standard coal. Then, all energy types are converted into the standard measure of heat energy in China, known as standard coal. Carbon emissions are calculated according to IPCC Carbon Emissions Calculation Guide using the following formula [17,18].

$$C = \sum_j E_j \times \delta_j \tag{10}$$

where C is the carbon emissions, E_j is the consumption of energy j, and δ_j is the conversion coefficient of energy j. Based on IPCC National Greenhouse Gas Emission Inventory Guidebook, we obtain the carbon emission conversion coefficient of various energy types (Table 1).

Table 1. Carbon emission coefficient of different sources of energy (unit: tc/tce).

Energy	Carbon Emission Coefficient	Energy	Carbon Emission Coefficient
Raw Coals	0.7559	Kerosene	0.5714
Cokes	0.8550	Diesel oils	0.5921
Crude oils	0.5857	Fuel oils	0.6185
Gasoline	0.5538	Natural gas	0.4483

Using the carbon emissions coefficient and the calculation formula of carbon emission, we obtain the carbon emissions of the 30 Chinese provinces. Then, we then obtain the corresponding carbon intensity of each province by dividing those values by each province's GDP. In this way, carbon emission and carbon intensity are determined.

2.2.2. Core Explanatory Variables

As mentioned above, financial development involves the expansion of financial scale and the enhancement of financial efficiency. Therefore, we use financial scale and financial efficiency to measure the level of financial development.

(1) Financial scale

Generally speaking, the Mckinnon Index and Goldsmith index are the most popular indexes used to measure financial scale; however, because China's regional financial assets and M2 (Broad money) statistics cannot be obtained from previous statistical data, we cannot use the Mckinnon Index and Goldsmith index to calculate financial scale in the paper. Thus, we use the deformation of Goldsmith index with deposits and loans used in place of financial assets, based on this, we use the proportion of deposits and loans in GDP to measure financial scale.

(2) Financial efficiency

We use the proportion of non-state enterprises' loans in GDP to measure financial efficiency. Presently, there are a large number of policy-oriented loans and non-performing loan in China's banking sector, and most of the credit resources are allocated to the state sector; indeed, less than 30% of the bank loans flow to the non-state sector. Meanwhile, the contribution of the non-state sector to the GDP is about 65% [19], and so the proportion of non-state enterprises' loans in total loans can better measure the efficiency of China's financial system.

Because China's existing statistical yearbooks do not list the values of non-state enterprises' loans, we estimate this indicator based on the method developed by Zhao [20]. We assume that the proportion of state-owned enterprises' loans in total loans and the proportion of state-owned enterprises' fixed assets in the whole society constitute a fixed proportion β. Then, "Total loans/GDP" can be decomposed and expressed as:

$$\begin{aligned}\frac{\text{Total loans}}{\text{GDP}} &= \frac{\text{State}-\text{owned enterprises}'\text{ loans}}{\text{GDP}} + \frac{\text{Nonstate}-\text{owned enterprises}'\text{ loans}}{\text{GDP}} \\ &= \beta\frac{\text{Total loans}}{\text{GDP}} \cdot \frac{\text{State}-\text{owned enterprise fixed assets investment}}{\text{Total fixed assets investment}} + \\ &\quad \frac{\text{Non}-\text{state enterprises}'\text{ loan}}{\text{GDP}}\end{aligned} \tag{11}$$

Then, we estimate the value of β based on the fixed effect. In this way, we obtain the value of financial efficiency.

2.2.3. Threshold Variable

Because the effect of financial development on carbon intensity is restricted by the level of economic development, we select per capita GDP (PGDP) as the threshold variable of carbon intensity affected by financial development.

2.2.4. Control Variables

(1) The level of openness

Considering that improving the level of openness is helpful to obtain advanced technology and management experience related to carbon reduction, we introduce the level of openness as one of the control variables in this paper. We mainly use the proportion of total value of imports and exports in GDP as the index to measure the level of openness [21].

(2) Industrial Structure

China is in a critical period of industrial structure transformation and adjustment. Because various industries' demand for energy is different, carbon emissions also differ in different industries. Therefore, we introduce industrial structure as another control variable, and it is measured by the share of added value of the tertiary industry in GDP.

2.3. Data Sources

Given the lack of relative data in Tibet, Hong Kong, Macau and Tainwan, we select the other 30 provinces as the research objective, and the study period is from 2005 to 2012. The data come from China Statistical Yearbook and China Energy Statistical Yearbook. Table 2 shows the descriptive statistics for the aforementioned variables. Among all the variables, the maximum values are much larger than the minimum values, indicating that there are huge heterogeneities across provinces in China.

Table 2. Descriptive statistics of variables (N = 240).

Variable	Definitions and Measures	Mean	Std. Dev	Min	Max
FD	total deposits and loans/GDP	2.51	1.04	1.02	7.30
FE	non-state enterprises' loans/GDP	0.58	0.34	0.05	2.05
PGDP	GDP/population	28,895.05	18,242.11	5052	93,173
OPEN	total value of imports and exports/GDP	4.61	5.29	0.61	22.91
IND	added value of the tertiary industry/GDP	40.28	7.79	22.49	75.9

3. Empirical Results and Analysis

3.1. Regional Carbon Intensity Measurement

We can calculate the carbon intensity values of 30 provinces in China from 2005 to 2012 according to Equation (10), and the results are listed in Table 3.

Table 3 indicates that the lowest carbon intensity values are found in the provinces of Beijing, followed by Guangdong and Shanghai. Two of the highest scoring provinces with respect to carbon intensity are Shanxi and Ningxia, and both of these two provinces obtain average carbon intensity values greater than 4.0 on average throughout the observed period. Regarding the regional distribution of carbon intensity in China, the eastern region has the lowest carbon intensity value and has an average carbon intensity value of 0.98 for the 8 years period. The central region has an average carbon intensity value of 1.54 during this period. Finally, the western region reports the highest carbon intensity value at 1.88. Clearly, carbon intensity in China increases as one moves inland from the coast. The reason for this may be related to geographical location and industrial structure. On one hand, most of the lower-carbon-intensity provinces are located in the eastern coastal region, which allows them to obtain the abundant capital, management personnel, technology and informational to improve their carbon intensity levels. In contrast, the distance from the inland areas to the coastal areas hinders the transfer of the advanced technology, capital and management [22]. These factors make the comprehensive level of carbon intensity in the central and western regions relatively higher. On the other hand, those regions with lower levels of carbon intensity also contain most of the areas that have more highly developed industrial structures and markets. Since the implementation of the "Reform and Opening" process, Beijing, Shanghai, Guangdong, Fujian and other coastal areas have gradually implemented the "going out" strategy, which has contributed to the adjustment and optimization of the industrial structure under an export-oriented economy. At the same time, these areas have established a highly standardized market system, and this has helped transit them away from resource-oriented economic growth. The level of industrial development in the west of the country is low and the basic industries are resource-oriented. Given the development of electric power, coal mining, oil and other specialized industries, the technological development in these areas is relatively poor [23]. Moreover, the high levels of carbon intensity in these areas have remained unchanged. We can also see that the carbon intensity of all the provinces in China have exhibited a downward trend in recent years, this indicates that China's implementation of an energy reduction strategy has resulted in substantial improvements [24].

3.2. Multi-Collinearity Test and Stationary Test

3.2.1. Multi-Collinearity Test

Considering that we introduce numerous explanatory variables in the econometric model, we carry on the multi-collinearity test to assess whether there is a multi-collinearity between explanatory variables. The results are shown in Table 4. It can be seen from Table 4 that the VIF value of each variable is below 10, that is to say, there is no multi-collinearity among all explanatory variables.

Table 3. Carbon Intensity of 30 provinces in China from 2005 to 2012.

Province	2005	2006	2007	2008	2009	2010	2011	2012
Anhui	1.90144	1.36451	1.18958	1.10726	1.09397	0.93355	0.82782	0.82952
Beijing	1.45569	0.55252	0.40509	0.30330	0.24163	0.20096	0.15946	0.13981
Chongqing	0.92751	0.94415	0.88350	0.91987	0.81199	0.68926	0.59926	0.50667
Fujian	1.35363	0.75283	0.64638	0.55118	0.53103	0.44041	0.44747	0.38235
Gansu	6.94534	2.38405	1.89134	1.45327	1.21831	1.15519	1.07602	0.99637
Guangdong	1.55875	0.62128	0.48035	0.39150	0.33149	0.30708	0.29690	0.26403
Guangxi	1.54945	0.87817	0.75054	0.59102	0.58898	0.55827	0.50909	0.47836
Guizhou	3.64512	3.66069	3.31357	2.63333	2.51270	2.25590	1.92486	1.72932
Hainan	5.88460	1.35323	0.87482	0.59004	0.47646	0.38644	0.35660	0.33750
Hebei	2.79879	1.88391	1.67803	1.45819	1.41319	1.33529	1.21284	1.16001
Heilongjiang	4.05115	1.88093	1.56659	1.37631	1.30290	1.20932	1.06992	1.02311
Henan	2.08490	1.64873	1.49686	1.22878	1.24076	1.07015	1.03017	0.87761
Hubei	2.06157	1.21532	1.01270	0.78297	0.72211	0.67893	0.63750	0.56490
Hunan	1.76974	1.11964	0.98802	0.75378	0.67246	0.58974	0.53322	0.46501
Jiangsu	1.79411	0.90178	0.73042	0.59991	0.53069	0.48686	0.47395	0.42985
Jiangxi	1.59919	0.97760	0.85529	0.72204	0.60309	0.56739	0.50250	0.44278
Jilin	3.35502	1.79084	1.41156	1.17851	1.07422	1.00141	0.95807	0.85643
Liaoning	7.08184	2.33665	1.66361	1.21257	1.05876	0.89305	0.77241	0.69305
InnerMongolia	3.18939	2.95333	2.57666	2.47261	2.10918	1.44130	2.22740	2.09400
Ningxia	7.27537	5.16797	4.81474	4.15534	3.81109	3.63708	3.80199	3.18811
Qinghai	2.62426	1.51255	1.30848	1.29453	1.27534	0.99087	1.05624	1.10665
Shaanxi	4.69738	1.94276	1.49322	1.23163	1.09761	1.04334	0.93498	0.96286
Shandong	3.18005	1.56106	1.29938	1.04829	0.97823	0.88589	0.80216	0.74262
Shanghai	2.06268	0.74217	0.53883	0.43768	0.36669	0.33923	0.30380	0.27157
Shanxi	5.49681	5.47125	4.77691	3.90441	3.65558	3.18935	3.01833	3.11805
Sichuan	1.18163	0.93797	0.86237	0.79653	0.77306	0.62439	0.51413	0.47465
Tianjin	3.06597	1.14203	0.88543	0.65727	0.55751	0.48651	0.42497	0.38496
Xinjiang	7.09467	2.32796	1.75726	1.43981	1.62031	1.35691	1.32471	1.40170
Yunnan	1.75297	1.72786	1.50856	1.30605	1.28461	1.15867	0.98748	0.88904
Zhejiang	1.92510	0.85781	0.70277	0.57596	0.52612	0.44134	0.39420	0.35477
East	2.92375	1.15503	0.90046	0.71145	0.63744	0.56392	0.51316	0.46914
Central	2.78998	1.93360	1.66219	1.38176	1.29564	1.15498	1.07219	1.02218
West	3.71664	2.22159	1.92366	1.66309	1.55484	1.35556	1.35965	1.25707
China	3.17880	1.75372	1.47876	1.23913	1.14934	1.01180	0.97261	0.90552

Table 4. Results of the multi-collinearity test.

Variable	lnFD	lnFE	lnOPEN	lnIND
lnFD	1			
lnFE	0.3690	1		
lnOPEN	0.4676	0.3495	1	
lnIND	0.5128	0.2277	0.3624	1
VIF value	1.63	1.21	1.38	1.39

Note: the above results are getted using command "estat vif" in stata13.

3.2.2. Stationary Test

Because we will conduct the empirical analysis using panel data method, it is necessary to carry out a stationary test on each variable. Given that this empirical research works on 30 provinces of China, the presence of cross-section dependence (CD) is probable [25]. We therefore first use Pesaran (2004) CD test to assess the cross-section dependence of each variable, and the results are shown in Table 5.

Table 5. Results of CD test.

Variable	Lag Order			
	0	1	2	3
lnFD	36.167 ***	34.943 ***	15.787 ***	6.501 ***
lnFE	52.533 ***	47.726 ***	29.263 ***	18.054 ***
lnQ	21.037 ***	9.707 ***	0.115 *	0.665 *
lnPGDP	6.912 ***	3.130 ***	3.382 ***	0.761 *
lnOPEN	38.268 ***	38.268 ***	31.226 ***	28.829 ***
lnIND	25.109 ***	13.263 ***	13.263 ***	16.422 ***

Note: (1) ***, ** and * denote rejection of the null hypothesis at the 1%, 5% and 10% levels, respectively; (2) The cross regressions include both the constant and trend as deterministic components; (3) the above results are getted using command "xtcsd" in stata13.

The results in Table 5 indicate that the null hypothesis of cross-section independence is rejected at different lag orders, and this provides strong evidence of the existence of the cross sectional dependence among all the variables of 30 provinces. We should therefore use a Pesaran (2007) panel unit root test (CIPS) to assess the existence of the unit root [26]. Tthe results are shown in Table 6. It can be seen from Table 6 that not all of the variables can refuse the unit root assumption. That is to say, all of the variables are stationary.

Table 6. Results of CIPS test.

Variable	lnFD	lnFE	lnQ	lnPGDP	lnOPEN	lnIND
CIPS	−2.673 *	−3.059 ***	−3.067 ***	−2.256 ***	−3.352 ***	−4.391 ***

Note: (1) ***, ** and * denote rejection of the null hypothesis at the 1%, 5% and 10% levels, respectively; (2) The Akaike information criterion (AIC) is used in selecting the lag lengths; (3) The cross regressions include both the constant and trend as deterministic components; (4) the above results are getted using command "pescadf" in stata13.

3.3. *Panel Threshold Effect Test*

3.3.1. Panel Threshold Effect Test

Hansen used the bootstrap method to find the number of threshold according to the principle of the minimum residual sum of squares. He successively conducted the significance test on the number of thresholds. Thus, we adopt the bootstrap method to conduct the threshold effect test. Table 7 shows the results.

Table 7. Results of threshold effect test.

Hypothesis Test	LR (0%, 5%, 1% Critical Points)	F	p
H0:No threshold; H1:Single threshold	7.3523 (4.4754, 5.9906, 8.9557)	30.2544	0.0000
H0: Single threshold; H1:Double-threshold	9.1527 (5.1598, 6.6852, 9.3089)	18.5894	0.0000
H0: Double-threshold; H1:Triple-threshold	11.1324 (5.2659, 7.1286, 10.7881)	6.6683	0.0630

Note: (1) the frequency of bootstrap is 1000; (2) the above results are getted using command "xthreg" in stata13.

In Table 7, it can be seen that both the single-threshold model and double-threshold model of the two core explanatory variables pass the significance test at the confidence level of 1%. In contrast, the triple-threshold model does not pass the significance test at the confidence level of 5%. However, the 95% confidence intervals in Table 8 show that the confidence interval of the double-threshold is contained in the single-threshold, so we select the single-threshold model. According to Table 8, lnPGDP, the value of the threshold variable, is 10.5297, therefore the value of per capita GDP is RMB 37,410.

Table 8. The values of thresholds and the confidence interval.

Model	Estimated Value of Threshold	95% Confidence Interval
Single-threshold Model	10.5297	(9.8660, 11.1324)
Double-threshold Model	10.3118	(10.2635, 10.8349)

Note: the above results are getted using command "xthreg" in stata13.

Based on the results of the aforementioned tests and analyses, we will use the linear model and single-threshold model to analyze the relationship between financial development and carbon intensity. We use the level of economic development as the threshold variable to study the influence of financial scale and financial efficiency on carbon intensity; as is shown in Table 9, the influence of the level of economic development on carbon intensity exhibits a significant single threshold effect.

Table 9. Results of linear model and single-threshold model.

Variable	Linear Model	Variable	Single-Threshold Model
lnFD	0.9482 *** (4.4059)	lnFD (lnPGDP < 10.5297)	1.6455 *** (6.0460)
		lnFD (lnPGDP ⩾ 10.5297)	0.5573 * (1.6813)
lnFE	−0.8283 *** (−4.0871)	lnFE (lnPGDP < 10.5297)	0.8645 *** (4.1699)
		lnFE (lnPGDP ⩾ 10.5297)	−0.8446 *** (−4.3830)
lnOPEN	−0.2022 *** (−4.4795)	lnOPEN	−0.2184 *** (−4.8296)
lnIND	−1.3732 *** (−4.6436)	lnIND	−1.3114 *** (−4.6240)
Hausman Test	23.78 ***		23.78 ***

Note: (1) ***, ** and * denote rejection of the null hypothesis at the 1%, 5% and 10% levels, respectively; (2) hausman test results show that the fixed effects model is selected; (3) the results of Linear Model is getted using command "xtreg" in stata13; (4) the results of Single-threshold Model is getted using command "xthreg" in stata13.

3.3.2. Analysis of the Core Explanatory Variables

According to the estimated results of single-threshold model (Table 9), the effects of carbon intensity influenced by financial scale and financial efficiency can be obtained as shown in Figure 1.

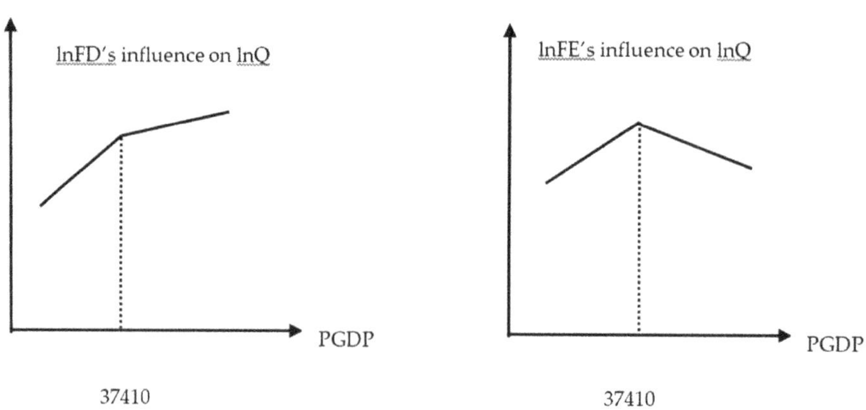

Figure 1. The threshold effect of financial development to carbon intensity.

The following can be seen in Figure 1:

(1) The direction of the effect of financial scale on carbon intensity before and after the threshold value does not change, but the influence degree does decreases beyond the threshold value. That is to say, the expansion of financial scale will increase carbon intensity, but the promotion effect becomes weaker upon reaching the threshold value.

(2) The direction of financial efficiency before and after the threshold value changes significantly, when the threshold value is not reached, the enhancement of financial efficiency will increase carbon intensity, while upon reaching the threshold, it will prompt the reduction in carbon intensity.

3.3.3. Analysis of the Threshold Variable

Based on the level of economic development and its value compared to the threshold value in each region, we divide the provinces into two zones. The low zone has per capita GDP that is lower than the threshold value, and the high zone has per capita GDP greater than the threshold value during the observation period. As Table 10 shows, 14 provinces, such as Shanghai, Beijing, Tianjin, Jiangsu, Zhejiang, Guangdong, Inner Mongolia, Shandong, Liaoning, Fujian, Jilin, Chongqing, Hubei, and Shaanxi, belong to the high zone. Most of these provinces that belong to the high zone are located in the eastern coastal areas, whereas the less developed provinces located in the Central and Western areas mainly belong to the low zone, in which the expansion of financial scale and the enhancement of financial efficiency will increase carbon intensity.

3.3.4. Analysis of the Control Variables

Table 9 also shows that the level of openness and industrial structure each has a significant negative relationship with carbon intensity; specifically, by raising the level of openness and improving the industrial structure, a region can significantly reduce the carbon intensity. The level of openness increasing by 1% reduces to the carbon intensity by 0.2022%, and the proportion of the tertiary industry in GDP increasing by 1% will reduce carbon intensity by 1.3732%.

Table 10. The threshold values of carbon intensity compared with regional GDP.

Year	Low Zone(GDP per Capita < 37,410 yuan)	High Zone(GDP per Capita > 37,410 yuan)
2005	Guizhou, Gansu, Yunnan, Anhui, Guangxi, Sichuan, Jiangxi, Shaanxi, Qinghai, Ningxia, Hunan, Hainan, Chongqing, Henan, Hubei, Shanxi, Jilin, Heilongjiang, Hebei, Inner Mongolia, Fujian, Liaoning, Shandong, Guangodng, Jiangsu, Zhejiang, Tianjin	Beijing, Shanghai
2009	Guizhou, Gansu, Yunnan, Anhui, Guangxi, Sichuan, Jiangxi, Hainan, Qinghai, Xinjiang, Hunan, Henan, Hebei, Shanxi, Ningxia, Heilongjiang, Chongqing, Jilin, Fujian, Liaoning, Shandong, Shaanxi	Inner Mongolia, Guangdong, Zhejiang, Jiangsu, Tianjin, Beijing, Shanghai
2012	Guizhou, Gansu, Yunnan, Anhui, Guangxi, Sichuan, Jiangxi, Hainan, Qinghai, Xinjiang, Hunan, Henan, Hebei, Shanxi, Ningxia, Heilongjiang	Shanghai, Beijing, Tianjin, Jiangsu, Zhejiang, Guangdong, Inner Mongolia, Shandong, Liaoning, Fujian, Jilin, Chongqing, Hubei, Shaanxi

3.4. Discussion

The aforementioned empirical results show that the impact of financial development on carbon emissions intensity does not indicate a simple linear relationship; changes occur in line with the threshold effect of the level of economic development. When the level of economic development is relatively low and per capita GDP is lower than the threshold value of RMB 37,410, the industrial structure tends to be based on high energy consumption and high pollution industries, and so the bank loans will mainly flow to these industries [27,28]. At this stage, the expansion of financial scale will increase carbon intensity. Meanwhile, when the economy develops to a certain degree so that per capita GDP exceeds the threshold value of RMB 37,410, the industrial structure will prioritize the service and high-tech industries, which feature low energy consumption and low pollution. When this occurs, bank loans will flow into these industries. Considering that the energy efficiency of non-state enterprises is higher than that of state-owned enterprises [29,30], while the loans to state-owned enterprise in China are much higher than those to the non-state-owned enterprises, the expansion of financial scale will also increase carbon intensity at this stage, while the promotion effect will attenuate compared to that seen in the previous stage. When the level of economic development is relatively low and per capita GDP is lower than the threshold value of RMB 37,410, although the enhancement of financial efficiency will promote the proportion of non-state financial loan in all loans, the industrial structure will be based on high energy consumption and high pollution industries, and so bank loans will mainly flow to those industries. At this stage, the enhancement of financial efficiency will increase carbon intensity. However, when the economy develops to a certain level where per capita GDP exceeds the threshold value of RMB 37,410, the industrial structure will prioritize service and high-tech industries, both of which are characterized by low energy consumption and low pollution, and so bank loans will flow to those industries. Considering that the energy efficiency of the non-state enterprises is higher than that of the state-owned enterprises, the increase of the share of the non-state-owned enterprises in the service and high-tech industries will gradually reduce carbon intensity [31].

We can also see that the per capita GDP of the provinces in the eastern coastal areas of China exceeds the threshold value, and so the decision-making departments of these areas should design effective financial policies that can support the non-state-owned enterprises [32]. In this way, they could create a positive interaction mechanism between financial development and carbon intensity. However, for the central and western areas of China, per capita GDP is less than the threshold value, and so the reasonable path in reducing the carbon intensity would be to become fully aware of the characteristics of the local industrial structure and design effective plans of regional industrial transition [33]. Meanwhile, they should also increase environmental awareness among financial institutions, so as to promote the decrease of the carbon intensity [34].

4. Conclusions and Policy Suggestions

We used the panel data from 2005 to 2012 to measure carbon emissions in various provinces in China based on the IPCC National Greenhouse Gas Emission Inventory Guidebook. On this basis, we tested the relationship between carbon intensity and the two indicators of financial development, namelyfinancial scale and financial efficiency, using the threshold model. We derived three key results as follows: (1) the relationships among carbon intensity, financial scale and financial efficiency are near simple linear, and they are more inclined to a piece-wise function with per capita GDP as a threshold variable that divides partition intervals. When per capita GDP is at a low level, the expansion of financial scale and the enhancement of financial efficiency will increase carbon intensity. When per capita GDP exceeds the threshold value, the expansion of financial scale will also increase carbon intensity, but this promotion effect will be weaker in the high zone, and the enhancement of financial efficiency will cause reduce carbon intensity. (2) Most of the provinces with per capita GDP greater than the threshold value (RMB 37,410) are located in the eastern coastal areas of China, whereas most of the provinces with per capita

GDP less than the threshold value are located in the central and western areas of China. (3) Increasing the level of openness and improving the industrial structure have significant positive effect on carbon intensity.

Based on the aforementioned empirical results, we should consider the following aspects by which to further reduce carbon intensity further: (1) The development style should be modified so as to focus on efficiency improvement rather than scale expansion; this modification is in keeping with a shift from the tendency of "aristocratic" in the past to the direction of "civilian" [35]. It is necessary to establish an effective credit mechanism for non-state-owned enterprises and develop financial products suitable for the non-state-owned enterprises. (2) We should accelerate the adjustment and upgrading of industrial structure, and continuously optimize the industrial structure and product structure of the heavy chemical industry; concurrently, it should enhance the proportion of high-tech industries and modern service industries in the national economy. (3) We should accelerate the pace of openness, while concurrently optimizing the open structure. It will be important to screen foreign investment and take advantage of international markets and international resources to introduce more advanced low-carbon industries and low-carbon technology; these steps are critical to bring about the win-win situation of concurrent economic development and energy conservation.

Acknowledgments: This paper is the stage achievement of National Nature Science Foundation Project (71303029), Liaoning province Social Science Fund Project (L13DGL033). The author is grateful for the support of the National Nature Science Foundation and Liaoning province Social Science Foundation. The contents of this paper are solely the responsibility of the authors and do not represent the official views of the aforementioned institutes and funding agencies.

Author Contributions: Xiongfeng Pan designed the research; Xiaoxue Peng and Qing Liu performed research and analyzed the data; Xiongfeng Pan and Yaobo Yan wrote the paper. All authors read and approved the final manuscript.

Conflicts of Interest: The authors declare no conflict of interest.

References

1. Miomir, J.; Ljiljana, K.; Aleksandra, D.; Vladimir, K. The Impact of Agro-Economic Factors on GHG Emissions: Evidence from European Developing and Advanced Economies. *Sustainability* **2015**, *7*, 16290–16310.
2. Vallati, A.; Grignaffini, S.; Romagna, M. A New Method to Energy Saving in a Micro Grid. *Sustainability* **2015**, *7*, 13904–13919.
3. Knight, K.W.; Schor, J.B. Economic Growth and Climate Change: A Cross-National Analysis of Territorial and Consumption-Based Carbon Emissions in High-Income Countries. *Sustainability* **2014**, *6*, 3722–3731.

4. Tamazian, A.; Chousa, J.P.; Vadlamannati, C. Dose Higher Economic and Financial Development Lead to Enviromental Degradation: Evidence from the BRIC Countries. *Energy Policy* **2009**, *37*, 246–253.
5. Tamazian, A.; Rao, B.B. A Model of Growth and Finance: FIML Estimates for India. *Energy Econ.* **2010**, *32*, 137–145.
6. Sadorsky, P. Financial Development and Energy Consumption in Central and Eastern European Frontier Economies. *Energ. Pol.* **2011**, *39*, 999–1006.
7. Shahbaz, M.; Hye, Q.M.A.; Tiwari, A.K. Economic Growth, Energy Consumption, Financial Development, International Trade and CO_2 Emissions in Indonesia. *Renew. Sustain. Energy Rev.* **2013**, *25*, 109–121.
8. Shahbaz, M.; Solarin, S.A.; Mahmood, H.; Arouri, M. Does Financial Development Reduce CO_2 Emissions in Malaysian Economy? A Time Series Analysis. *Econ. Model.* **2013**, *35*, 145–152.
9. Omri, A. CO_2 Emissions, Energy Consumption and Economic Growth Nexus in MENA Countries: Evidence from Simultaneous Equations Models. *Energy Econ.* **2013**, *40*, 657–664.
10. Boutabba, M.A. The Impact of Financial Development, Income, Energy and Trade on Carbon Emissions: Evidence from the Indian Economy. *Econ. Model.* **2014**, *40*, 33–41.
11. Ziaei, S.M. Effects of Financial Development Indicators on Energy Consumption and CO_2 Emission of European, East Asian and Oceania countries. *Renew. Sustain. Energy Rev.* **2015**, *42*, 752–759.
12. Al-mulali, U.; Che Sab, C.N.B. The Impact of Energy Consumption and CO_2 Emissions on the Economic Growth and Financial Development in the Sub Saharan African Contries. *Energy* **2012**, *39*, 180–186.
13. Zhang, Y. The Impact of Financial Development on Carbon Emissions : An Empirical Analysis in China. *Energy Policy* **2011**, *39*, 2197–2203.
14. Salahuddin, M.; Gow, J.; Ozturk, I. Is the Long-run Relationship between Economic Growth, Electricity Consumption, Carbon Dioxide Emissions and Financial Development in Gulf Cooperation Council Countries robust? *Renew. Sustain. Energy Rev.* **2015**, *51*, 317–326.
15. Abbasi, F.; Riaz, K. CO_2 Emissions and Financial Development in an Emerging Economy: An Augmented VAR Approach. *Energy Policy* **2016**, *90*, 102–114.
16. Hansen, B.E. Threshold Effects in Non-dynamic Panels: Estimation, Testing, and Inference. *J. Econ.* **1993**, *93*, 345–368.
17. Zhang, N.; Wei, X. Dynamic Total Factor Carbon Emissions Performance Changes in the Chinese Transportation Industry. *Appl. Energy* **2015**, *146*, 409–420.
18. Zhang, N.; Zhou, P.; Kung, C. Total-factor Carbon Emission Performance of the Chinese Transportation Industry: A Bootstrapped Non-radial Malmquist Index Analysis. *Renew. Sustain. Energy Rev.* **2015**, *41*, 584–593.
19. Li, J.; Wei, P. Financial Development and Total Factor Productivity Growth: Empirical Research Based on China's Provincial Panel Data. *Econ. Theory Bus. Manag.* **2015**, *35*, 47–64.

20. Zhao, Y.; Lei, D. Financial Development and Economic Growth: Productivity Promotion *versus* Capital Formation. *J. World Econ.* **2010**, *33*, 37–50.
21. Pan, X.; Yang, Y. Research on Interactive Relationship between Foreign Trade and Carbon Emission: Based on Simultaneous Equation Model. *Oper. Res. Manag. Sci.* **2013**, *01*, 132–137.
22. Zhang, N.; Choi, Y. Environmental Energy Efficiency of China's Regional Economies: A Non-oriented Slacks-based Measure Approach. *Soc. Sci. J.* **2013**, *50*, 225–234.
23. Liu, L.; Liang, Q.; Wang, Q. Accounting for China's Regional Carbon Emissions in 2002 and 2007: Production-based *versus* Consumption-based Principles. *J. Clean Prod.* **2015**, *103*, 384–392.
24. Huang, G.; Ouyang, X.; Yao, X. Dynamics of China's regional carbon emissions under gradient economic development mode. *Ecol. Indic.* **2015**, *51*, 197–204.
25. Pesaran, M.H. *General Diagnostic Tests for Cross Section Dependence in Panels*; Cambridge Working Papers in Economics, No. 0435; University of Cambridge: Cambridge, UK, 2004.
26. Pesaran, M.H. A Simple Panel Unit Root Test in the Presence of Cross-section Dependence. *J. Appl. Econ.* **2007**, *22*, 265–312.
27. Yang, Y. Impact of Non-state Economic Sector on China's Industrial Enterprises Technical Efficiency. *Econ. Res. J.* **1998**, *12*, 29–35.
28. Luo, X. The Effects of Privatization on Industrial Performance in Chinese Economic Transition—Analysis of the 2001 National Industrial Census data. *Econ. Res. J.* **2004**, *8*, 16–26.
29. Zhang, N.; Choi, Y. Total-factor Carbon Emission Performance of Fossil Fuel Power Plants in China: A Metafrontier Non-radial Malmquist Index Analysis. *Energy Econ.* **2013**, *40*, 549–559.
30. Zhang, N.; Kong, F.; Choi, Y.; Zhou, P. The Effect of Size-control Policy on Unified Energy and Carbon Efficiency for Chinese Fossil Fuel Power Plants. *Energy Policy* **2014**, *70*, 193–200.
31. Guo, W.; Sun, T.; Dai, H. Effect of Polution Structure Change on Cabon Emission in China. *Sustainability* **2016**, *8*, 225–244.
32. Liu, Y.; Xiao, H.; Zhang, N. Industrial Carbon Emissions of China's Region: A Spatial Econometric Analysis. *Sustainbility* **2016**, *8*, 210–223.
33. Zhang, X.; Karplus, V.; Qi, T.; Zhang, D.; He, J. Carbon Emissions in China: How Far Can New Efforts Bend the Curve? *Energy Econ.* **2016**, *54*, 388–395.
34. Wang, Q.; Chen, X. Energy Policies for Managing China's Carbon Emission. *Renew. Sustain. Energy Rev.* **2015**, *50*, 470–479.
35. Fang, J. Ownership, Institutional Environment and Capital Allocation. *Econ. Res. J.* **2007**, *12*, 82–92.

Analysis on Impact Factors of Water Utilization Structure in Tianjin, China

Conglin Zhang, Leihua Dong, Yu Liu and Haijuan Qiao

Abstract: Water is an essential foundation for socio-economic development and environmental protection. As such, it is very critical for a city's sustainable development. This study analyzed the changes in water utilization structure and its impact factors using water consumption data for agricultural, industrial, domestic and ecological areas in the city of Tianjin, China from 2004 to 2013. On this base, the evolution law and impact factors of water utilization structure were depicted by information entropy and grey correlation respectively. These analyses lead to three main results. First, the total amount of water consumption in Tianjin increased slightly from 2004 to 2013. Second, the information entropy and equilibrium degree peaked in 2010. From 2004 to 2010, the water utilization structure tended to be more disordered and balanced. Third, the economic and social factors seemed to influence the water utilization structure, while the main impact factors were industrial structure, per capita green area, cultivated area, effective irrigation area, rural electricity consumption, animal husbandry output, resident population, per capita domestic water etc.

Reprinted from *Sustainability*. Cite as: Zhang, C.; Dong, L.; Liu, Y.; Qiao, H. Analysis on Impact Factors of Water Utilization Structure in Tianjin, China. *Sustainability* **2016**, *8*, 241.

1. Introduction

As a kind of strategic resources, water resources are not only an important condition of economic and social sustainable development, but also a precondition of improving ecological environment. Water utilization structure is one of the most important elements in water resources and effective water resources management.

Analysis of water utilization structure was prevalent in some scientific literatures. Liu *et al.* [1] qualitatively analyzed the change of water utilization structure of Beijing from 1980 to 2000. Jenerette [2] asserted that the water utilization form depended on population, livestock, climate change, and ecological water supply condition etc. This author judged that there would be continuous water supply crisis in most cities, so attention must be paid to sustainable water resources management. Su *et al.* [3], Ma *et al.* [4] and Zheng *et al.* [5] discussed the water utilization structure of Guanzhong area using information entropy. Lv *et al.* [6] used grey correlation degree to analyze the driving forces of water utilization structure change in Zhengzhou. Chen [7] used information entropy and grey correlation degree to analyze the change of water utilization structure and its impact factors respectively in Xiamen. Furthermore, the impact factors of water utilization structure were chosen by principal component analysis to satisfy the requirement of water resources management. On this basis, Yun *et al.* [8] built a compositional data linear regression model based on partial least-squares regression, and demonstrated the connection coherence between water utilization and industrial structure. Hereafter, Liu *et al.* [9] used a variable stability test, the co-integration and Granger causality test to study the relationship between water utilization structure and industrial sustainable development. Zhai *et al.* [10] not only considered industrial, agricultural and domestic water consumption, but also explored the influence of ecological water consumption on water utilization structure. Indeed, ecological water consumption is very important for sustainable ecological development. Paola *et al.* [11,12] presented a multi-objective approach for the automatic partitioning of a water distribution network into District Metering Areas (DMAs). On this base, Paola *et al.* [13] presented a model for valve setting in water distribution networks (WDNs), with the aim of reducing the level of leakage. Focusing on an African case, Paola *et al.* [14] analyzed the problem of the sustainable design and development of urban storm-water systems in response to the climate change and anthropic modification. Finally, Nunez *et al.* [15] presented a system of two-stage processing for the detection of acoustic emissions and the localization of the sources, which is also instructive.

Water consumption and its impact factors were discussed in some literatures. In general, water consumption can be divided into four categories: industrial, agricultural, domestic and ecological. Marios [16] supposed that the impact factors of water consumption included population growth, sustainable economic development, technical progress, land use pattern, urbanization progress etc. Through logarithmic mean Divisia index Method (LMDI), Liu *et al.* [17] analyzed the industrial water consumption of Anhui province by considering economic scale, industrial structure and water quota. On this basis, Zhang *et al.* [18] added water intensity and industrial water saving effect to those impact factors, and explained the change of industrial water in Anhui Province. Arbues *et al.* [19] calculated the weight of domestic water consumption. Cui *et al.* [20] proposed that the impact factors of domestic water consumption mainly included resident income and water price. Adding drinking water and family population, Xu [21] built a regression model to research the impact factors of domestic water consumption in Beijing.

Above results were meaningful, but the impact factors of water utilization structure should be further perfected. The impact factors from different literatures were complementary. Based on these impact factors and the data we could get (mainly from *Tianjin Statistical Yearbook 2005–2014* [22]), the impact factors were filtered. Besides, in *Tianjin Statistical Yearbook 2005–2014* [22], some indexes did not appear in above literatures, but they had influence on the water consumption of Tianjin. These kinds of indexes also could be used as impact factors. This study took the water consumption of Tianjin from 2004 to 2013 as sample [22]. Tianjin is located in the northeast of North China Plain and in the center of Bohai economic circle. As the largest port city of North China, Tianjin is an important communication hub, directly connecting Beijing to Northeast China, East China and the Pacific Ocean. In this study, the change of water utilization structure in Tianjin was analyzed, and its main impact factors were identified afterwards. These impact factors of water utilization structure are essential for us to conduct effective water resources management and sustainable development in Tianjin.

The study period was from 2004 to 2013. Due to the limitation of water quantity monitoring and resources management abilities, agricultural, industrial, domestic and ecological water consumption data of Tianjin before 2004 was unavailable.

2. Material and Methods

2.1. Research Area

The annual precipitation of Tianjin is 560 mm–720 mm on average, with the precipitation from June to August comprising 75% of total annual precipitation. With the rapid development of economy and society, the water shortage of Tianjin became more and more serious. In fact, water resources had become a key restrictive factor of the sustainable development in Tianjin. After the construction of Luanhe-Tianjin water diversion project, Luanhe River became an important water source for Tianjin. In the past decade, the annual water resources quantity of Tianjin was 1.177 billion m^3, and it was larger in the northern area than in the southern area. From 2000, four times the water supply from Yellow River to Tianjin were transported to ease the water shortage of Tianjin. After the completion of the first-phase of the middle-line of the South-to-North Water Transfer Project, although the water supply quantity is increased, the water shortage of Tianjin was also severe(Figure 1). Water resources remained a critical issue in Tianjin.

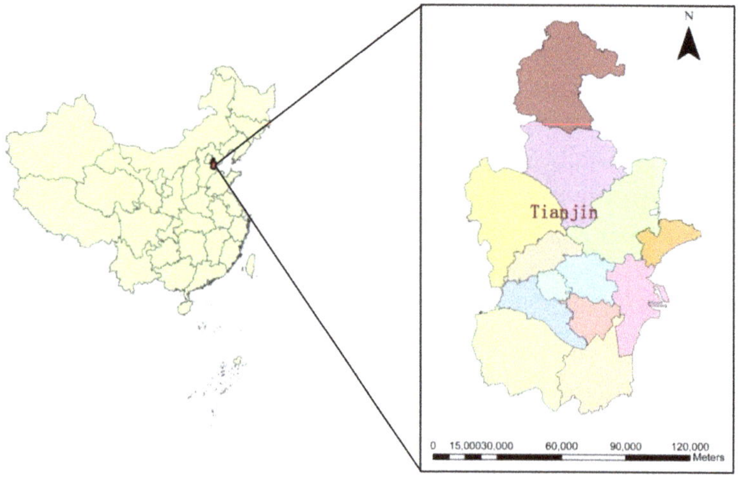

Figure 1. The map of the research area.

2.2. Data Source

Water consumption is affected by many factors. Through literature review, the impact factors of water consumption were summarized as the follows (Table 1).

Table 1. The impact factors of water consumption in literatures.

Water consumption	Impact factor
Agricultural water consumption	canal lining [23], canal facilities [23], water management [23], planting structure adjustment [23], rainfall [23], cultivated area [4,24], climate factors [4] , grain yield [25], water-saving effect [26], water price [26], irrigation area [26,27]
Industrial water consumption	industrial water saving [18], economic scale [17,18], industrial structure [17,18], water intensity [17,18], quota effect [17], population effect [24,28], technology effect [29], water reuse efficiency in industry [24,28], gross industrial output value [24,30], GDP [24], fixed assets investment [24,31]
Domestic water consumption	population [32], water saving model [22], price [32], economy development [32], regional difference [32], municipal infrastructure [32], the lowest temperature [28], the maximum and minimum humidity [32], green coverage of built-up area [24], living space per capita [33], consciousness of water saving [34], family income [21], water price [4,35]
Ecological water consumption	per capita green area [8,36]

Data on water consumption and impact factors were derived from the *2005–2014 Tianjin Statistical Yearbook* [22]. The specific impact factors for agricultural, industrial, domestic, and ecological water consumption are presented in Table 1.The impact factors of agricultural water consumption included cultivated area, effective irrigation area, effective water saving irrigation area, agriculture forestry, animal husbandry, and fishery output value, and rural electricity consumption. The impact factors of industrial water consumption included industrial output value, water consumption of 10,000 Chinese Yuan (CNY) industrial added value, water consumption of 10,000 CNY industrial output value, output value of light industry, heavy industry and high-tech industry. The impact factors of domestic water consumption included resident population, urban per capita income, per capita domestic water consumption. The impact factors of ecological water consumption included per capita green area, garden green area and park area.

3. Analysis Methods

3.1. Information Entropy Anasysis

Entropy is a concept of thermodynamics. Shannon introduced it into information theory, and defined it as information entropy to describe the confusion or disorder degree of system or material [37]. The evolution of water utilization structure has some stages, and they are irreversible. So information entropy could be used to analyze the water utilization structure. This index is useful in water resources sustainable management.

Set the total water consumption equals to Q, and the water consumption is $X_i\,(i = 1,2,3\cdots,n)$. The proportion of each water consumption type is p_i, and $\sum_{i=1}^{n} p_i = 1\,(p_i \neq 0)$. According to Shannon formula, $H = -\sum_{i=1}^{n} p_i \ln p_i$, where H is information entropy. The concept of equilibrium degree is J, and $J = H/H_{\max}$, where $H_{\max}$ is the maximum of information entropy, $H_{\max} = \ln(n)$. With the increase of H, the disorder degree of the system increases, and the structure of the system tends to be more balanced. With the increase of J, the equilibrium degree and stability of the system increase.

3.2. Grey Correlation Analysis Method

A grey system is a system containing both known and unknown information [38,39]. The grey correlation analysis model belongs to one multi-element statistical analysis methods. Based on the sample data, this model descripts the relevance size and order of factors through grey correlation. Since the relationship between water utilization structure and its impact factors is grey, this study used the grey correlation analysis method to analyze these impact factors.

$X_i(t)$ and $X_j(t)$ are the coordinates of the gravity centers of element i and j, respectively. So $\Delta_{i,j}(t) = |X_i(t) - X_j(t)|$ represents the absolute value of the difference between these two coordinates. Set $\Delta(\max)$ and $\Delta(\min)$ are the maximum and minimum values of these differences, then $0 \leqslant \frac{\Delta(\min)}{\Delta(\max)} \leqslant \frac{\Delta_{ij}(t)}{\Delta(\max)} \leqslant 1$. Obviously, with the increase of $\frac{\Delta_{ij}(t)}{\Delta(\max)}$, the consistency of change between (X_i) and (X_j) becomes weaker. In order to limit the standardized data between 0 and 1, the value could be taken as $\frac{\Delta(\min)/\Delta(\max)}{\Delta_{ij}(t)/\Delta(\max)}$.

For convenience, the grey correlation degree was classified into three groups: 0–0.60 implies weak correlation degree, 0.60–0.80 implies moderate correlation degree, 0.80–1.0 implies strong correlation degree [40].

4. The Water Consumption and Its Impact Factors

The total water consumption of Tianjin increased from 2.206 billion m^3 in 2004 to 2.337 billion m^3 in 2007. Its annual growth rate was 1.94% on average. The total water consumption of Tianjin in 2008 and 2009 fluctuated slightly. From 2010 to 2013, the annual growth rate was 1.85% on average (Figure 2).

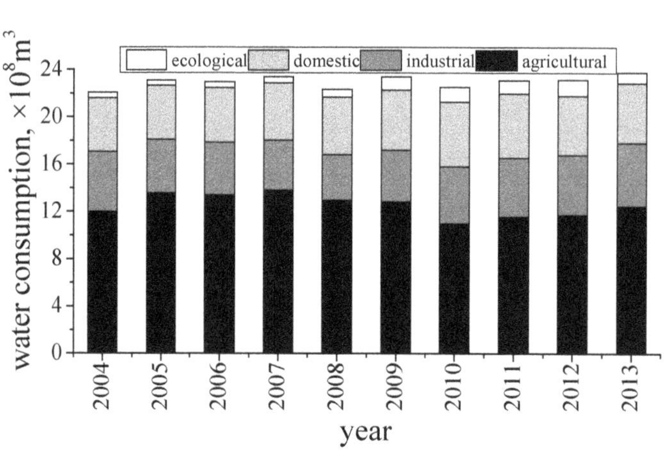

Figure 2. The water consumption of Tianjin from 2004 to 2013.

4.1. Agricultural Water Consumption

From 2004 to 2013, the agricultural water consumption of Tianjin accounted for 48%–59% of the total water consumption, and it was the largest water user in Tianjin. The change of agricultural water consumption could be divided into three stages. From 2004 to 2007, it increased from 1.198 billion m^3 to 1.384 billion m^3, and the annual growth rate was 4.93% on average; after 2007, it decreased to

1.097 billion m^3, and the annual growth rate was 7.45% on average; hereafter, it increased slightly.

The agricultural water consumption changed because the decrease of cultivated area and effective irrigation area and the increase of efficient water saving irrigation area. Indeed, the cultivated area of Tianjin decreased from 0.42 million hm^2 in 2004 to 0.39 million hm^2 in 2013, and its annual change rate was −0.63% on average; the effective irrigation area decreased from 0.35 million hm^2 in 2004 to 0.31 million hm^2 in 2013, and its annual change rate was −1.5% on average; efficient water saving irrigation area increased from 0.19 million hm^2 in 2004 to 0.29 million hm^2 in 2013, and its annual growth rate was 5.16% on average. In addition, the agricultural electricity consumption increased from 4.84 billion kWh in 2004 to 6.92 billion kWh in 2013, and its annual growth rate was 4.05% on average; the output value of primary industry increased from 22.14 billion CNY in 2004 to 40.20 billion CNY in 2013, and its annual growth rate was 6.86% on average.

Fishery output value had weak correlation with agricultural water consumption. Other impact factors had moderate correlation with agricultural water consumption (Table 2). It appeared that the adjustment of agricultural structure and the improvement of water saving irrigation technology influenced the agricultural water consumption. In these years, Tianjin City conducted the strategic layout of taking planting as a foundation and taking breeding industry as a sustainable development emphases. With the promotion of water saving irrigation technology, unreasonable irrigation methods were eliminated. On this basis, supposing that, the agricultural water of Tianjin developed in a sustainable direction.

Table 2. Grey correlation degree between the agricultural water and its impact factors in Tianjin.

	Correlation Degree		Correlation Degree
Cultivated area	0.7357	Forestry output value	0.6921
Effective water saving irrigation area	0.6571	Animal husbandry output value	0.7087
Effective irrigation area	0.6699	Fishery output value	0.4901
Agricultural output value	0.6642	Rural electricity consumption	0.7071

4.2. Industrial Water Consumption

From 2004 to 2013, the industrial water consumption of Tianjin accounted for 17%–23% of total water consumption. Its change could be divided into two stages: from 2004 to 2008, it reduced from 0.507 billion m^3 to 3.81 billion m^3, and its annual change rate was 6.89% on average; from 2008 to 2013, it increased from 0.381 billion m^3 to 0.537 billion m^3, and its annual growth rate was 7.11% on average.

The change of industrial water consumption was significantly influenced by the policy and regulation. In order to achieve sustainable water resources management, Tianjin City proposed "regulations on water saving of Tianjin" in 2002, which was the first local regulation on water saving in China. From 2004 to 2013, the industrial output value of Tianjin increased from 168.59 billion CNY to 727.55 billion CNY, while the water consumption of 10,000 CNY industrial output value decreased from 30.07 m^3 to 7.38 m^3. The water saving technology was improved and the industrial structure was adjusted. Consequently, from 2004 to 2008, the industrial water consumption decreased. However, in 2009, the administrative division of the New Coastal Region of Tianjin was reformed. The rapid development of the New Coastal Region of Tianjin accelerated the industrial economic development of Tianjin. At this time, the industrial output value of Tianjin was 398.78 billion CNY, and the New Coastal Region of Tianjin accounted for only 9.2% of Tianjin; in 2013, the industrial output value of Tianjin was 727.55 billion CNY, and the New Coastal Region of Tianjin accounted for 70.6% of Tianjin. The supernormal development of the New Coastal Region of Tianjin was an important explanation for the increase of the industrial water consumption of Tianjin after 2008. As a large industrial city, controlling total water consumption while ensuring the economic development was a challenge for Tianjin.

All of the impact factors had moderate correlation with the industrial water consumption of Tianjin (Table 3). Thus, the industrial economic quality had a significant influence on total water consumption in Tianjin. This finding suggested that improving industrial economic quality would promote the improvement of water resources management.

Table 3. Grey correlation degree between the industrial water and its impact factors in Tianjin.

	Industrial output value	Water consumption of ten thousand CNY industrial output value	Water consumption of ten thousand CNY industrial added value	Light industry output value	Heavy industry output value	High technology industry output value
Correlation degree	0.6768	0.7144	0.7520	0.7056	0.6645	0.6786

4.3. *Domestic Water Consumption*

From 2004 to 2013, the domestic water consumption of Tianjin accounted for 20%–25% of the total water consumption. The change of the domestic water consumption could be divided into two stages: In the first stage, the domestic water consumption increased steadily, while in the second stage, it reduced with fluctuation. Specifically, from 2004 to 2010, it increased from 0.453 billion m^3 in 2004 to 0.548 billion m^3 in 2010, and its annual growth rate was 3.22% on average; and from 2010 to 2013, it decreased from 0.548 billion m^3 to 0.515 billion m^3, and its annual change rate was 2.69% on average.

The basic explanation for the change of the domestic water consumption of Tianjin was population growth. The resident population increased from 10.24 million in 2004 to 14.72 million in 2013, and the annual growth rate was 4.12% on average. With the simultaneous improvement in living standard, per capita domestic water consumption increased from 123.6 m^3 in 2005 to 142.34m^3 in 2013.

All of the impact factors had moderate correlation with the domestic water consumption of Tianjin (Table 4). With the adjustment of fertility policy and the improvement of income level, the domestic water consumption of Tianjin would be increased in the future. Green lifestyle could be very important in forming sustainable water resources management.

Table 4. Grey correlation degree between the domestic water and its impact factors in Tianjin.

	Resident population	Urban per capita income	Per capita domestic water consumption
Correlation degree	0.7375	0.6898	0.6806

4.4. Ecological Water Consumption

Given the importance of ecological environment protection and sustainability, the ecological water consumption of Tianjin was also studied. It increased from 48 million m^3 in 2004 to 136 million m^3 in 2012, and its annual growth rate was 13.9% on average; in 2013, it reduced by 33.8%. Certain factors may explain this change in ecological water consumption. For example, the per capita green area of Tianjin increased from 13.87 m^2 in 2004 to 15.76 m^2 in 2013, and its growth rate was 13.6% on average; the garden green area increased from 14238.1 hm^2 in 2004 to 23916 hm^2 in 2013, and its annual growth rate was 5.93% on average; the park area increased from 2424 hm^2 in 2004 to 7279 hm^2 in 2013, and its annual growth rate was 13% on average.

Garden green area had a moderate correlation degree with the ecological water consumption. Park area had a weak correlation degree with the ecological water consumption (Table 5). Ecological environment is essential for sustainable development.

Table 5. Grey correlation degree between the ecological water and its impact factors in Tianjin.

	Per capita green area	Garden green area	Park area
Correlation degree	0.4946	0.7052	0.6414

5. The Evolution of Water Utilization Structure and Its Impact Factors

5.1. Water Utilization Structure Analysis

The change of water utilization structure of Tianjin could be divided into two stages. In the first stage, the information entropy increased from 1.078 in 2004 to 1.183 in 2010, and its annual average rate was 1.56%. Through relative literatures, if the annual average rate was larger than 2%, the change of information entropy was considered to be obvious or rapid; otherwise, it was considered to be inconspicuous or slow [3,5,7,41]. Therefore, when the information entropy increased from 1.078 to 1.183, it increased slowly. The water utilization structure of Tianjin tended to be more disordered and balanced. This result implied that that the dominance of a single water consumption type decreased generally. The water utilization structure of Tianjin developed in a sustainable direction. It also demonstrated that the water resources management tended to be more sustainable. In the second stage, the information entropy fluctuated between 1.128 and 1.165. The relatively high information entropy meant that the water utilization structure of Tianjin was relatively mature, and influence of single factor reduced continuously (Figure 3). Similar results could be obtained by considering the equilibrium degree.

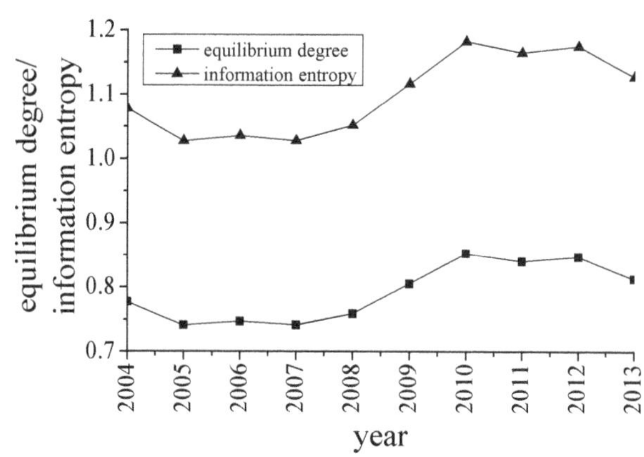

Figure 3. The change of equilibrium degree and information entropy of water utilization structure in Tianjin from 2004 to 2013.

5.2. *The Impact Factors Analysis of Water Utilization Structure*

Rejecting the impact factors whose correlation degree was less than 0.6, other impact factors were used to do correlation analysis with the information entropy of water utilization structure. Besides, considering industrial structure had obvious influence on water utilization structure (Table 1), relative indexes were added. The information entropy of the water utilization structure was taken as reference sequence X_0, the originally selected impact factors included: the proportion of primary industry X_1, the proportion of secondary industry X_2, the proportion of tertiary industry X_3, cultivated area X_4, effective irrigation area X_5, effective water saving irrigation area X_6, agricultural output value X_7, forestry output value X_8, animal husbandry output value X_9, rural electricity consumption X_{10}, water consumption of 10,000 CNY industrial added value X_{11}, water consumption of 10,000 CNY industrial output value X_{12}, light industry output value X_{13}, heavy industry output value X_{14}, high technology industry output value X_{15}, resident population X_{16}, urban per capita income X_{17}, per capita domestic water consumption X_{18}, garden green area X_{19}, park area X_{20}.

After calculation, the correlation degrees of the impact factors were: $X_{10}>X_4>X_{19}>X_7>X_{12}>X_{13}>X_{11}>X_{16}>X_{14}>X_9>0.7>X_{20}>X_{15}>X_1>X_{17}>X_{18}>X_5>X_6>X_8>X_2>X_3>0.6$. They all had moderate correlation with the water utilization structure of Tianjin (Table 6).

Table 6. Grey correlation degree between the water utilization structure and its impact factors in Tianjin.

	X_1	X_2	X_3	X_4	X_5	X_6	X_7	X_8	X_9	X_{10}
Correlation degree	0.6787	0.6326	0.6295	0.7478	0.6636	0.6420	0.7434	0.6417	0.7024	0.7726
	X_{11}	X_{12}	X_{13}	X_{14}	X_{15}	X_{16}	X_{17}	X_{18}	X_{19}	X_{20}
Correlation degree	0.7201	0.7383	0.7301	0.7126	0.6804	0.7151	0.6690	0.6653	0.7446	0.6934

6. Conclusions

Tianjin has an important position in China and an undeniably good foundation of water quantity monitoring in China, making it an ideal study site. Through information entropy analysis and grey correlation analysis method, the changing process of water consumption was studied, thereafter the evolution law and the impact factors of the water utilization structure were discussed. Compared with previous studies, the impact factors of the water utilization structure in this study were more comprehensive. The results of this study were meaningful for adjusting water consumption and optimizing water utilization. The water utilization structure reflected the level of water resources management sustainable development to some extent.

From 2004 to 2010, the water utilization structure of Tianjin tended to be disordered and balanced. It meant that the dominance of a single water consumption type decreased and the equilibrium degree increased generally. In general, the water utilization structure of Tianjin developed in a sustainable direction. It also demonstrated that the water resources management tended to be more sustainable. From 2011 to 2013, the relatively high information entropy implied that the water utilization structure of Tianjin was relatively mature, and influence of single factors reduced continuously.

From 2004 to 2013, rural electricity consumption, cultivated area, garden green area, agricultural output value, water consumption of 10,000 CNY industrial output value, light industry output value, per capita green area, per capita green area, water consumption of 10,000 CNY industrial added value, resident population, heavy industry value and animal husbandry output value had strong correlation with the water utilization structure of Tianjin. These results could be used to adjust water consumption and further optimize water utilization structure.

In the future, if the water consumption data is plentiful and the related technologies are mature, the water utilization structure could be predicted accurately, and its impact factors will be identified effectively.

Acknowledgments: This research is supported by the National Natural Science Foundation of China (No. 91325302 and No. 71503245).

Author Contributions: As the first author, Conglin Zhang made substantial contributions to the acquisition, analysis and interpretation of the data as well as the drafting and revising of article; Yu Liu and Leihua Dong provided some ideas for the article, and gave important advices for the final version; Haijuan Qiao helped to revise the manuscript.

Conflicts of Interest: The authors declare no conflict of interest.

References

1. Liu, B.Q.; Yao, Z.J.; Gao, Y.C. Trend and driving forces of water consumed structure changes in Beijing. *Resour. Sci.* **2003**, *25*, 38–43. (In Chinese)
2. Jenerette, G.D. A global perspective on changing sustainable urban water supplies. *Glob. Planet. Chang.* **2006**, *50*, 202–211.
3. Su, X.L.; Zhao, C.; Ma, L.H. Water consumed structure evolvement and its driving forces for Guanzhong Region in the past 20 years. *J. Irrig. Drain.* **2008**, *27*, 71–73. (In Chinese)
4. Ma, L.H.; Kang, S.Z.; Su, X.L. Study on evolution and its driving forces of water utilization structure of Shiyang River Basin in northwest arid areas. *Agric. Res. Arid Areas* **2008**, *26*, 125–130. (In Chinese)

5. Zheng, A.Q.; Wang, W.K.; Duan, L. Water consumption structure evolvement and its driving forces for Guanzhong Basin. *J. Arid Land Resour. Environ.* **2011**, *25*, 75–79. (In Chinese)
6. Lv, C.M.; Wu, Z.N.; Hu, C.H. Grey relational degree analysis on the major driving forces of water use structure. *Water Sav. Irrig.* **2008**, *2*, 39–41. (In Chinese)
7. Chen, X.F. Analysis of grey relation degree based on water utilization evolution and driving force of information entropy in Xiamen. *J. Water Resour. Water Eng.* **2013**, *24*, 188–191. (In Chinese)
8. Yun, Y.; Zou, Z.H.; Wang, H.W. A regression model based on the compositional data of Beijing's water consumed structure and industrial structure. *Syst. Eng.* **2008**, *26*, 67–71. (In Chinese)
9. Liu, X.X.; Xie, J.C. The interactive relationship between water consumption structure and industrial structure of Shanxi Province. *Syst. Eng.* **2011**, *29*, 45–51. (In Chinese)
10. Zhai, Y.Z.; Wang, J.S.; Zheng, J.Q.; Huan, H. Evolution and driving forces of water consumed structure in Beijng during the past 30 years. *J. Nat. Resour.* **2011**, *26*, 635–643. (In Chinese)
11. Paola, F.D.; Fontana, N.; Galdiero, E.; Giugni, M.; Savic, D.; Uberti, G.S. Automatic multi-objective sectorization of a water distribution network. *Procedia Eng.* **2014**, *89*, 1200–1207.
12. Paola, F.D.; Fontana, N.; Galdiero, E.; Giugni, M.; Uberti, G.S.; Vitaletti, M. Optimal design of district metered areas in water distribution networks. *Procedia Eng.* **2014**, *70*, 449–457.
13. Paola, F.D.; Galdiero, E.; Giugni, M. A jazz-based approach for optimal setting of pressure reducing valves in water distribution networks. *Eng. Optim.* **2015**.
14. Paola, F.D.; Galdiero, E.; Giugni, M.; Pugliese, F. Sustainable development of storm-water systems in African cities considering climate change. *Procedia Eng.* **2015**, *119*, 1181–1191.
15. Nunez, B.; Roman, J.E.P.; Souto, J.A.G. Decision support system for the optimal design of district metered areas. *J. Hydroinf.* **2015**.
16. Marios, S. Global and regional water availability and demand prospects for the future. *Nat. Resour. Res.* **2004**, *13*, 61–75.
17. Liu, C.; Bai, M.G. Change in industrial water use in Anhui Province based on LMDI. *Resour. Sci.* **2012**, *34*, 2299–2305. (In Chinese).
18. Zhang, L.B.; Xu, Y.J.; Jin, J.L.; Wu, C.G. Analysis of influence factors of regional industry water use in Anhui Province. *J. Hydraulic Eng.* **2014**, *45*, 837–843. (In Chinese)
19. Arbues, F.; Barberan, R.; Villanua, I. Price impact on urban residential water demand: A dynamic panel data approach. *Water Resour. Res.* **2004**, *40*, 1–9.
20. Cui, H.S.; Deng, Y.Q. Factors influencing residential water consumption. *Water Resour. Prot.* **2009**, *25*, 83–85. (In Chinese)
21. Xu, H. *Investigation on Influencing Factors of Domestic Water Use in Beijing*; Capital Normal University: Beijing, China, 2012. (In Chinese)
22. Tianjin Municipal Bureau of Statistics. *Tianjin Statistical Yearbook 2005–2014*; China Statistic Press: Beijing, China, 2005–2014. (In Chinese)

23. Zhao, L.R.; Wu, J.W.; Yang, X.; Huang, Z.Q.; Li, H.Q. Analysis of influence factors of irrigation water quantity of Hetao Irrigation District based on analysis of variance. *Water Sav. Irrig.* **2009**, *7*, 1–3. (In Chinese)
24. Liu, Z.X. *Analysis on Water Demand Structure Change and Water Consumption Forecasting in Baotou*; Northwest A & F University: Xi'an, China, 2013. (In Chinese)
25. Xie, H.T. *The Research of Water Resources Utilization and Its Influencing Factor Analysis in Hunan Province*; Hunan University: Hunan, China, 2011. (In Chinese)
26. Qin, H.Y. *Performance Evaluation and Influence Factor Analysis of Agricultural Water of Ganzhou District: A Perspective of Farmer Water User Associations*; Lanzhou University: Lanzhou, China, 2014. (In Chinese)
27. Harun, R.; Muresan, I.C.; Arion, F.H. Analysis of factors that influence the willingness to pay for irrigation water in the Kurdistan Regional Government, Iraq. *Sustainability* **2015**, *7*, 9574–9586.
28. Mai, Z.Y.; He, Z.J.; Wang, X.H. Urban hourly water demand prediction based on principal factors analysis. *J. Zhejiang Univ. (Eng. Sci.)* **2012**, *46*, 1968–1974. (In Chinese)
29. Liu, Y.F.; Kong, W. Analysis of industry water utilization change in Beijing based on factor decomposition model. *Water Resour. Power* **2013**, *31*, 26–29. (In Chinese)
30. Wu, F.; Zhan, J.Y.; Zhang, Q. Evaluating impacts of industrial transformation on water consumption in the Heihe River Basin of Northwest China. *Sustainability* **2014**, *6*, 8283–8296.
31. Shi, Q.L.; Chen, S.Y.; Shi, C.C. The impact of industrial transformation on water use efficiency in Northwest Region of China. *Sustainability* **2015**, *7*, 56–74.
32. Sun, Y.; Xu, Z.X. Impact factors classification urban water consumption forecast and priority weight analysis for forecasting target. *Energy Environ. Prot. (S1)* **2008**, *34*, 114–117. (In Chinese)
33. Tian, S.Y.; Li, X.C.; Yang, B.Z.; Qiu, Y.Q.; Wang, C.Y.; Du, J.K. Xi'an urban residents living water demand factors analysis. *China Rural Water Hydropower* **2014**, *2*, 35–37. (In Chinese)
34. Cheng, Z.L.; Zuo, J.L.; Li, H.; Xu, Y.T.; Wang, L.; Wang, Z.P. Analysis of affecting factors for water consumption of new rural residents. *Water Sav. Irrig.* **2011**, *10*, 51–59. (In Chinese)
35. Ma, D.C.; Xian, C.F.; Zhang, J. The evaluation of water footprints and sustainable water utilization in Beijing. *Sustainability* **2015**, *7*, 13206–13221.
36. Yan, J.J.; Sha, J.H.; Chu, X. Dynamic evaluation of water quality improvement based on effective utilization of stockbreeding biomass resource. *Sustainability* **2014**, *6*, 8218–8236.
37. Shannon, C.E. A mathematical theory of communication. *Mob. Comput. Commun. Rev.* **1948**, *5*, 3–55.
38. Zhang, S.L.; Zhang, G.L. Comparison of grey correlation calculation methods and its problem. *Syst. Eng.* **1996**, *14*, 45–49. (In Chinese)
39. Deng, J.L. *Foundation of Grey System*; Huazhong University of Science and Technology: Wuhan, China, 2002; pp. 122–209. (In Chinese)

40. Fan, J.G. The grey analysis on grain output and its main input elements in Shaanxi Province from 1983 to 2004. *Agric. Res. Arid Areas* **2007**, *25*, 209–212. (In Chinese)
41. Lei, H.; Zhang, X. Anslysis on water use structure change and its driving forces during last 10 years in Shaanxi Province. *J. Water Reour. Water Eng.* **2012**, *23*, 125–128. (In Chinese)

Green Development Performance in China: A Metafrontier Non-Radial Approach

Ke Li and Malin Song

Abstract: This paper proposes a green development growth index (GDGI) for measuring the changes in sustainable development over time. This index considers a wide range of pollutants, and allows for the incorporation of group heterogeneity and non-radial slack in the conventional green development index. The GDGI is calculated based on a non-radial directional distance function derived by several data envelopment analysis (DEA) models, and was decomposed into an efficiency change (EC) index, a best-practice gap change (BPC) index and a technology gap change (TGC) index. The proposed indices are employed to measure green development performance in 30 provinces in China from 2000 to 2012. The empirical results show that China has a low level of green development, with a 2.58% increase per year driven by an innovation effect. China's green development is mainly led by the eastern region, and the technology gaps between the eastern region and the other two regions (the central and western regions) have become wider over the years. The group innovative provinces have set a target for resource utilization of non-innovative provinces in order to catch-up with the corresponding groups, while the metafrontier innovative provinces provide targets for the technology levels of other provinces to improve their green development performance.

Reprinted from *Sustainability*. Cite as: Li, K.; Song, M. Green Development Performance in China: A Metafrontier Non-Radial Approach. *Sustainability* **2016**, *8*, 219.

1. Introduction

Based on the concept of sustainable development, green growth has attracted increasing attention by the international community. The formal definition of green growth was introduced by the Organization for Economic Cooperation and Development (OECD): "green growth means fostering economic growth and development while ensuring that natural assets continue to provide the resources and environmental services on which our well-being relies" [1]. Achieving green growth is an important target for the Chinese government, and it is an effective strategy to follow given increasing energy demand/supply restraints, huge environmental costs and backward energy use technologies [2]. In this paper, we adopted recently developed techniques to measure green development performance based on four main environmental pollutants in 30 provinces of China over the period 2000–2012.

The approach adopted allows for the incorporation of heterogeneity across provinces and non-radial slack in the conventional green development index.

China's economic growth is usually described as "unstable, unbalanced, uncoordinated and ultimately unsustainable" [3]. According to Ho and Jorgenson [4], China's total environmental damage accounted for approximately 2%–8% of total GDP in 1997. Faced with increasingly serious environmental costs, green development has become a priority for the Chinese government. Recently, green development became one of China's development philosophies in the 13th Five-Year Plan (13th FYP, corresponding the period 2016–2020).

Measuring the green development performance of China has become an important topic attracting attention from both academia and government policy-makers. Traditionally, the growth of total factor productivity (TFP) has been used as a signal of development performance. However, conventional TFP studies are seriously misleading and do not provide a reliable indicator of green development performance because they neglect both the energy inputs required for economic growth and their environmental impacts [3,5]. Taking environmental pollutants as undesirable outputs (bads), Chung *et al.* [6] presented the directional distance function (DDF). The rationale is that it strives to increase the "goods" while simultaneously decreasing the "bads" under the framework of data envelope analysis (DEA). Under their framework, Zhang and Choi [7] and Zhang *et al.* [8] proposed the total-factor CO_2 emission performance index (TCPI) and the metafrontier non-radial Malmquist CO_2 emission performance index (MNMCPI) for changes in TCPI over the years.

Our measure of green development performance is the extended form of TCPI and MNMCPI. In our opinion, green development is a comprehensive concept, and more pollutants should be considered. Thus, we adopt four main pollutants, including waste water, sulphur dioxide emission, solid wastes and CO_2 emissions rather than only one pollutant to measure its performance. Furthermore, we evaluate green development performance considering the heterogeneities across China's provinces under the framework of metafrontier analysis. Based on this augmented methodology, the parameters efficiency change, best-practice gap change and technology gap change present us with rich information about China's green development performance.

The rest of the paper is organized as follows. Section 2 is the literature review. Section 3 is the methodology of the study. Section 4 describes the data. Section 5 provides the results and discussion. Section 6 concludes this paper.

2. Literature

Economic theory measuring a "sustainable"/"green" economic growth model was developed more than half a century ago. As early as 1957, Solow [9] argued that sustainable economic growth only depends on technological progress in the long term. In other words, a rising contribution of total factor productivity (TFP) to economic growth is the only way to achieve sustainable growth [3]. Many studies analyze the sustainability of China's economic growth based on TFP via growth accounting [5,10–12] or non-parametric methods [13]. In essence, these studies are based on the framework of productivity improvement.

However, the conventional TFP studies ignore the environmental costs of economic growth (such as CO_2 emissions). It is widely accepted that China's economic growth is accompanied by huge environmental costs. In this sense, these studies may overestimate the true contribution of TFP to output growth, and thereby may lead to a wrong result about China's green development. Furthermore, this bias might direct our attention to less efficient use of environmental friendly abatement technologies or send wrong signals to policy-makers [5]. Thus, green development has been introduced, which is derived from the integration of productivity improvement and environmental protection, and it is a strategy for enhancing productivity and environmental performance for overall socio-economic development [1,14,15]. Obviously, measuring green development performance should take undesirable outputs into consideration.

Considering undesirable outputs, a number of studies evaluate green development performance via DEA models [14,16,17]. Zhang and Choi [7] introduced TCPI and MNMCPI to measure the CO_2 emission performance. These two indices are based on the non-radial DDF model, and incorporate the metafrontier approach. Further, a bootstrapping approach is conducted to introduce statistical inferences into these indices [8]. As an application, Zhang, Zhou and Kung [8] found that the total-factor carbon emission performance of the Chinese transportation industry as a whole decreased by 32.8% over the period 2002–2010, and this reduction was primarily caused by technological decline. Yu-Ying Lin *et al.* [18] measures green (environmental) productivity in 70 countries over the period 1981–2007 by combining the directional distance function (DDF) model and the metafrontier analysis.

The Malmquist-Luenberger productivity index (MLPI), which is introduced by Chung, Färe and Grosskopf [6] and calculated by using the estimated DDF scores, is widely used to measure productivity change or green TFP because it incorporates undesirable output production [3], though it has an inconsistency problem [19]. Chen and Golley [3] used MLPI to evaluate the changing patterns of "green" TFP growth of 38 Chinese industrial sectors during the period 1980–2010. Song, *et al.* [20] used a non-radial DEA model with a slacks-based measure to evaluate the green development of Chinese transportation sectors across provinces in 2003–2012. The empirical result indicated that performance was poor and the average environmental efficiency was 0.45.

Acknowledging the shortcomings of the "discriminating power problem" and "technical regress" associated with the conventional DEA models in measuring green development performance, Li and Lin [21] proposed the SS-ML index, which improves the estimation method for DEA models with the combination of super-efficiency and the sequential DEA method, to measure the green productivity growth of Chinese industrial sectors under the constraints of energy consumption and CO_2 emissions. Combining the SS-ML index and the meta-frontier analysis, Li and Lin [22] found that eastern China has the highest level of green development, followed by western China and central China.

Due to data limitations, studies on green development at the micro-level are rare. Li *et al.* [23] adopted an improved DEA model to measure the green development performance of the 24 Chinese electric power supply companies of the State Power Grid (SGCC). Wei *et al.* [24] assessed the abatement potential of energy and associated CO_2 emissions from coal-fired power enterprises in China's Zhejiang province.

Nevertheless, the existing literature aiming at evaluating green development performance focuses primarily on CO_2 emissions or SO_2 emissions [25], and rarely considers other environmental pollutants. Therefore, they may not give an overall prospect of green development. This paper argues that measuring green development should include the main environmental pollutants, and the non-radial DDF model is a suitable method as it can accurately determine the efficiency when non-zero slacks exist. Zhou *et al.* [26] and Zhang, Zhou and Kung [8] applied this method. However, their research only takes CO_2 emissions as the environmental pollutant. This paper improves their method so as to incorporate more environmental pollutants, and presents a new index to evaluate the green development performance of China.

3. Methodology

3.1. Non-Radial Directional Distance Function

It is well known that desirable and undesirable outputs are jointly produced during the production process. Generally speaking, the production process strives to increase the desirable outputs while simultaneously decreasing the undesirable ones under a given input set and technology. In other words, considering the desirable outputs while ignoring the undesirable outputs to evaluate green development is misleading. Färe *et al.* [27] examined environmental technology that incorporated weak disposability of outputs and null-jointness. This framework is very useful to investigate environmental performance. Assume that $x = (x_1, x_2, \cdots, x_N) \in \mathbb{R}_N^+$, $y = (y_1, y_2, \cdots, y_M) \in \mathbb{R}_M^+$ and $b = (b_1, b_2, \cdots, b_J) \in \mathbb{R}_J^+$ the vectors of inputs, desirable outputs and undesirable outputs respectively. Then, the production process can be described as using inputs x to produce desirable outputs y while also emitting undesirable environmental pollutants b. This production technology can also be conceptually defined as T = {(x, y, b): x can produce (y, b)}. Mathematically, the technology set T is usually assumed to be a closed, bounded, and convexity set. Additionally, it must satisfy the following properties [21,27,28]:

(1) If $(x, y, b) \in T$, and $b = 0$, then $y = 0$;
(2) If $(x, y, b) \in T$, and $x' > x$, then $T(x) \subseteq T(x')$;
(3) If $(x, y, b) \in T$, and $y' < y$, then $(x, y', b) \in P(x)$;
(4) If $(x, y, b) \in T$, and $0 \leqslant \theta \leqslant 1$, then $(\theta y, \theta b) \in P(x)$.

The production technology T has been well-defined conceptually to reflect green development as it considers both the desirable and undesirable outputs, and the abatement of undesirable outputs is not free but costly in terms of a proportional reduction in desirable outputs. However, it cannot be applied in empirical analysis directly.

In fact, the directional distance function (DDF) method introduced by Chung, Färe and Grosskopf [6] is a suitable framework to measure the above production technology T and its properties. Originally, the DDF is defined as follows:

$$\vec{D}_o(x, y, b; g) = \sup\{\beta : (y, b) + g \times \beta \in T\} \tag{1}$$

where $g = (g_y - g_b) \in \mathbb{R}^+_M \times \mathbb{R}^+_J$ is a direction vector and it implies that the desirable outputs and the undesirable outputs can increase and decrease simultaneously. As the DDF reflects the essence of green development, it is widely used in empirical studies to evaluate environmental performance. Examples of such studies include Picazo-Tadeo *et al.* [29], Färe *et al.* [27], Watanabe and Tanaka [30], Macpherson *et al.* [31], Yuan *et al.* [32], Li and Lin [21] and Li and Lin [22].

However, Equation (1) has a rigid assumption that the expansion of desirable outputs and the reduction of the undesirable outputs are at the same rate. In this sense, it is a radical efficiency measure, and when there are non-zero slacks it may overestimate the environmental performance. Another limitation of Equation (1) is that it cannot evaluate the performance of a given input [33]. In order to overcome these limitations, Zhou, Ang and Wang [26] presented a non-radial directional distance function (NDDF) in order to model green performance considering both the desirable and undesirable outputs. Zhou, Ang and Wang [26] believed that NDDF relaxes the assumption of the proportional adjustments of inputs and (desirable and undesirable) outputs, and has higher discriminating power than DDF. It can be defined as follows [26]:

$$\vec{ND}(x, y, b; g) = \sup\{\mathrm{w}^T\beta : (x, y, b) + g \times diag(\beta) \in T\} \tag{2}$$

where $\beta = (\beta_x, \beta_y, \beta_b)^T \geqslant 0$ is a vector of scaling factors which measure the distance of the real inputs/outputs from the optimal state; $\mathrm{w}^T = (w_x, w_y, w_b)$

denotes the weights of the inputs/outputs; $g = (g_x, g_y, g_b)$ is a directional vector, while *diag* implies the diagonal matrices. One of the advantages of NDDF is that $w^T = (w_x, w_y, w_b)$ and $g = (g_x, g_y, g_b)$ can be set in different ways according to the research goals. Furthermore, Model (2) is related to Fukuyama and Weber [34] (FW hereafter) with the difference that FW did not consider bad outputs. However, as Pastor and Aparicio [35] pointed out, the directional slacks-based measure of FW is a weighted additive-type measure in the sense that it maximizes a specific weighted sum of slacks. In fact, Model (2) is mathematically equivalent to a weighted additive-type model. Similar to DDF, NDDF also cannot be employed in empirical analysis directly.

Most studies use the piecewise convex combinations (DEA-type) of the observed data to characterize the production technology and the related distance function. As a nonparametric method, DEA constructs the best-practice frontier using linear programming. This paper applies DEA to calculate the NDDF to evaluate the green development performance of China. Specifically, each province is regarded as a DMU, and it uses capital stock (*K*), labor force (*L*), and energy (*E*) to generate the desirable output—gross regional product (*Y*). Meanwhile, we assume the production process generates four undesirable environmental pollutants as byproducts, which include waste water (*W*), sulphur dioxide emission (*S*), solid wastes (*D*) and carbon dioxide emissions (*C*). To accurately measure the green development performance, it is appropriate to assume the non-energy inputs are fixed in the NDDF model, as capital stock (*K*) and labor force (*L*) do not emit pollutants directly [8]. In other words, the weight vector and the directional vector are set as $w^T = (0, 0, \frac{1}{3}, \frac{1}{3}, \frac{1}{12}, \frac{1}{12}, \frac{1}{12}, \frac{1}{12})$ and $g = (0, 0, -E, Y, -W, -S, -D, -C)$, respectively. The assumption of weight vector $\left(0, 0, \frac{1}{3}, \frac{1}{3}, \frac{1}{12}, \frac{1}{12}, \frac{1}{12}, \frac{1}{12}\right)$ is similar to Wang *et al.* [36], Zhang, Zhou and Kung [8] and Lin and Du [37], and it means that energy reduction, desirable output expansion and pollutants reduction are assigned the same importance, namely 1/3. For pollutants, the Chinese government introduced a series of regulations and measures to control the main pollutants in 2006. In other words, it is difficult to present the accurate weight for the different pollutants. Thus, we give the weight $\frac{1}{12}\left(= \frac{1}{3} \times \frac{1}{4}\right)$ to the pollutants. This setup also assigns the same importance to reduction of waste water (*W*), sulphur dioxide emission (*S*), solid wastes (*D*) and carbon dioxide emissions (*C*) (Note that different choices of the weight vector might lead to different performance scores. We thank a reviewer for pointing out this.). Although it is a naive choice, it is a common one and has been widely used in previous studies. Furthermore, as the effects of capital and labor are eliminated, it places emphasis on the effective consumption of energy inputs. Based on the above analysis, the value of NDDF of a specific DMU (a province in China) can be calculated by solving the following linear programming,

$$\begin{aligned}
&\overrightarrow{ND}(K,L,E,Y,W,S,D,C;g) = \max\left\{\frac{1}{3}\beta_E + \frac{1}{3}\beta_Y + \frac{1}{12}\beta_W + \frac{1}{12}\beta_S + \frac{1}{12}\beta_D + \frac{1}{12}\beta_C\right\} \\
&s.t.\ \sum_{t=1}^{T}\sum_{i=1}^{N}\lambda_{it}K_{it} \leqslant K \\
&\sum_{t=1}^{T}\sum_{i=1}^{N}\lambda_{it}L_{it} \leqslant L \\
&\sum_{t=1}^{T}\sum_{i=1}^{N}\lambda_{it}E_{it} \leqslant E - \beta_E g_E \\
&\sum_{t=1}^{T}\sum_{i=1}^{N}\lambda_{it}Y_{it} \geqslant Y + \beta_Y g_Y \\
&\sum_{t=1}^{T}\sum_{i=1}^{N}\lambda_{it}W_{it} = W - \beta_W g_W \\
&\sum_{t=1}^{T}\sum_{i=1}^{N}\lambda_{it}S_{it} = S - \beta_S g_S \\
&\sum_{t=1}^{T}\sum_{i=1}^{N}\lambda_{it}D_{it} = D - \beta_D g_D \\
&\sum_{t=1}^{T}\sum_{i=1}^{N}\lambda_{it}C_{it} = C - \beta_C g_C \\
&\lambda_{it} \geqslant 0\ (i = 1,2,\cdots,N; t = 1,2,\cdots,T) \\
&\beta_j \geqslant 0\ (j = E,Y,W,S,D,C)
\end{aligned} \tag{3}$$

It is worth noting that we employ the global DEA method, which uses the whole sample to construct the best-practice frontier. That implies a single global frontier is used as a benchmark technology for all DMUs over the whole period. Thus, the estimate values can be compared over time.

Suppose that $\beta^* = \left(0,0,\beta_E^*,\beta_Y^*,\beta_W^*,\beta_S^*,\beta_D^*,\beta_C^*\right)^T \geqslant 0$ are the optimal solutions to Equation (3), which implies that, under the best-practice scenario, the energy input, desirable output and undesirable environmental pollutants will be $E_{it} - \beta_{E,it}^* \times E_{it}$, $Y_{it} + \beta_{Y,it}^* \times Y_{it}$ and $U_{it} - \beta_{U,it}^* \times U_{it}(U = W,S,D,C)$. Obviously, $\beta^* = 0$ means that the DMU on the best-practice frontier. Following Zhou, Ang and Wang [26], Zhang *et al.* [38] and Lin and Du [37], the green development performance index (GDPI) can be defined as follows:

$$\begin{aligned}
GDPI_{it} &= \frac{1}{2}\left[\frac{\left(E_{it} - \beta_{E,it}^* \times E_{it}\right)/\left(Y_{it} - \beta_{Y,it}^* \times Y_{it}\right)}{E_{it}/Y_{it}}\right] + \\
&\frac{1}{2}\left[\frac{1}{4}\sum_{J=W,S,D,C}\frac{\left(J_{it} - \beta_{J,it}^* \times J_{it}\right)/\left(Y_{it} - \beta_{Y,it}^* \times Y_{it}\right)}{J_{it}/Y_{it}}\right]
\end{aligned} \tag{4}$$

It should be noted that the values of GDPI lie between zero and unity, and the higher the GDPI, the better the green development performance.

3.2. Green Development Growth Index and Its Decomposition

In the above section, we define an index, *i.e.*, GDPI to measure green development performance. However, it assumes all provinces have the same level of technology, or it does not account for the technology heterogeneities across provinces. In fact, the different regions may have various technical conditions or technical constraints. For example, the eastern provinces have a relatively high level of economic development, and have advanced technologies to combat undesirable environmental pollutants. Furthermore, the GDPI index is a static index, or it cannot measure the change of green development over years. In order to investigate the changes in GDPI over time while also considering the region heterogeneity, we first combine the GDPI with a meta-frontier analysis.

Suppose N DMUs (provinces of China in this study, $N = 30$) can be divided into h ($h = 3$, including the eastern, central and western China, respectively) independent groups, and each group includes N_h ($h = 1, 2, 3$) DMUs and uses different production technologies. Thus, we can define three benchmark technology sets: a contemporaneous benchmark technology, an intertemporal benchmark technology and a global benchmark technology [7,39,40].

For a given group h, the contemporaneous benchmark technology (denoted as T_h^C) can be reflected in the following linear programming,

$$
\begin{aligned}
&\overrightarrow{ND}_h^C(K,L,E,Y,W,S,D,C;g) = \max\left\{\frac{1}{3}\beta_E + \frac{1}{3}\beta_Y + \frac{1}{12}\beta_W + \frac{1}{12}\beta_S + \frac{1}{12}\beta_D + \frac{1}{12}\beta_C\right\} \\
&s.t. \quad \sum_{i=1}^{N_h} \lambda_{it} K_{it} \leqslant K \\
&\qquad \sum_{i=1}^{N_h} \lambda_{it} L_{it} \leqslant L \\
&\qquad \sum_{i=1}^{N_h} \lambda_{it} E_{it} \leqslant E - \beta_E g_E \\
&\qquad \sum_{i=1}^{N_h} \lambda_{it} Y_{it} \geqslant Y + \beta_Y g_Y \\
&\qquad \sum_{i=1}^{N_h} \lambda_{it} W_{it} = W - \beta_W g_W \\
&\qquad \sum_{i=1}^{N_h} \lambda_{it} S_{it} = S - \beta_S g_S \\
&\qquad \sum_{i=1}^{N_h} \lambda_{it} D_{it} = D - \beta_D g_D \\
&\qquad \sum_{i=1}^{N_h} \lambda_{it} C_{it} = C - \beta_C g_C \\
&\qquad \lambda_{it} \geqslant 0 \; (i = 1,2,\cdots,N_h; t = 1,2,\cdots,T; h = 1,2,3) \\
&\qquad \beta_j \geqslant 0 \; (j = E,Y,W,S,D,C)
\end{aligned}
\tag{5}
$$

Essentially, T_h^C constructs the best-practice frontier based on the observations for the specific group h only for that time period t. Based on $\hat{\beta}_j (j = E, Y, W, S, D, C)$ calculated by Equation (5), the GDPI given by Equation (4) can be denoted as $GDPI_h^C$.

An intertemporal benchmark technology for a given group h is defined as $T_h^I = T_h^1 \cup T_h^2 \cup \cdots \cup T_h^T$. Obviously, the intertemporal benchmark technology develops a single best-practice frontier for a given group h over the whole period. Because we have h (= 3) independent groups, there are h (= 3) distinct intertemporal technologies. Based on T_h^I, the result of GDPI can be denoted as $GDPI_h^I$ and can be calculated by Equation (6) and Equation (4).

$$\begin{aligned}
&\overrightarrow{ND}_h^I(K, L, E, Y, W, S, D, C; g) = \max\left\{\frac{1}{3}\beta_E + \frac{1}{3}\beta_Y + \frac{1}{12}\beta_W + \frac{1}{12}\beta_S + \frac{1}{12}\beta_D + \frac{1}{12}\beta_C\right\} \\
&s.t. \quad \sum_{t=1}^{T}\sum_{i=1}^{N_h} \lambda_{it} K_{it} \leqslant K \\
&\sum_{t=1}^{T}\sum_{i=1}^{N_h} \lambda_{it} L_{it} \leqslant L \\
&\sum_{t=1}^{T}\sum_{i=1}^{N_h} \lambda_{it} E_{it} \leqslant E - \beta_E g_E \\
&\sum_{t=1}^{T}\sum_{i=1}^{N_h} \lambda_{it} Y_{it} \geqslant Y + \beta_Y g_Y \\
&\sum_{t=1}^{T}\sum_{i=1}^{N_h} \lambda_{it} W_{it} = W - \beta_W g_W \\
&\sum_{t=1}^{T}\sum_{i=1}^{N_h} \lambda_{it} S_{it} = S - \beta_S g_S \\
&\sum_{t=1}^{T}\sum_{i=1}^{N_h} \lambda_{it} D_{it} = D - \beta_D g_D \\
&\sum_{t=1}^{T}\sum_{i=1}^{N_h} \lambda_{it} C_{it} = C - \beta_C g_C \\
&\lambda_{it} \geqslant 0 \ (i = 1, 2, \cdots, N_h; t = 1, 2, \cdots, T; h = 1, 2, 3) \\
&\beta_j \geqslant 0 \ (j = E, Y, W, S, D, C)
\end{aligned} \tag{6}$$

Finally, a global benchmark technology can be defined as $T^G = \text{Convex}\{T_1^I \cup \cdots \cup T_h^I\}$. T^G is the convex envelope of all the contemporaneous technologies. It can be inferred that T^G constructs only one best-practice frontier based on the whole observation (including observations for all groups over the entire time period) [41]. Obviously, T^G can be reflected in Equation (3), and the corresponding GDPI is denoted as $GDPI_h^I$.

Similar to a global Malmquist productivity index introduced by Pastor and Lovell [42], a new index, namely the green development growth index (GDGI), is introduced to measure the changes in GDPI over time,

$$\begin{aligned} GDGI_{i,t}^{t+1} &= \frac{GDPI^{G}(K_{i,t+1}, L_{i,t+1}, E_{i,t+1}, Y_{i,t+1}, W_{i,t+1}, S_{i,t+1}, D_{i,t+1}, C_{i,t+1})}{GDPI^{G}(K_{i,t}, L_{i,t}, E_{i,t}, Y_{i,t}, W_{i,t}, S_{i,t} D_{i,t}, C_{i,t})} \\ &\triangleq \frac{GDPI^{G}(\cdot^{t+1})}{GDPI^{G}(\cdot^{t})} \end{aligned} \tag{7}$$

Since $GDPI_{h}^{I}$ is calculated by a global DEA method (Equation 3), in other words there is only one global benchmark technology, and its scores can be compared over the years [43], so there is no need to resort to the geometric mean convention when defining the global index [41,42].

As developed by Oh [40] and Zhang and Choi [7], the GDGI can be decomposed into different components as follows:

$$\begin{aligned} GDGI_{i,t}^{t+1} &= \left[\frac{GDPI^{C}(\cdot^{t+1})}{GDPI^{C}(\cdot^{t})}\right] \times \left[\frac{GDPI^{I}(\cdot^{t+1})/GDPI^{C}(\cdot^{t+1})}{GDPI^{I}(\cdot^{t})/GDPI^{C}(\cdot^{t})}\right] \times \left[\frac{GDPI^{G}(\cdot^{t+1})/GDPI^{I}(\cdot^{t+1})}{GDPI^{G}(\cdot^{t})/GDPI^{I}(\cdot^{t})}\right] \\ &\triangleq \frac{TE^{t+1}}{TE^{t}} \times \frac{BPR^{t+1}}{BPR^{t}} \times \frac{TGR^{t+1}}{TGR^{t}} \\ &\triangleq EC \times BPC \times TGC \end{aligned} \tag{8}$$

In Equation (8), the green development performance at time period s (= t, $t + 1$) based on the contemporaneous benchmark technology (T_{h}^{C}) is measured by $GDPI^{C}(\cdot^{S})$ and abbreviated as TE^{S}. So, the EC term in Equation (8) measures how close a DMU for a specific group moves toward the contemporaneous benchmark technology at time period $t + 1$ compared to time period t, which indicates a "catch-up" effect. $EC >$ (or $<$) 1 implies that a DMU moves close to (or far away from) the best-practice frontier at time period $t + 1$ compared to time period t. $BPR = GDPI^{I}/GDPI^{C}$ measures the best-practice gap ratio between a contemporaneous benchmark technology and an intertemporal benchmark technology, and it changes over time measured by BPC in Equation (8). $BPC >$ (or $<$) 1 means the contemporaneous technology frontier shifts toward (or far away from) the intertemporal technology frontier, and is also considered to reflect the innovation effect [7]. $TGR = GDPI^{G}/GDPI^{I}$ is the technology gap ratio between an intertemporal benchmark technology and a global benchmark technology, and it changes over time measured by TGC in Equation (8). $TGC >$ (or $<$) 1 indicates a decrease (increase) in the technology gap between the intertemporal technology for a specific group and the global technology. Thus, TGC reflects the technology leadership effect for a given group.

4. Data

As stated in Section 3, data on the inputs, desirable output and undesirable environmental pollutants are collected to evaluate the green development performance across 30 Chinese provinces from 2000 to 2012. Tibet is excluded because of data unavailability. The variables and data sources are discussed as follows.

(1) Inputs: include capital stock (K), labor force (L) and energy consumption (E). The data on capital stock (constant 2000 prices) and labor force can be found in Li and Lin [22]. The data on capital stock is estimated by the perpetual inventory approach, and each province has different depreciation rates of gross fixed capital formation [44]. Energy consumption is measured by 10^4 tons of standard coal equivalent (10^4 tce), and the data is obtained from the *China Energy Statistical Yearbook* [45].

(2) Desirable output: it is measured by the gross regional product (GRP) at 100 million Yuan (Y). It has been converted into 2000 prices with GRP deflators. Data are collected from the *China Energy Statistical Yearbook* [45].

(3) Undesirable environmental pollutants: include waste water (W), sulphur dioxide emission (S), solid wastes (D) and carbon dioxide (CO_2) emissions (C). According to the *China Statistical Yearbook* and *China Statistical Yearbook of Environment*, the China's environmental protection puts more emphasis on the status quo and the recovery of three wastes (waste water, waste gas and industrial residue). Correspondingly, the main environmental indicators of this study are waste water (W), sulphur dioxide emissions (S), carbon dioxide emissions (C) and solid wastes (D). Here, sulphur dioxide emissions (S) and carbon dioxide emissions (C) are two main types of waste gas and are closely related to air quality. In *China Statistical Yearbook*, the solid wastes (D) is the main form of industrial residue. From the perspective of statistical indicators released by the Chinese government, using only one pollutant to measure the performance of green development is not comprehensive. The data on waste water (W), sulphur dioxide emissions (S) and solid wastes (D) are collected from the *China Statistical Yearbook* [46]. Carbon dioxide is emitted from fossil fuel usage, and it can be estimated by using the fuel-based carbon calculation model introduced by IPCC (2006) [47]. This paper uses eight fuel types (including coal, coke, crude oil, gasoline, kerosene, diesel oil, fuel oil, natural gas) to estimate CO_2 emissions, while their consumption levels are collected from the *China Energy Statistical Yearbook* [45].

Table 1 shows the descriptive statistics for the input and output variables. Among all the variables, the maximum values are much larger than the minimum values, indicating that there are huge heterogeneities across provinces in China. Furthermore, Table 2 indicates the average share of outputs for each province of the country. Overall, the eastern region has a high share of desirable output and a relatively low share of undesirable environmental pollutants, while the central and the western regions are found to have the opposite results. Taking Beijing for example; its GRP accounts for about 3.088% of the whole country, but its waste water, sulphur dioxide emissions, solid wastes and CO_2 emissions only account for about 0.557%, 0.494%, 0.698% and 1.428%, respectively. For the western provinces such as Yunnan, Gansu and Qinghai, their shares of undesirable environmental pollutants are higher than their shares of desirable outputs. In other words, the environment cost of economic growth in eastern provinces is lower than that of the western provinces.

Table 1. Descriptive statistics of variables (N = 390).

Variable (Unit)	Mean	Std.dev	Minimum	Maximum
Capital stock (100 million RMB)	20,067.44	17,952.33	1390.72	108,951.70
Labor (ten thousand workers)	2321.64	1540.87	239.47	6242.54
Energy consumption (10^4 tce)	9570.24	7005.88	479.95	38,899.00
Gross Regional Product (100 million RMB)	7211.28	6938.54	263.68	42,865.93
Waste water (10^4 ton)	75,186.57	63,500.79	3453.00	296,318.00
Sulphur dioxide emission (10^4 ton)	62.74	39.23	1.93	171.50
Solid wastes (10^4 ton)	5725.87	5876.08	75.00	45,576.00
CO_2 emissions (10^4 ton)	26,934.84	21,364.15	445.25	122,379.40

In order to further describe the green development of China, the 30 provinces of China are categorized into three groups based on geographical closeness and regional development strategies, namely eastern China, central China and western China [22]. The average values and growth rates for the variables in the three groups for the entire period are presented in Table 3. It can be seen that the eastern provinces use much more inputs than the central and western provinces, and its average GRP is also much higher. Also, heterogeneities across groups can be found among undesirable environmental pollutants. For example, the average values of sulphur dioxide emissions, solid wastes and CO_2 emissions for the eastern provinces are much higher than those for the western provinces, but they are very close to those for the central provinces. However, the average value of waste water for the eastern provinces is about 1.64 times larger than that of the central provinces. These differences in patterns of outputs/inputs across groups imply that different regions have varied production technologies, and it provides justification for developing a metafrontier to measure and decompose green growth using the GDGI index.

Table 2. The average share of outputs for each province to the whole country (2000–2012) (unit: %).

Province	GRP	W	S	D	C
(E) Beijing	3.088	0.557	0.494	0.698	1.428
(E) Fujian	3.739	5.132	1.806	2.775	1.919
(E) Guangdong	11.313	8.022	5.454	2.115	5.311
(E) Guangxi	2.062	5.846	4.166	2.619	1.518
(E) Hainan	0.501	0.304	0.127	0.105	0.371
(E) Hebei	4.809	5.116	6.271	11.798	7.495
(E) Jiangsu	9.144	11.463	5.898	3.796	6.150
(E) Liaoning	4.715	4.038	4.513	8.248	6.604
(E) Shandong	8.849	6.704	8.059	6.540	9.269
(E) Shanghai	4.617	2.321	1.590	1.160	2.832
(E) Tianjin	2.048	0.948	1.136	0.686	1.690
(E) Zhejiang	6.305	8.076	3.657	1.753	4.014
(C) Anhui	2.808	2.984	2.401	3.612	2.964
(C) Heilongjiang	2.958	2.006	1.894	2.405	3.378
(C) Henan	5.012	5.638	5.988	4.693	5.544
(C) Hubei	3.444	4.253	2.954	2.694	3.416
(C) Hunan	3.471	4.706	3.520	2.547	2.760
(C) Inner Mongolia	2.162	1.228	5.653	5.859	5.027
(C) Jilin	1.991	1.685	1.469	1.751	2.436
(C) Jiangxi	2.012	2.631	2.422	4.429	1.595
(C)Shanxi	1.904	1.679	5.736	8.296	7.158
(W) Chongqing	1.655	3.046	3.247	1.188	1.315
(W) Gansu	0.980	0.817	2.276	1.794	1.731
(W) Guizhou	0.986	0.735	3.811	3.103	2.415
(W) Ningxia	0.287	0.730	1.508	0.738	1.108
(W) Qinghai	0.271	0.286	0.487	1.494	0.352
(W) Shaanxi	1.924	1.753	3.814	2.796	2.758
(W) Sichuan	3.959	4.774	5.221	4.631	2.919
(W) Xinjiang	1.216	0.939	2.154	1.451	2.327
(W) Yunnan	1.770	1.581	2.274	4.226	2.197
Eastern China	61.191	57.211	43.698	50.396	58.526
Central China	25.761	26.374	33.628	30.973	26.812
Western China	13.048	16.415	22.674	18.631	14.662

Note: E—the eastern China; C—the central China; W—the western China.

Table 3. Average values and growth rates for variables in three groups (2010–2012).

Group	Num	K	L	E	Y	W	S	D	C
Eastern	156	28,701.88 (14.56%)	2536.28 (2.92%)	12,057.56 (9.19%)	11,031.55 (12.19%)	110,009.60 (1.69%)	67.72 (0.27%)	6054.16 (11.57%)	32,726.45 (10.48%)
Central	117	17,641.95 (16.72%)	2602.41 (1.45%)	9880.52 (9.75%)	6192.39 (12.62%)	67,196.26 (1.92%)	67.00 (4.34%)	6925.58 (11.87%)	30,775.59 (9.32%)
Western	117	10,980.35 (14.30%)	1754.68 (1.67%)	5943.55 (10.66%)	3136.46 (12.05%)	36,746.17 (1.87%)	51.85 (6.18%)	4088.44 (19.31%)	15,371.94 (11.84%)

Note: The growth rates in parentheses.

5. Results and Discussion

5.1. Staticistal Analysis: the Green Development Performance Index under the Global Benchmark Technology

We use R3.2 to solve the Equations (3) and (4), and the estimation results of GDPI under the global benchmark technology across China's 30 provinces during the period 2000–2012 are reported in Table 4. The average score of GDPI in China during the sample period is only 0.398, and Table 4 shows that only 85 of the 390 estimated scores are found to be higher than 0.5, indicating that China is at a low level of green development. It is worth noting that during the first 10 years (2000–2009), there are 58 estimated scores higher than 0.5. However, during the past three years (2010–2012), there are 27 scores. Obviously, the average proportion of the development during 2010–2012 per year is higher than during 2000–2009 (27/90 > 58/300). This result indicates that green development performance has made significant progress in recent years. Additionally, only a few scores of the GDPI are equal to unity, and most of the unity scores correspond to developed provinces, such as Beijing, Tianjin and Guangdong. Except for the year 2005, Guangdong has unity scores of GDPI during 2000–2012, indicating its high performance in green development.

Compared to the scores of GDPI in 2000, the scores of GDPI in 26 provinces increased in 2012. Four provinces, namely Yunnan, Shaanxi, Ningxia and Xinjiang, decreased. These four provinces are in western China, indicating its low level of green development. An important finding is that the GDPIs of eight provinces in 2009 are lower than the level in 2008. The possible reason is that large-scale economic stimulus plan adopted by the government to address the global crisis worsened the performance.

Figure 1 plots the trends of the average scores of the GDPI in the three regions (the eastern, central and western China). It can be found that the scores of GDPI in all regions show upward trends, with the average scores of 0.525, 0.330 and 0.297 in eastern, central and western China respectively. Furthermore, eastern China not only has the highest GDPI, but also shows the fastest growth rate. Specifically, its average score across provinces increases from 0.443 in 2000 to 0.653 in 2012, an increase of 47.321%. Figure 1 also indicates that the scores of GDPI in western China are lower than in central China, and both have slower growth rates than eastern China. Thus, the green development performance gap between eastern China and the other two regions becomes widened. In this sense, enhancing the green development performance in central and western China is crucial to China's sustainable development.

Table 4. Estimation results of green development performance index (GDPI) in China.

Province	2000	2001	2002	2003	2004	2005	2006	2007	2008	2009	2010	2011	2012
Beijing	0.351	0.388	0.434	0.464	0.493	0.533	0.601	0.711	0.836	0.903	1.000	1.000	1.000
Tianjin	0.294	0.308	0.334	0.366	0.382	0.382	0.412	0.461	0.537	0.582	0.610	0.714	1.000
Hebei	0.236	0.261	0.238	0.238	0.231	0.235	0.249	0.265	0.288	0.314	0.330	0.246	0.250
Shanxi	0.189	0.172	0.165	0.175	0.176	0.183	0.176	0.196	0.212	0.224	0.228	0.228	0.232
Inner Mongolia	0.238	0.242	0.231	0.235	0.235	0.239	0.249	0.257	0.264	0.189	0.213	0.206	0.252
Liaoning	0.221	0.246	0.275	0.292	0.292	0.300	0.320	0.340	0.339	0.363	0.384	0.367	0.381
Jilin	0.277	0.299	0.298	0.302	0.310	0.327	0.346	0.298	0.295	0.317	0.334	0.342	0.375
Heilongjiang	0.288	0.321	0.349	0.350	0.364	0.373	0.391	0.425	0.446	0.473	0.486	0.494	0.484
Shanghai	0.394	0.411	0.436	0.469	0.518	0.538	0.583	0.667	0.737	0.793	1.000	0.913	1.000
Jiangsu	0.477	0.497	0.514	0.515	0.486	0.444	0.460	0.493	0.528	0.557	0.571	0.577	0.607
Zhejiang	0.499	0.530	0.493	0.491	0.490	0.491	0.501	0.526	0.556	0.581	0.602	0.624	0.661
Anhui	0.298	0.310	0.328	0.345	0.361	0.372	0.357	0.371	0.387	0.402	0.427	0.442	0.461
Fujian	0.580	0.637	0.561	0.536	0.524	0.497	0.517	0.534	0.552	0.555	0.576	0.597	0.622
Jiangxi	0.371	0.417	0.378	0.377	0.378	0.385	0.358	0.369	0.396	0.412	0.423	0.428	0.451
Shandong	0.371	0.429	0.363	0.360	0.353	0.331	0.341	0.354	0.373	0.387	0.390	0.394	0.405
Henan	0.324	0.338	0.340	0.336	0.311	0.313	0.323	0.340	0.306	0.302	0.311	0.320	0.350
Hubei	0.285	0.322	0.323	0.316	0.308	0.318	0.331	0.354	0.325	0.343	0.351	0.352	0.363
Hunan	0.435	0.408	0.393	0.375	0.354	0.321	0.344	0.363	0.391	0.410	0.426	0.426	0.455
Guangdong	1.000	1.000	0.955	1.000	1.000	0.924	1.000	1.000	1.000	1.000	1.000	1.000	1.000
Guangxi	0.367	0.393	0.386	0.369	0.345	0.340	0.353	0.360	0.381	0.339	0.343	0.356	0.371
Hainan	0.532	1.000	0.504	0.504	0.520	0.543	0.522	0.512	0.516	0.547	0.579	0.519	0.544
Chongqing	0.303	0.287	0.325	0.333	0.321	0.281	0.289	0.302	0.317	0.290	0.310	0.329	0.365
Sichuan	0.409	0.466	0.498	0.530	0.630	0.632	0.732	1.000	0.601	0.717	0.879	1.000	1.000
Guizhou	0.178	0.191	0.186	0.176	0.172	0.198	0.198	0.212	0.226	0.230	0.239	0.241	0.245
Yunnan	0.300	0.321	0.306	0.300	0.286	0.284	0.290	0.306	0.332	0.249	0.262	0.262	0.277
Shaanxi	0.314	0.304	0.299	0.299	0.293	0.285	0.300	0.311	0.336	0.274	0.284	0.298	0.313
Gansu	0.207	0.226	0.229	0.228	0.241	0.238	0.252	0.271	0.280	0.297	0.219	0.299	0.230
Qinghai	0.206	0.218	0.246	0.229	0.220	0.204	0.204	0.211	0.205	0.214	0.226	0.210	0.212
Ningxia	0.169	0.210	0.187	0.149	0.156	0.136	0.139	0.132	0.131	0.129	0.130	0.127	0.139
Xinjiang	0.261	0.269	0.273	0.278	0.271	0.266	0.263	0.264	0.268	0.255	0.243	0.215	0.191

For individual provinces, the GDPI scores show a huge difference. Among the 30 provinces, only eight provinces show a relatively strong performance, with the average scores higher than 0.5 (Figure 2). Guangdong shows the best performance, and its average scores is 0.991; while Ningxia shows the lowest score (0.149).

Corresponding to Figure 1, Figure 2 also indicates that the eastern provinces performed better than the central and western provinces.

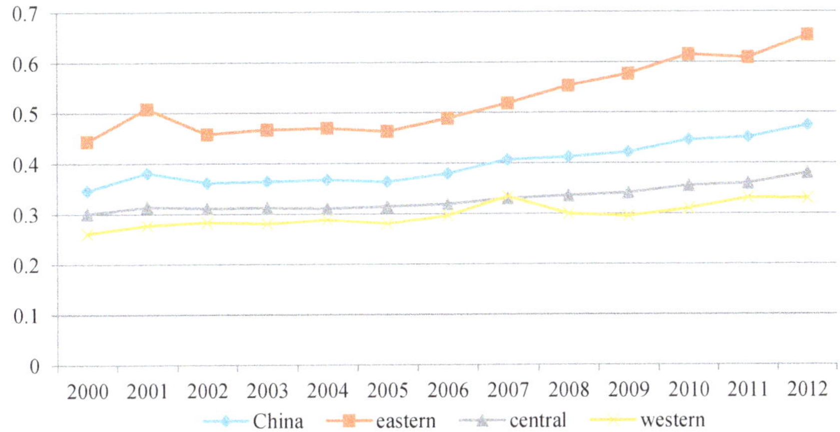

Figure 1. The average green development performance index (GDPI) in 2000–2012.

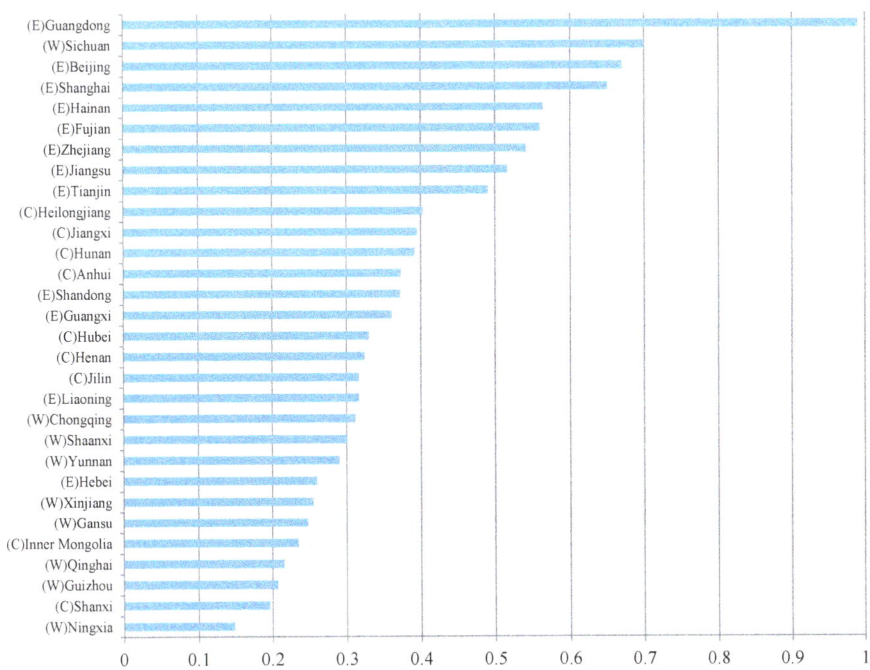

Figure 2. The average green development performance index (GDPI) across provinces. (Note: E—the eastern China; C—the central China; W—the western China.)

5.2. The Green Development Growth Index

To assess the changes in green development performance in China, the green development growth index (GDGI) is calculated based on Equations (7) and (8). Table 5 shows the average results of GDGI and its decomposition during the period 2000–2012. Recall that GDGI values greater (less) than unity indicate improvement (deterioration) in green development performances. Thus, Table 5 indicates that the growth rate of green development performance in the study sample is 2.58% per year, and it is mainly driven by the innovation effect (BPC). Furthermore, heterogeneities across groups/regions can be confirmed in green development. The average annual growth rate in green development of the eastern region is higher than those of the central and the western regions. Innovation effect (BPC) is the main contributor to green development for all three regions. Additionally, "catch-up" and technical leadership effects cannot be found in China.

Table 5. Average values of GDGI and its decomposition in 2000–2012.

Group	GDGI	EC	BPC	TGC
Eastern	1.0375	0.9868	1.0535	1.0000
Central	1.0214	0.9995	1.0322	0.9992
Western	1.0146	0.9897	1.0570	0.9902
China	1.0258	0.9915	1.0482	0.9968

We examine the cumulative GDPI and its decomposition by setting the values in year 2000 equal to 1, and the results are shown in Table 6. It can be seen that for China, the green development performance increased by 35.12% in 2012 compared with the year 2000. It further indicates that the innovation effect increased by 73.76% during the period 2000–2012. In other words, the innovation effect is the main contributor to green development in China. This conclusion is applicable to all three regions.

Table 6. The cumulative values of GDGI and its decomposition in 2000–2012.

Group	GDGI	EC	BPC	TGC
Eastern	1.5357	0.8517	1.8451	1.0000
Central	1.2862	0.9918	1.4393	0.9800
Western	1.1828	0.8654	1.8634	0.8742
China	1.3512	0.9011	1.7376	0.9588

Figure 3 shows the changes in green development performance and the decomposed sources in China and the three regions. For the period 2001–2002 and 2004–2005, the values of GDGI for China is less than unity, indicating a decrease in green development performance. After 2005, the GDGI of China shows values higher than unity. In other words, China's green development performance has shown improvement since 2005. This result implies that the policy of "energy-saving and emission-reduction" since 2006 does matter for China's green development. The eastern region shows larger fluctuations in GDGI than the central and western regions. For example, due to the economic crisis, the GDGI of the eastern region during the period 2010–2011 sharply declined from 1.059 to 0.984. Interestingly, the GDGI of the western region during the same period increased to 1.045. The decomposed results show that the main reason is that the value of BPC increased to the highest point (1.306), which implies that the contemporaneous technology frontier of the western region significantly shifted toward an intertemporal technology frontier. In terms of EC index, the values of EC in China during 2005–2012 are less than unity, implying there is no "catch-up" effect. Because *TE* measures the efficiency given the technology level, these results imply the poor performance of allocative efficiency. It also reflects the dilemma and hysteresis of reform in factor markets. In 2010–2011, the values of EC for the western region sharply declined to 0.831, indicating efficiency deterioration under the constraint of maintaining economic growth through huge investment. The values of BPC show opposite trends to EC. It seems that the "innovation effect" is driven by the "poor" performance of the intertemporal technology frontier rather than the "better" performance of the contemporaneous technology frontier. The values of TGC for the eastern region are always unity, indicating that the technology in the eastern region is of the highest level in China. Since 2006, the values of TGC for the central and western regions are less than unity, indicating a decrease in the technology leadership effect, or the technology gaps becoming wider rather than narrower.

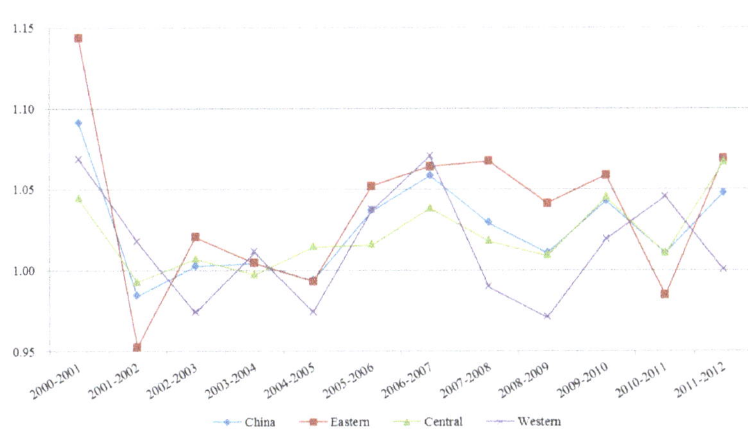

(a)

Figure 3. *Cont.*

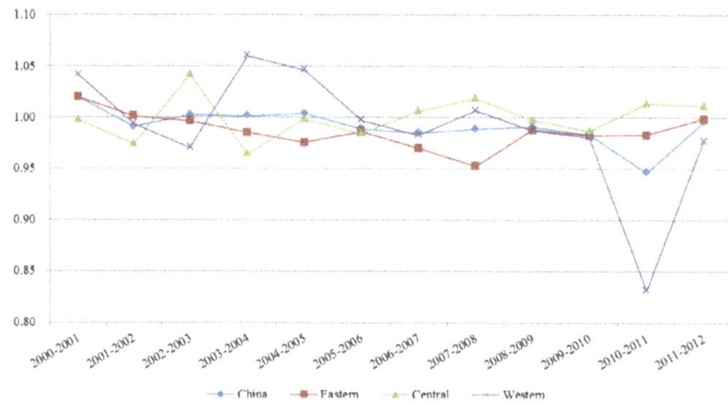

(**b**)

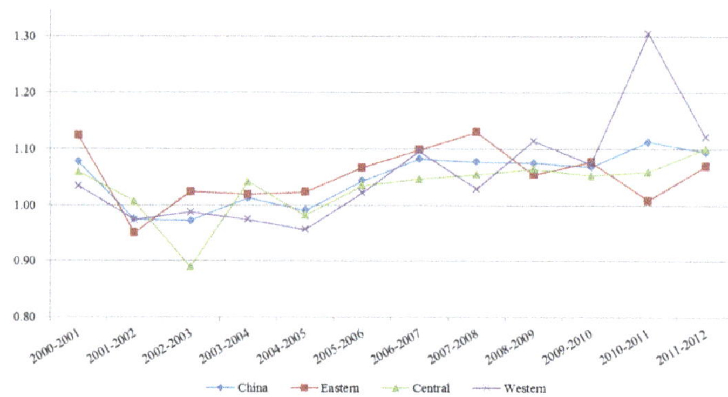

(**c**)

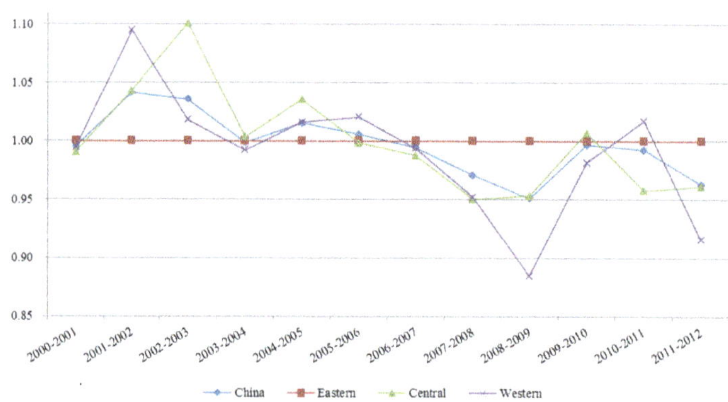

(**d**)

Figure 3. Trends in the green development growth index (GDGI) and its decomposition. (**a**) green development growth index (GDGI); (**b**) efficiency change (EC); (**c**) best-practice gap change (BPC); (**d**) technology gap change (TGC) index.

Figure 4 shows the cumulative changes in green development performance and the decomposed sources based on the GDPI. The GDPI shows values larger than unity, indicating an increase in green development performance. It is found that the sample provinces as a whole show an increase in green development performance by approximately 35.12% from 2000 to 2012. Among the three regions, the eastern region shows the highest GDPI (53.57%), followed by the central region and the western region (28.62% and 18.28%, respectively). This finding indicates that all the regions experienced green development performance gains. For efficiency changes (EC), Figure 4b indicates that the eastern region and the western region show upward trends during the period 2001–2004 and 2000–2010, respectively. However, all three regions experienced efficiency loss over the whole period (2000–2012), which implies there has been a deterioration in efficiency in recent years. Regarding the innovation effect (BPC) in Figure 4c, all the regions show increasing trends in 2000–2012. The eastern and western regions have a similar growth rate at 84.51% and 86.34%, respectively, while the central region increased by 43.94%. Figure 4d further confirms that the technology in the eastern region is of the highest level in China.

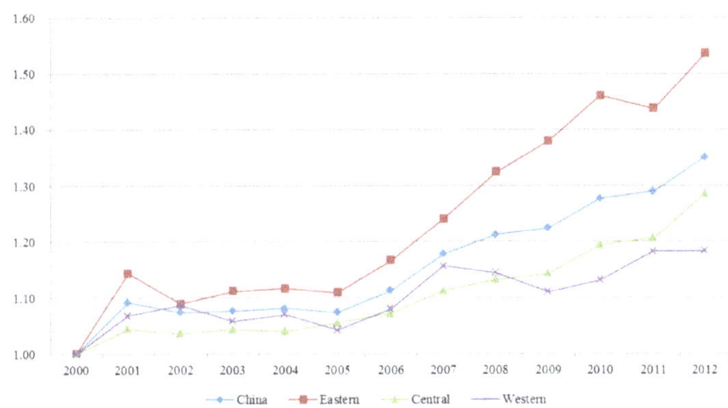

(a)

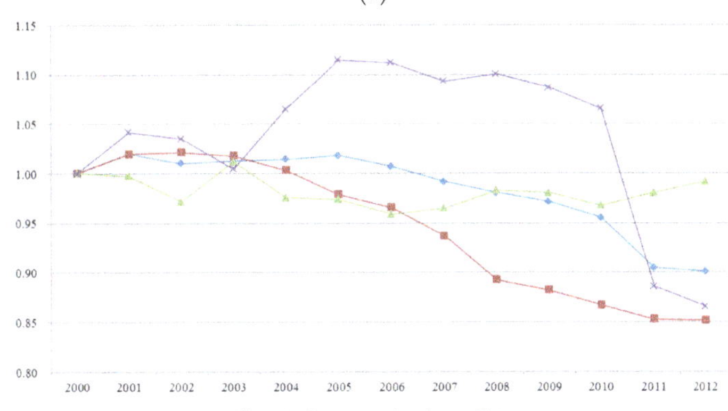

(b)

Figure 4. *Cont.*

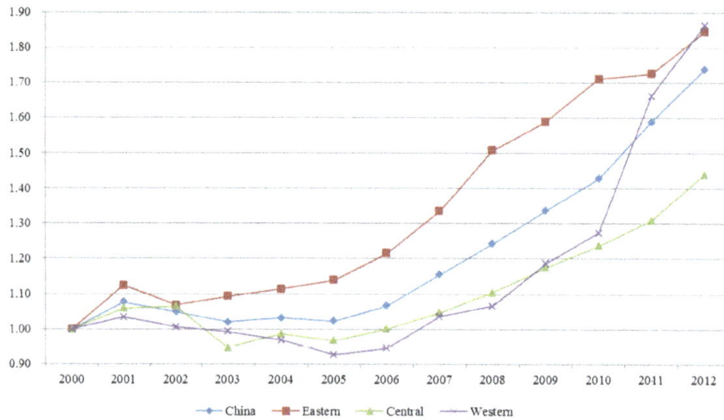

(c)

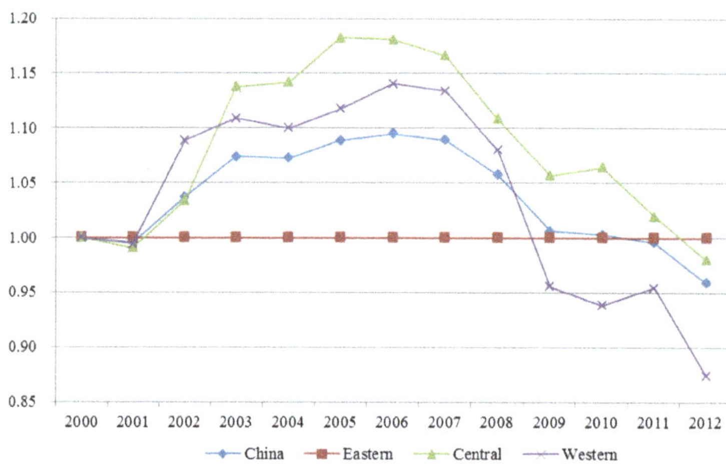

(d)

Figure 4. Trends of cumulative GDGI and its decomposition (2000–2012). (**a**) green development growth index (GDGI); (**b**) efficiency change (EC); (**c**) best-practice gap change (BPC); (**d**) technology gap change (TGC) index.

Table 7 shows the empirical results for the average GDGI for the 2000–2012 period and its decomposition for each province. Although most provinces have larger GDGI than unity, three provinces (Yunnan, Ningxia and Xinjiang) are observed to have GDGI lower than unity. Tianjin has the highest GDGI (average = 11.1%), indicating an 11.1% increase in green development performance. For EC, seven provinces show "catch-up" effects (their values of EC are larger than unity), whereas 12 provinces show no "catch-up effects". The BPCs of all provinces are larger than unity, implying there are innovation effects. The TGC of all eastern provinces and two other provinces (Guangxi and Qinghai) are at unity, implying high level of technology.

Table 7. Average green development growth (GDGI), efficiency change (EC), best practice gap change (BPC), and technical gap ratio change (TGC) of provinces in 2000–2012.

Province	GDGI	EC	BPC	TGC
Beijing	1.093	1.000	1.093	1.000
Tianjin	1.111	1.033	1.084	1.000
Hebei	1.010	0.967	1.046	1.000
Shanxi	1.019	0.988	1.058	0.978
Inner Mongolia	1.012	1.000	1.024	1.034
Liaoning	1.047	1.004	1.044	1.000
Jilin	1.027	1.002	1.042	0.990
Heilongjiang	1.045	1.000	1.039	1.015
Shanghai	1.083	1.000	1.083	1.000
Jiangsu	1.021	0.987	1.035	1.000
Zhejiang	1.024	0.989	1.037	1.000
Anhui	1.037	1.005	1.036	0.998
Fujian	1.007	0.964	1.050	1.000
Jiangxi	1.018	1.000	1.021	1.002
Shandong	1.010	0.969	1.042	1.000
Henan	1.008	0.990	1.027	0.995
Hubei	1.022	1.011	1.040	0.976
Hunan	1.005	1.000	1.002	1.005
Guangdong	1.001	1.000	1.001	1.000
Guangxi	1.002	0.973	1.031	1.000
Hainan	1.039	0.956	1.098	1.000
Chongqing	1.018	1.000	1.045	0.975
Sichuan	1.096	1.000	1.063	1.039
Guizhou	1.028	1.032	1.086	0.979
Yunnan	0.998	0.973	1.053	0.982
Shaanxi	1.002	1.000	1.047	0.960
Gansu	1.021	1.008	1.063	0.982
Qinghai	1.004	0.939	1.080	1.000
Ningxia	0.990	0.956	1.053	0.988
Xinjiang	0.975	1.000	1.022	1.006

5.3. Innovative Provinces

Although TGC indicates technology leadership, it cannot present information on which provinces are leading the way. In other words, it cannot specify which provinces serve the role of innovators. Because we have three benchmark technology sets, there are two ways to define innovators: group and metafrontier innovative provinces. The former refers to the outstanding provinces within a given group, while the latter refers to the innovative provinces from an integrated perspective. According to Zhang and Choi [7] and Li and Lin [21], three conditions are used for determining group innovative provinces,

$$BPC > 1 \tag{9}$$

$$GDPI_h^{C,t}(K^{t+1}, L^{t+1}, E^{t+1}, Y^{t+1}, W^{t+1}, S^{t+1}, D^{t+1}, C^{t+1}) > 1 \tag{10}$$

$$GDPI_h^{C,t+1}(K^{t+1}, L^{t+1}, E^{t+1}, Y^{t+1}, W^{t+1}, S^{t+1}, D^{t+1}, C^{t+1}) = 1 \tag{11}$$

Equation (9) suggests that the contemporaneous technology frontier should shift toward the intertemporal technology frontier, or the group innovators should have an innovation effect. Equation (10) implies that for group innovators, the technology in period t cannot produce the required quantity of outputs in period $t + 1$. Equation (11) specifies that the innovators must be located on the best practice frontier.

Because the metafrontier innovative provinces is based on global benchmark technology, we add two additional conditions to choose metafrontier innovative provinces,

$$TGC > 1 \tag{12}$$

$$GDPI^G(K^{t+1}, L^{t+1}, E^{t+1}, Y^{t+1}, W^{t+1}, S^{t+1}, D^{t+1}, C^{t+1}) = 1 \tag{13}$$

Equation (12) means that a metafrontier innovative province should be among the technologically leading provinces. Equation (13) suggests that a metafrontier innovative province should be located along the global technology frontier.

Table 8 shows the innovative provinces for every period. In the eastern group, Guangdong is found to be an innovator six times. The province is also shown to perform the highest in green development (Figure 2). Heilongjiang is an important innovator in the central group, and it is registered as an innovator six times. For the western group, Sichuan is found to be an innovator four times. Guangdong is a metafrontier innovator twice, Beijing and Tianjin are metafrontier innovators once. The group innovative provinces set targets for non-innovative ones to catch-up with the corresponding groups, while the metafrontier innovative provinces can be regarded as targets of other provinces to improve their green development performance.

Table 8. Group and metafrontier innovators.

Year	Group Innovator			Metafrontier Innovator
	Eastern	Central	Western	
2000–2001	Hainan	Inner Mongolia, Jiangxi	Xinjiang	-
2001–2002	-	Heilongjiang	-	-
2002–2003	Guangdong	-	-	-
2003–2004	Guangdong	Heilongjiang	-	Guangdong
2004–2005	-	Heilongjiang	-	-
2005–2006	Guangdong	-	-	-
2006–2007	-	-	Sichuan	-
2007–2008	-	Heilongjiang, Hunan	-	-
2008–2009	Guangdong	Heilongjiang	Sichuan	Guangdong
2009–2010	Beijing, Shanghai	Hunan	-	Beijing
2010–2011	Guangdong	-	Sichuan, Xinjiang	-
2011–2012	Tianjin, Shanghai, Guangdong	Inner Mongolia, Jiangxi, Heilongjiang, Anhui, Hubei, Hunan	Chongqing, Sichuan, Shaanxi, Xinjiang	Tianjin

6. Conclusions

Although many studies measure the green development performance of China, most of them only consider one environmentally harmful by-product, for example CO_2 emissions. In theory, green development is quite encompassing and other major pollutants need to be considered in the case of China. This paper uses four main pollutants, namely waste water, sulphur dioxide emission, solid wastes and CO_2 emissions, to measure green development performance. In terms of methodology, the green development performance index (GDPI) is presented, which is based on the total-factor production efficiency framework, and considers non-radial slacks. In order to incorporate *ex ante* group heterogeneities, the above GDPI is combined with the metafrontier analysis. With this augmented methodology, the green development growth index (GDGI) is decomposed into EC, BPC and TGC indices. All the above indices can be calculated by solving several non-radial DEA-type models.

The proposed approach is employed to analyze the changes in green development performance of 30 provinces in China during the period 2000–2012. The main results are: (1) the average score of GDPI in China during the sample period is 0.398, indicating a low level of green development, but it has improved in recent years; (2) mainly driven by BPC, China's green development performance improves by 2.58% each year; (3) green development is mainly led by the eastern region. It should be noted that this result is relative. In other words, it only implies that the eastern region's performance is better than the central and the western regions. In fact, the average score of GDPI for the eastern region is 0.525, implying a low level of green development.

There are some limitations for this study. Methodologically, the paper adopts geographical closeness as a criterion for grouping provinces. In theory, heterogeneity is main reflected by production technology. In this regard, an in-depth analysis of the heterogeneity is more effective in reflecting different characteristics of provinces in China. Furthermore, in our case, it needs to be recomputed when a new time period is added to the data set. In this sense, biennial technology rather than global technology may be more suitable [48]. Additionally, we do not provide statistical inferences on green development performance. By doing this, the bootstrapping method should be used to perform the statistical inference. Finally, a regression analysis on the determinants of green development performance may be an interesting empirical extension.

Acknowledgments: This work was supported by Major Projects in Philosophy and Social Science Research of the Ministry of Education of China (Grant No. 14JZD031); the National Natural Science Foundation of China (Grant Nos. 71171001 & 71471001); and the Fundamental Research Funds for the Central Universities (No. 20720151028).

Author Contributions: Ke Li and Malin Song conceived and designed the research; Ke Li performed research, analyzed the data and wrote the paper; Ke Li and Malin Song revised the paper together. All authors read and approved the final manuscript.

Conflicts of Interest: The authors declare no conflict of interest.

References

1. Organisation for Economic Co-operation and Development (OECD). *Towards Green Growth*; OECD: Paris, France, 2011.
2. Li, K.; Lin, B. The efficiency improvement potential for coal, oil and electricity in China's manufacturing sectors. *Energy* **2015**, *86*, 403–413.
3. Chen, S.; Golley, J. 'Green' productivity growth in China's industrial economy. *Energ. Econ.* **2014**, *44*, 89–98.
4. Ho, M.; Jorgenson, D. Sector allocation of emissions and damage and Policies to control air pollution damages. In *Clearing the Air: The Health and Economic Damages of Air Pollution in China*; Ho, M.S., Nielsen, C.P., Eds.; MIT Press: Cambridge, UK, 2007; pp. 279–372.
5. Cao, J. Measuring Green Productivity Growth for China's Manufacturing Sectors: 1991–2000. *Asian Econ. J.* **2007**, *21*, 425–451.
6. Chung, Y.H.; Färe, R.; Grosskopf, S. Productivity and Undesirable Outputs: A Directional Distance Function Approach. *J. Environ. Manag.* **1997**, *51*, 229–240.
7. Zhang, N.; Choi, Y. Total-factor carbon emission performance of fossil fuel power plants in China: A metafrontier non-radial Malmquist index analysis. *Energ. Econ.* **2013**, *40*, 549–559.

8. Zhang, N.; Zhou, P.; Kung, C.-C. Total-factor carbon emission performance of the Chinese transportation industry: A bootstrapped non-radial Malmquist index analysis. *Renew. Sustain. Energ. Rev.* **2015**, *41*, 584–593.
9. Solow, R.M. Technical Change and the Aggregate Production Function. *Rev. Econ. Stat.* **1957**, *39*, 312–320.
10. Li, K.-W. China's total factor productivity estimates by region, investment sources and ownership. *Economic Systems* **2009**, *33*, 213–230.
11. You, K.; Sarantis, N. Structural breaks, rural transformation and total factor productivity growth in China. *J. Prod. Anal.* **2013**, *39*, 231–242.
12. Chen, P.C.; Yu, M.M.; Chang, C.C.; Hsu, S.H. Total factor productivity growth in China's agricultural sector. *China Econ. Rev.* **2008**, *19*, 580–593.
13. Yang, Q.; Wan, X.; Ma, H. Assessing Green Development Efficiency of Municipalities and Provinces in China Integrating Models of Super-Efficiency DEA and Malmquist Index. *Sustainability* **2015**, *7*, 4492–4510.
14. Ahmed, E.M. Green TFP Intensity Impact on Sustainable East Asian Productivity Growth. *J. Econ. Anal. Pol.* **2012**, *42*, 67–78.
15. Choi, Y. Intermediary Propositions for Green Growth with Sustainable Governance. *Sustainability* **2015**, *7*, 14785–14801.
16. Sueyoshi, T.; Goto, M. DEA environmental assessment in a time horizon: Malmquist index on fuel mix, electricity and CO_2 of industrial nations. *Energ. Econ.* **2013**, *40*, 370–382.
17. Rashidi, K.; Saen, R.F. Measuring eco-efficiency based on green indicators and potentials in energy saving and undesirable output abatement. *Energ. Econ.* **2015**, *50*, 18–26.
18. Lin, E.Y.-Y.; Chen, P.-Y.; Chen, C.-C. Measuring green productivity of country: A generlized metafrontier Malmquist productivity index approach. *Energy* **2013**, *55*, 340–353.
19. Aparicio, J.; Pastor, J.T.; Zofio, J.L. On the inconsistency of the Malmquist–Luenberger index. *Eur. J. Oper. Res.* **2013**, *229*, 738–742.
20. Song, X.; Hao, Y.; Zhu, X. Analysis of the Environmental Efficiency of the Chinese Transportation Sector Using an Undesirable Output Slacks-Based Measure Data Envelopment Analysis Model. *Sustainability* **2015**, *7*, 9187–9206.
21. Li, K.; Lin, B. Measuring green productivity growth of Chinese industrial sectors during 1998–2011. *China Econ. Rev.* **2015**, *36*, 279–295.
22. Li, K.; Lin, B. Metafroniter energy efficiency with CO_2 emissions and its convergence analysis for China. *Energ. Econ.* **2015**, *48*, 230–241.
23. Li, J.; Li, J.; Zheng, F. Unified Efficiency Measurement of Electric Power Supply Companies in China. *Sustainability* **2014**, *6*, 779–793.
24. Wei, C.; Löschel, A.; Liu, B. Energy-saving and emission-abatement potential of Chinese coal-fired power enterprise: A non-parametric analysis. *Energ. Econ.* **2015**, *49*, 33–43.
25. Tu, Z.; Shen, R. Can China's Industrial SO_2 Emissions Trading Pilot Scheme Reduce Pollution Abatement Costs? *Sustainability* **2014**, *6*, 7621–7645.

26. Zhou, P.; Ang, B.W.; Wang, H. Energy and CO_2 emission performance in electricity generation: A non-radial directional distance function approach. *Eur. J. Oper. Res.* **2012**, *221*, 625–635.
27. Färe, R.; Grosskopf, S.; Pasurka, C.A., Jr. Environmental production functions and environmental directional distance functions. *Energy* **2007**, *32*, 1055–1066.
28. Wu, P.-I.; Chen, C.T.; Liou, J.-L. The meta-technology cost ratio: An indicator for judging the cost performance of CO_2 reduction. *Econ. Model.* **2013**, *35*, 1–9.
29. Picazo-Tadeo, A.J.; Reig-Martínez, E.; Hernández-Sancho, F. Directional distance functions and environmental regulation. *Resource Energ. Econ.* **2005**, *27*, 131–142.
30. Watanabe, M.; Tanaka, K. Efficiency analysis of Chinese industry: A directional distance function approach. *Energ. Policy* **2007**, *35*, 6323–6331.
31. Macpherson, A.J.; Principe, P.P.; Smith, E.R. A directional distance function approach to regional environmental–economic assessments. *Ecol. Econ.* **2010**, *69*, 1918–1925.
32. Yuan, P.; Cheng, S.; Sun, J.; Liang, W. Measuring the environmental efficiency of the Chinese industrial sector: A directional distance function approach. *Math. Comput. Model.* **2013**, *58*, 936–947.
33. Chang, T.-P.; Hu, J.-L. Total-factor energy productivity growth, technical progress, and efficiency change: An empirical study of China. *Appl. Energ.* **2010**, *87*, 3262–3270.
34. Fukuyama, H.; Weber, W.L. A directional slacks-based measure of technical inefficiency. *Soc. Econ. Plan. Sci.* **2009**, *43*, 274–287.
35. Pastor, J.T.; Aparicio, J. A note on "A directional slacks-based measure of technical inefficiency". *Soc. Econ. Plan. Sci.* **2010**, *44*, 174–175.
36. Wang, K.; Lu, B.; Wei, Y.-M. China's regional energy and environmental efficiency: A Range-Adjusted Measure based analysis. *Appl. Energ.* **2013**, *112*, 1403–1415.
37. Lin, B.; Du, K. Energy and CO_2 emissions performance in China's regional economies: Do market-oriented reforms matter? *Energy Policy* **2015**, *78*, 113–124.
38. Zhang, N.; Kong, F.; Choi, Y.; Zhou, P. The effect of size-control policy on unified energy and carbon efficiency for Chinese fossil fuel power plants. *Energ. Policy* **2014**, *70*, 193–200.
39. Tulkens, H.; Eeckaut, P.V. Non-parametric efficiency, progress and regress measures for panel data: Methodological aspects. *Eur. J. Oper. Res.* **1995**, *80*, 474–499.
40. Oh, D.-H. A metafrontier approach for measuring an environmentally sensitive productivity growth index. *Energ. Econ.* **2010**, *32*, 146–157.
41. Afsharian, M.; Ahn, H. The overall Malmquist index: A new approach for measuring productivity changes over time. *Ann. Oper. Res.* **2014**, *226*, 1–27.
42. Pastor, J.T.; Lovell, C.A.K. A global Malmquist productivity index. *Econ. Lett.* **2005**, *88*, 266–271.
43. Liou, J.-L.; Wu, P.-I. Will economic development enhance the energy use efficiency and CO_2 emission control efficiency? *Expert Syst. Appl.* **2011**, *38*, 12379–12387.
44. Wu, Y. *Productivity, Efficiency and Economic Growth in China*; Palgrave Macmillan: London, UK, 2008.

45. National Bureau of Statistics of the People's Republic of China. *The China Energy Statistical Yearbook*; 2001–2014. National Bureau of Statistics of the People's Republic of China: Beijing, China, 2001–2014.
46. National Bureau of Statistics of the People's Republic of China. *The China Statistical Yearbook*; 2001–2015. National Bureau of Statistics of the People's Republic of China: Beijing, China, 2001–2015.
47. Intergovernmental Panel on Climate Change (IPCC) National Greenhouse Gas Inventories Programme. *IPCC Guidelines for National Greenhouse Gas Inventories*; Eggleston, S., Buendia, L., Miwa, K., Ngara, T., Tanabe, K., Eds; Institute for Global Environmental Strategies (IGES): Hayama, Japan, 2006.
48. Pastor, J.T.; Asmild, M.; Lovell, C.A.K. The biennial Malmquist productivity change index. *Soc. Econ. Plan. Sci.* **2011**, *45*, 10–15.

Industrial Carbon Emissions of China's Regions: A Spatial Econometric Analysis

Yu Liu, Hongwei Xiao and Ning Zhang

Abstract: This paper proposes an extended Stochastic Impacts by Regression on Population, Affluence, and Technology (STIRPAT) model to investigate the factors driving industrial carbon emissions in China. In the first stage, a spatial Durbin model is applied to investigate the determinants of regional industrial carbon emissions. In the second stage, a geographically and temporally weighted regression is applied to investigate temporal and spatial variations in the impacts of these driving factors on the scale and intensity of regional industrial carbon emissions. The empirical results suggest that the provinces with low carbon emissions act as exemplars for those with high carbon emissions and that driving factors impact carbon emission both directly and indirectly. All of the factors were investigated, except energy intensity, energy price, and openness, significantly impact carbon emissions. Overall, the results suggest that spatial correlation, heterogeneity, and spillover effects should be taken into account when formulating policies aiming at reducing industrial carbon emissions. The paper concludes with relevant policy recommendations taking full account of the regional industrial carbon emissions, heterogeneity and spillover.

Reprinted from *Sustainability*. Cite as: Liu, Y.; Xiao, H.; Zhang, N. Industrial Carbon Emissions of China's Regions: A Spatial Econometric Analysis. *Sustainability* **2016**, *8*, 210.

1. Introduction

For the past two decades, China has been one of the world's fastest growing economies, with an average annual Gross Domestic Product (GDP) growth rate of 10.4% between 1990 and 2010. This, combined with the fact that China is the world's most populous country, has led to unprecedented growth in energy demand, especially that for fossil fuels. The 2013 *World Energy Outlook* reports that China and India accounted for about 10% of total world energy consumption in 1990 but around 24% in 2010. China's economy rapidly developed between 1990 and 2010, and the country's carbon dioxide emissions more than tripled during this period. By 2010, China's carbon emissions constituted 24% of global emissions. After gains in energy efficiency coupled with deployment of cleaner energy, enabled by economic development, China's carbon intensity (measured in CO_2 emitted per unit of GDP), declined by 15% between 2005 and 2011. Projections indicate that China and India will continue to lead not only future global economic growth but also future growth in energy demand, making up 34% of total world energy consumption in 2040.

Electricity demand is the main driver of China's emissions growth: coal-powered electricity generation was responsible for almost all emissions growth between 1990 and 2010, despite notable improvements in the emissions performance of coal-fired power generating processes. In addition, the demand for large-scale infrastructure investments and Chinese products has grown in tandem with the economy. This has motivated increased energy consumption—mainly of fossil fuels—by the industrial sector, which accounts for around 70% of China's total carbon emissions. The government has recognized the need to prioritize reducing industrial-sector carbon emissions. This is reflected, for example, in the country's 12th Five-Year Plan, which aims to lower CO_2 emissions per unit GDP by 17% between 2010 and 2015. Provincial and local governments are central to implementing this plan, which requires a systematic understanding of China's industrial energy use and CO_2 emissions at sub-national levels. In particular, a spatial understanding—at the provincial and, where possible, local government levels—would help policymakers identify the areas of the greatest energy-saving and emissions-reduction potential.

Most past studies have investigated this issue at the national or aggregate level. The development of spatial econometrics, however, has presented researchers with an opportunity to incorporate spatial dynamics into their analyses of the trends and drivers of China's carbon emissions. Most of these studies empirically test the environmental Kuznets curve (EKC) hypothesis at the regional or industry level. The EKC argument hypothesizes that there is a relationship between various indicators of environmental degradation and economic development measured in terms of income per capita. It argues that initially, environmental degradation rises as a country industrializes; as the country develops economically and is able to invest in combating environmental degradation and the economy shifts from industry to services, however, environmental outcomes improve. Prior work using spatial panel data models has shown that China's regional economic development is spatially correlated with environmental quality [1–7]. By estimating a spatial panel data model with fixed effects, Zhu *et al.* [8] (pp. 65–74) empirically investigated the EKC of China's regional emissions of industrial pollutants. They found that regions' industrial pollutant emissions are strongly correlated and that the spatial panel data model is more robust than the traditional panel model. Wang *et al.* [9] (pp. 818–825) also applied a dynamic spatial panel data model to examine the relationship between China's environmental pollution and its economic growth. The results revealed that environmental pollution is spatially correlated, supporting the hypothesis that the EKC does exist. Yao and Ni [10] (pp. 1432–1438) and Xu and Deng [4] (pp. 30–43) applied spatial panel data models to analyze the relationship between China's carbon intensity and foreign direct investment (FDI). The results showed that FDI decreases China's regional carbon intensity due to FDI's technology spillover effect, which reduces CO_2 emissions and associated carbon intensity. Yu [11] (pp. 93–101) also applied a spatial panel data model to investigate the causes of low energy efficiency. That study found that both increases in total factor productivity (TFP) and changes in industrial structure increased China's energy efficiency, with the former playing a more important role than the latter. Wei *et al.* [12] (pp. 478–488) conducted a comparative study on China's energy efficiency among countries and confirmed China's low energy efficiency. Wei *et al.* [13] (pp. 552–565), Choi *et al.* [14] (pp. 198–208) and Wang *et al.* [7] (pp. 2584–2600) investigated the regional carbon emission efficiency for Chinese Provinces. However, previous studies did not consider the spatial patterns and drivers of carbon emissions for China.

This paper seeks to contribute to the spatial understanding of patterns and drivers of industrial carbon emissions in China. Such analysis is important due to the substantial nature of the cross-regional variation in industrial carbon emissions. As such, for policies aimed at reducing carbon emissions to be effective in China, they must be anchored in a comprehensive understanding of spatial dynamics and reflect the variations in emissions reduction potential across regions. To provide such insights in the form of more effective estimates, this paper applies a spatial econometric model, incorporating both temporal and spatial effects, to investigate the factors influencing regional industrial carbon emissions. Based on the estimations, the paper provides practical and effective policy recommendations for reducing industrial carbon emissions in China.

2. Materials and Methods

2.1. The Extended STIRPAT Model

Ehrlish *et al.* [15] (pp. 1212–1217) were among the pioneers proposing the IPAT (Impact, Population, Affluence, Technology) accounting identity for the impact of human activities on the environment. The IPAT equation, born out of a need to identify sources of human-induced environmental change, posits that interactions between population size, affluence or economic growth, and technological change impact the environment as follows:

$$I = PAT \tag{1}$$

where I represents environmental impact, P is population size, A is affluence or wealth per capita, and T is the technology level. Though simple, the IPAT model emphasizes that environmental degradation (or improvements) result from multiple factors acting together to have a compound effect on the environment. The Intergovernmental Panel on Climate Change (IPCC) has used the IPAT model to assess the contributions of population size, affluence, and technology on greenhouse gas (GHG) emissions. Over time, empirical estimation of the equation has been strengthened by adding additional socio-political and technical factors, thereby improving the equation's predictive power. Some of studies that have used the IPAT equation include Harrison [16], Raskin [17] (pp. 225–233), York [18] (pp. 18–34), Shi [1] (pp. 29–42), Cole [19] (pp. 5–21), and Rosa [20] (pp. 509–512).

The major limitation of the IPAT equation is that, as an accounting identity, it does not allow hypothesis testing because it assumes a proportional functional relationship between the IPAT factors. Based on this limitation, numerous researchers have proposed alternative versions of the equation. Notably, Dietz and Rosa [21] (pp. 277–300) present the IPAT equation in a stochastic manner and propose a Stochastic Impacts by Regression on Population, Affluence, and Technology (STIRPAT) model. In this approach, carbon emissions are modeled stochastically by regressing them on population, wealth, and technology. The STIRPAT model is expressed as follows:

$$I = aP^{b}A^{c}T^{d}e \tag{2}$$

where a represents the model coefficient; b, c, and d are the coefficients on population, wealth, and technology, respectively; and e is the error term. Environmental pressure is represented by I, population size by P, wealth by A, and technology by T. Taking the natural logarithm of both sides leads to Equation (3).

$$\ln I = \ln a + b\left(\ln P\right) + c\left(\ln A\right) + d\left(\ln T\right) + \ln e \tag{3}$$

The elasticity between the independent variables and the dependent variable is reflected in the coefficients, which express the percentage change in the dependent variable arising from a 1% change in the dependent variable, all else held constant. The STIRPAT model builds on the IPAT equation by making it possible to test hypotheses, allowing estimation of the coefficients and decomposition of the influencing factors, and allowing for the inclusion of other factors. Studies based on the model and its adjustment include Dietz and Rosa [21] (pp. 277–300) and York *et al.* [22] (pp. 351–365).

This paper extends the STIRPAT model to investigate the various impacts of the factors driving regional industrial carbon emissions in China. At the regional industrial level, the level or scale of investment is considered a more important determinant of carbon emissions than is population size. As a result, we place population size with the scale of investment in our model. Additional variables, such as energy consumption structure, energy price, and openness (Openness: measured by the percentage of FDI in total investment by industrial enterprises, unit: %.), are also included. The resulting extended STIRPAT model is expressed as in Equations (4) and (5).

$$\begin{aligned} \ln CS &= \ln a + \beta_1 (\ln SI) + \beta_2 (\ln IE) + \beta_3 (\ln EI) + \beta_4 (\ln ECS) + \beta_5 (\ln EP) + \beta_6 (\ln OPEN) + \ln e \\ &= \alpha + \beta_1 (\ln SI) + \beta_2 (\ln IE) + \beta_3 (\ln EI) + \beta_4 (\ln ECS) + \beta_5 (\ln EP) + \beta_6 (\ln OPEN) + \varepsilon \end{aligned} \tag{4}$$

$$\begin{aligned} \ln CI &= \ln a + \beta_1 (\ln SI) + \beta_2 (\ln IE) + \beta_3 (\ln EI) + \beta_4 (\ln ECS) + \beta_5 (\ln EP) + \beta_6 (\ln OPEN) + \ln e \\ &= \alpha + \beta_1 (\ln SI) + \beta_2 (\ln IE) + \beta_3 (\ln EI) + \beta_4 (\ln ECS) + \beta_5 (\ln EP) + \beta_6 (\ln OPEN) + \varepsilon \end{aligned} \tag{5}$$

where *CS* denotes the level or scale of carbon emissions (measured in 10,000 tons of industrial carbon emitted), *CI* is carbon intensity (CO_2 emissions per unit of industrial value-added, measured in tons/10,000 Yuan), *SI* represents the scale of investment (measured by total assets of industrial enterprises above a designated size, in 100 million Yuan), *IE* denotes industrial economic intensity (measured by industrial added-value per employee), *EI* is energy intensity (measured by energy consumption per unit of industrial added-value, in tons of standard coal per 10,000 Yuan), *ECS* is energy consumption structure (measured by the percentage of total energy consumption that is industrial coal consumption), *EP* is energy price (measured by producer price index for manufactured products), and *OPEN* is openness (measured by the percentage of FDI in total investment by industrial enterprises).

2.2. *Spatial Econometric Model with Carbon Emissions*

Starting from the improved STIRPAT, we build a spatial econometric model by taking into account the fact that carbon emissions are heterogeneous and spatially correlated among regions and industries. As highlighted in Equations (4) and (5), the two dependent variables used in the analysis are industrial: the level of carbon emissions and carbon intensity. The spatial panel data econometric model, which integrates spatial econometrics (spatial effects) and panel data (time effects), this paper utilizes a space and time fixed-effects in a spatial econometric framework. This makes spatial econometric analysis more efficient. The spatial panel data econometric model includes three basic models: the spatial lag panel data model, the spatial error panel data model, and the spatial Durbin panel data model. These models are discussed in detail below.

2.2.1. Spatial Lag Panel Data Model (SLPDM)

The spatial lag panel data model (SLPDM) used in the paper is represented as follows:

$$\begin{aligned}\text{lnCS}_{it} &= \delta \sum_{j=1}^{N} w_{ij}\text{lnCS}_{it} + \text{ln}a + \beta_1 (InSI_{it}) + \beta_2 (InIE_{it}) + \beta_3 (InEI_{it}) + \beta_4 (InECS_{it}) + \beta_5 (InEP_{it}) \\ &+\beta_6 (InOPEN_{it}) + \mu_i + \lambda_t + \text{ln}e_{it} \\ &= \delta \sum_{j=1}^{N} w_{ij}\text{lnCS}_{it} + \alpha + \beta_1 (InSI_{it}) + \beta_2 (InIE_{it}) + \beta_3 (InEI_{it}) + \beta_4 (InECS_{it}) + \beta_5 (InEP_{it}) \\ &+\beta_6 (InOPEN_{it}) + \mu_i + \lambda_t + \varepsilon_{it}\end{aligned} \tag{6}$$

$$\begin{aligned}\text{ln}CI_{it} &= \delta \sum_{j=1}^{N} w_{ij}\text{ln}CI_{it} + \text{ln}a + \beta_1 (InSI_{it}) + \beta_2 (InIE_{it}) + \beta_3 (InEI_{it}) + \beta_4 (InECS_{it}) + \beta_5 (InEP_{it}) \\ &+\beta_6 (InOPEN_{it}) + \mu_i + \lambda_t + \text{ln}e_{it} \\ &= \delta \sum_{j=1}^{N} w_{ij}\text{ln}CI_{it} + \alpha + \beta_1 (\text{ln}SI_{it}) + \beta_2 (\text{ln}IE_{it}) + \beta_3 (InEI_{it}) + \beta_4 (InECS_{it}) + \beta_5 (InEP_{it}) \\ &+\beta_6 (InOPEN_{it}) + \mu_i + \lambda_t + \varepsilon_{it}\end{aligned} \tag{7}$$

where CS_{it} and CI_{it} denote industrial carbon emissions scale and intensity of the region i at time t, respectively. $\sum w_{ij}\text{lnCS}_{it}$ and $\sum w_{ij}\text{ln}CI_{it}$ represent the spatial correlation between $\text{ln}CS_{it}$ and $\text{ln}CI_{it}$ of region i and that of its adjacent regions. $\text{ln}SI_{it}$, $\text{ln}IE_{it}$, $\text{ln}EI_{it}$, $\text{ln}ECS_{it}$, $\text{ln}EP_{it}$, and $\text{ln}OPEN_{it}$ are independent variables corresponding to region I at time $t.\delta$ is the spatial auto-correlation index, w_{ij} is an element of the spatial weight matrix representing the spatial relations between regions i and j, α is the constant term, the βs are coefficients to be estimated, μ_i is the individual (region) fixed effect and λ_t is the time fixed effect.

2.2.2. Spatial Error Panel Data Model (SEPDM)

The spatial error panel data model (SEPDM) is specified as follows:

$$\begin{aligned}\text{lnCS}_{it} &= \text{ln}a + \beta_1 (InSI_{it}) + \beta_2 (InIE_{it}) + \beta_3 (InEI_{it}) + \beta_4 (InECS_{it}) + \beta_5 (InEP_{it}) + \beta_6 (InOPEN_{it}) \\ &+\mu_i + \lambda_t + \text{ln}e_{it} \\ &= \alpha + \beta_1 (\text{ln}SI_{it}) + \beta_2 (\text{ln}IE_{it}) + \beta_3 (\text{ln}EI_{it}) + \beta_4 (InECS_{it}) + \beta_5 (InEP_{it}) + \beta_6 (InOPEN_{it}) \\ &+\mu_i + \lambda_t + \varnothing_{it} \\ \varnothing_{it} &= \rho \sum_{j=1}^{N} w_{ij}\varnothing_{it} + \varepsilon_{it}\end{aligned} \tag{8}$$

$$\begin{aligned}\text{ln}CI_{lt} &= \text{ln}a + \beta_1 (InSI_{it}) + \beta_2 (InIE_{it}) + \beta_3 (InEI_{it}) + \beta_4 (InECS_{it}) + \beta_5 (InEP_{it}) + \beta_6 (InOPEN_{it}) \\ &+\mu_i + \lambda_t + \text{ln}e_{it} \\ &= \alpha + \beta_1 (\text{ln}SI_{it}) + \beta_2 (\text{ln}IE_{it}) + \beta_3 (\text{ln}EI_{it}) + \beta_4 (InECS_{it}) + \beta_5 (InEP_{it}) + \beta_6 (InOPEN_{it}) \\ &+\mu_i + \lambda_t + \varnothing_{it} \\ \varnothing_{it} &= \rho \sum_{j=1}^{N} w_{ij}\varnothing_{it} + \varepsilon_{it}\end{aligned} \tag{9}$$

where $\text{ln}CS_{it}$, $\text{ln}CI_{it}$, α, β, μ_i, λ_t, ε_{it}, $\text{ln}SI_{it}$, $\text{ln}IE_{it}$, $\text{ln}EI_{it}$, $\text{ln}ECS_{it}$, $\text{ln}EP_{it}$, and $\text{ln}OPEN_{it}$ are defined as in Equations (6) and (7). φ_{it} denotes spatial error auto-correlation, and ρ is the spatial auto-correlation index.

2.2.3. Spatial Durbin Panel Date Model (SDPDM)

The spatial Durbin panel data model (SDPDM) is specified as follows:

$$\begin{aligned}
\ln CS_{it} &= \delta \sum_{j=1}^{N} w_{ij}\ln CS_{it} + \ln a + \beta_1 (lnSI_{it}) + \beta_2 (lnIE_{it}) + \beta_3 (lnEI_{it}) + \beta_4 (lnECS_{it}) + \beta_5 (lnEP_{it}) \\
&+\beta_6 (lnOPEN_{it}) + \theta_1 \sum_{j=1}^{N} w_{ij}\ln SI_{ijt} + \theta_2 \sum_{j=1}^{N} w_{ij}\ln IE_{ijt} + \theta_3 \sum_{j=1}^{N} w_{ij}\ln EI_{ijt} \\
&+\theta_4 \sum_{j=1}^{N} w_{ij}\ln ECS_{ijt} + \theta_5 \sum_{j=1}^{N} w_{ij}\ln EP_{ijt} + \theta_6 \sum_{j=1}^{N} w_{ij}\ln OPEN_{ijt} + \mu_i + \lambda_t + \ln e_{it} \\
&= \delta \sum_{j=1}^{N} w_{ij}\ln CS_{it} + \ln a + \beta_1 (lnSI_{it}) + \beta_2 (lnIE_{it}) + \beta_3 (lnEI_{it}) + \beta_4 (lnECS_{it}) + \beta_5 (lnEP_{it}) \\
&+\beta_6 (lnOPEN_{it}) + \theta_1 \sum_{j=1}^{N} w_{ij}\ln SI_{ijt} + \theta_2 \sum_{j=1}^{N} w_{ij}\ln IE_{ijt} + \theta_3 \sum_{j=1}^{N} w_{ij}\ln EI_{ijt} \\
&+\theta_4 \sum_{j=1}^{N} w_{ij}\ln ECS_{ijt} + \theta_5 \sum_{j=1}^{N} w_{ij}\ln EP_{ijt} + \theta_6 \sum_{j=1}^{N} w_{ij}\ln OPEN_{ijt} + \mu_i + \lambda_t + \varepsilon_{it}
\end{aligned} \tag{10}$$

$$\begin{aligned}
\ln CI_{it} &= \delta \sum_{j=1}^{N} w_{ij}\ln CS_{it} + \ln a + \beta_1 (lnSI_{it}) + \beta_2 (lnIE_{it}) + \beta_3 (lnEI_{it}) + \beta_4 (lnECS_{it}) \\
&+\beta_5 (lnEP_{it}) + \beta_6 (lnOPEN_{it}) + \theta_1 \sum_{j=1}^{N} w_{ij}\ln SI_{ijt} + \theta_2 \sum_{j=1}^{N} w_{ij}\ln IE_{ijt} \\
&+\theta_3 \sum_{j=1}^{N} w_{ij}\ln EI_{ijt} + \theta_4 \sum_{j=1}^{N} w_{ij}\ln ECS_{ijt} + \theta_5 \sum_{j=1}^{N} w_{ij}\ln EP_{ijt} \\
&+\theta_6 \sum_{j=1}^{N} w_{ij}\ln OPEN_{ijt} + \mu_i + \lambda_t + \ln e_{it} \\
&= \delta \sum_{j=1}^{N} w_{ij}\ln CS_{it} + \ln a + \beta_1 (lnSI_{it}) + \beta_2 (lnIE_{it}) \\
&+\beta_3 (lnEI_{it}) + \beta_4 (lnECS_{it}) + \beta_5 (lnEP_{it}) + \beta_6 (lnOPEN_{it}) \\
&+\theta_1 \sum_{j=1}^{N} w_{ij}\ln SI_{ijt} + \theta_2 \sum_{j=1}^{N} w_{ij}\ln IE_{ijt} + \theta_3 \sum_{j=1}^{N} w_{ij}\ln EI_{ijt} \\
&+\theta_4 \sum_{j=1}^{N} w_{ij}\ln ECS_{ijt} + \theta_5 \sum_{j=1}^{N} w_{ij}\ln EP_{ijt} + \theta_6 \sum_{j=1}^{N} w_{ij}\ln OPEN_{ijt} + \mu_i \\
&+\lambda_t + \varepsilon_{it}
\end{aligned} \tag{11}$$

where $\ln SI_{it}$, $\ln IE_{it}$, $\ln EI_{it}$, $\ln ECS_{it}$, $\ln EP_{it}$, $\ln OPEN_{it}$, α, μ_i, λ_t are defined as in Equations (6) and (7). θ is a vector of coefficients to be estimated. We test the joint hypotheses H_0: $\theta = 0$ and H_0: $\theta + \delta\beta = 0$; rejection of the hypotheses indicates that the SDPDM fits the data optimally.

Our extension of the STIRPAT model not only allows examination of the impacts of the above-mentioned independent variables on the scale and intensity of a region's industrial carbon emissions but also makes it possible to examine the impacts of the adjacent regions' independent variables on a given region's industrial carbon emissions scale and intensity. It also allows measurement of the impact of the adjacent region's carbon emissions scale and intensity on a given region's industrial carbon emissions scale and intensity.

2.2.4. Spatial Weight Matrix

A spatial weight matrix must be constructed to reflect the spatial correlation among regions, and an appropriate spatial weight matrix is essential for obtaining a sound spatial econometric result. We opt for both geographic and economic spatial weight matrixes. The former is constructed via the inverse distance method.

$$W^{GS}{}_{ij} = \begin{cases} \dfrac{1}{d_{ij}{}^{\alpha}} & i \neq j \\ 0 & i = j \end{cases} \tag{12}$$

where d_{ij} is the distance between regions i and j, which is calculated using their longitudes and latitudes.

The economic spatial weight matrix W^* is a product of W and the economic weight matrix, E.

$$E_{ij} = \begin{cases} \dfrac{1}{\left|\overline{G}_i - \overline{G}_i\right|^{\alpha} + m} & i \neq j \\ 0 & i = j \end{cases} \qquad \overline{G}_i = \frac{1}{5}\sum_{t=2006}^{2010} G_{it} \tag{13}$$

where G_{it} denotes per capita industrial value-added, representing the actual industrial output per capita of region i at time t (Deflated by the price index in 2006). Thus, W^* incorporates economic development into the weight matrix.

Regional industrial carbon emissions are obtained from energy types via the stable carbon emissions factors for electricity and thermo, such as raw coal, washed coal, other washed coal, briquette, coke, coke oven gas, other gases, other coking products, crude oil, gasoline, kerosene, diesel oil, fuel oil, liquefied petroleum gas, refinery gas, other petroleum products, and natural gas. The data is from *China Energy Statistical Yearbook 2007–2011* [23].

3. Empirical Investigation

3.1. Model Specification

Two Lagrange Multiplier tests (LM-Lag and LM-Error tests) are applied to choose which model of those described in Sections 3.1–3 best fits the data. In classical panel data models, there are four options for fixed effects, namely individual fixed effects, time fixed effects, individual and time fixed effects, and no fixed effects. We test these four options via the LM test. Tables 1 and 2 show the LM test statistics for Models 1 and 2, with dependent variables Log (CS) and Log (CI), respectively.

Table 1 shows that in Model 1 the LM statistic for the spatial error model with time fixed effects is significant at the 5% level. The LM statistics for the spatial lag model under no spatial effects and time fixed effects are significant (Table 2). Thus, the hypothesis that spatial correlation does not exist is rejected. In addition, the likelihood ratio (LR) tests reject the hypotheses that individual fixed effects and time fixed effects do not exist, indicating that the individual and time fixed effect model outperforms its alternatives. We further determine which model (SLPDM, SEPDM, or SDPDM) best fits the data using Wald and LR tests.

Table 1. LM test for Model 1 (dependent variable: Log (*CS*)).

Variable	Pooled OLS	Individual Fixed Effects	Time Fixed Effect	Individual and Time Fixed Effect
Constant	−5.6650 *** (−5.2619)	–	–	–
Log(SI)	1.1638 *** (28.2449)	0.8632 *** (30.1539)	1.1750 *** (29.4582)	1.0276 *** (9.4509)
Log(IE)	0.4286 *** (4.1127)	0.0181 (0.3679)	0.5213 *** (4.9142)	0.0247 (0.4977)
Log(EI)	1.1953 *** (13.7546)	0.8343 *** (22.9815)	1.1824 *** (14.1376)	0.8428 *** (23.1001)
Log(ECS)	0.5700 *** (6.2444)	0.0313 (1.1421)	0.6205 *** (6.9625)	0.0639 ** (2.3508)
Log(EP)	0.8240 ** (2.1345)	0.1033 ** (2.4452)	2.1933 *** (2.9781)	−0.0040 (−0.0492)
Log(OPEN)	0.0857 ** (1.9728)	−0.0176 (−1.1628)	0.1130 *** (2.6721)	−0.0157 (−1.0903)
σ^2	0.0163	0.0002	0.0148	0.0001
R^2	0.8821	0.9201	0.8914	0.8447
Adjusted R^2	0.8771	0.9173	0.8877	0.8393
Durbin-Watson	2.1067	1.8217	2.2265	2.0198
Log-likelihood	99.3054	438.8255	106.1706	450.8258
LM spatial Lag	0.0480 (0.827)	0.1109 (0.739)	0.0084 (0.927)	0.3449 (0.557)
Robust LM spatial Lag	0.4311 (0.511)	0.0212 (0.884)	1.4950 (0.221)	2.2990 (0.129)
LM spatial error	2.9681 (0.085)	0.8291 (0.363)	5.1309 (0.024)	1.3763 (0.241)
Robust LM spatial error	3.3513 (0.067)	0.7394 (0.390)	6.6176 (0.010)	3.3304 (0.068)
Joint test of significance LR	Fixed-effects	Statistics	df	P-value
	Individual-fixed effects	689.3104	30	0.0000
	Time-fixed effects	24.0007	5	0.0002

Note: t or z-values are in the parentheses. P-values in the parentheses under the coefficients of the LM tests. * represents significance at 10%, ** 5%, and *** 1% respectively.

3.2. *Industrial Carbon Emissions SDPDM*

Three Durbin models are considered: (1) individual and time fixed effects (*Model* 3); (2) both time and individual effects, with bias correction borrowed from (*Model* 4) [23]; and (3) individual random effect and time fixed effect (*Model* 5). The estimation results using Log (*CS*) and Log (*CI*) as dependent variables are reported in Tables 3 and 4 respectively. In Model3 and Model 4, the coefficients of *SI*, *IE*, *EI*, *ECS*, *EP*, *OPEN*, and σ^2 changed slightly after bias correction, and the coefficients of the spatially lagged dependent and independent variables are also sensitive to bias correction. Thus, bias correction is necessary for the spatial Durbin model with both individual and time fixed effects. The SDPDM has two hypotheses: H_0: $\theta = 0$ and H_0:

$\theta + \delta\beta = 0$; rejection of both indicates that the SDPDM fits the data best. Both Wald and LR tests reject the two hypotheses, suggesting that neither the SLPDM nor the SEPDM is appropriate. We thus opt for the SDPDM. Meanwhile, the Hausman test lends support to Model 4, the coefficients in Model 4 align with our expectations, and its goodness of fit is greater than those of the alternatives.

Table 2. LM test for Model 2 (dependent variable: Log (CI)).

Variable	Pooled OLS	Individual Fixed Effects	Time Fixed Effect	Individual and Time Fixed Effects
Constant	−1.5254 *** (−2.8568)	–	–	–
Log(SI)	0.0705 *** (3.4491)	−0.1369 *** (−4.7811)	0.0771 *** (3.7957)	0.0277 (0.2544)
Log(IE)	0.1595 ***(3.0865)	0.0182 (0.3685)	0.1973 *** (3.6527)	0.0247 (0.4986)
Log(EI)	1.1555 ***(26.8095)	0.8343 *** (22.9802)	1.1596 ***(27.2301)	0.8428 ***(23.0986)
Log(ECS)	0.1957 ***(4.3232)	0.0314 (1.1439)	0.2158 *** (4.7543)	0.0640 ** (2.3528)
Log(EP)	0.2765(1.4439)	0.1033 ** (2.4443)	0.2896 (0.7723)	−0.0041 (−0.0498)
Log(OPEN)	0.0063 (0.2940)	−0.0175 (−1.1607)	0.0103 (0.4765)	−0.0157 (−1.0880)
σ^2	0.0040	0.0002	0.0038	0.0001
R^2	0.9411	0.9309	0.9418	0.8007
Adjusted R^2	0.9386	0.9285	0.9398	0.7937
Durbin-Watson	2.0015	1.8220	2.0977	2.0200
Log-likelihood	204.4955	438.8169	207.4130	450.8174
LM spatial Lag	7.9070 (0.005)	2.6476 (0.104)	9.1879 (0.002)	1.3206 (0.250)
Robust LM spatial Lag	9.3611 (0.002)	1.8238 (0.177)	9.6301 (0.002)	5.3170 (0.021)
LM spatial error	0.0046 (0.946)	0.8303 (0.362)	0.1920 (0.661)	1.3742 (0.241)
Robust LM spatial error	1.4587 (0.227)	0.0065 (0.936)	0.6342 (0.426)	5.3706 (0.020)
Joint test of significance (LR)	Fixed-effects	Statistics	df	P-value
	Individual fixed effects	486.8087	30	0.0000
	Time fixed effects	24.0011	5	0.0002

Note: t or z-values are in the parentheses. P-values in the parentheses under the coefficients of the LM tests. * represents significance at 10%, ** 5%, and *** 1% respectively.

3.2.1. Discussion

Since Model 4 fits the data best, we discuss only the results of Model 4, as shown in the third columns of Tables 3 and 4 respectively, for the two dependent variables. The coefficients on most of the independent variables are significant and have the expected signs. The coefficients for the spatially lagged dependent variables are negative and insignificant in both models (line 2, column 3 in Tables 3 and 4), indicating that carbon emissions are correlated among regions. A region's industrial carbon emissions are estimated to decrease by 0.05% if its neighboring regions' scale and intensity of industrial carbon emissions increase by 1%. This indicates that under the Chinese government's 11th Five-Year Plan, the provinces managed to optimize industrial structures, innovate in industrial technology, and encourage energy savings and reduced emissions in industrial enterprises. Since the successful provinces 'exemplary achievements play an important role in adjusting economic structures nationwide, strategic planning of regional industrial carbon emissions is necessary to further optimize the industrial structure and reduce industrial carbon emissions. Below we discuss the impacts of the independent variables on regional industrial carbon emissions.

Table 3. Spatial Durbin model with both individual and time fixed effect (dependent variable: Log (CS)).

Variables	Individual and Time Fixed Effect	Individual and Time Fixed Effect (Bias Corrected)	Individual Random Effect and Time Fixed Effect
W*Log(CS)	−0.0930 (−0.9547)	−0.0484 (−0.4917)	−0.0810 (−0.8257)
Log(SI)	1.1804 *** (11.0696)	1.1860 *** (9.7884)	1.1141 *** (17.0034)
Log(IE)	0.0679 (1.4516)	0.0695 (1.3071)	0.0612 (1.1652)
Log(EI)	0.7823 *** (21.5919)	0.7809 *** (18.9591)	0.8461 *** (22.5372)
Log(ECS)	0.0557 ** (2.1734)	0.0561 * (1.9237)	0.0530 * (1.8354)
Log(EP)	−0.0100 (−0.1353)	−0.0078 (−0.0931)	−0.0133 (−0.1565)
Log(OPEN)	−0.0062 (−0.4694)	−0.0061 (−0.4066)	−0.0127 (−0.8649)
W*Log(SI)	−0.6954 ** (−2.5700)	−0.7466 ** (−2.4707)	0.0194 (0.1272)
W*Log(IE)	−0.2031 ** (−2.3151)	−0.2069 ** (−2.0754)	−0.1202 (−1.2563)
W*Log(EI)	0.1863 * (1.8330)	0.1502 (1.3894)	0.1395 (1.2992)
W*Log(ECS)	−0.1110 ** (−1.9561)	−0.1142 * (−1.7711)	−0.1014 (−1.5774)
W*Log(EP)	−0.1751 (−1.2580)	−0.1734 (−1.0952)	−0.2302 (−1.4528)
W*Log(OPEN)	0.0088 (0.5443)	0.0094 (0.5144)	0.0088 (0.4831)
teta	–	–	0.0369 *** (5.4795)
σ^2	0.0001	0.0002	0.0002
R^2	0.9991	0.9991	0.9988
Square correlation coefficient	0.8713	0.8714	0.8237
Log likelihood	465.4752	465.4752	345.2436
Wald test spatial Lag	31.9461 (0.000)	24.0161 (0.000)	14.8409 (0.0215)
LR test spatial Lag	28.9664 (0.000)	28.9664 (0.000)	NA
Wald test spatial error	30.7361 (0.000)	24.1035 (0.000)	14.7719 (0.0221)
LR test spatial error	27.7109 (0.000)	27.7109 (0.000)	NA
Hausman test	Statistics	df	P-value
	23.7970	13	0.0330

Note: t or z-values are in the parentheses. P-values in the parentheses under the coefficients of the LM and Wald tests. * represents significance at 10%, ** 5%, and *** 1% respectively.

Table 4. Spatial Durbin model with both individual and time fixed effect (dependent variable: Log (CI)).

Variables	Individual and Time Fixed Effect	Individual and Time Fixed Effect (Bias Corrected)	Individual Random Effect and Time Fixed Effect
W*Log(CI)	−0.0920 (−0.9407)	−0.0470 (−0.4761)	−0.0520 (−0.5448)
Log(SI)	0.1805 * (1.6928)	0.1861 (1.5360)	−0.0061 (−0.1588)
Log(IE)	0.0680 (1.4530)	0.0696 (1.3086)	0.0457 (0.8805)
Log(EI)	0.7823 *** (21.5900)	0.7809 *** (18.9574)	0.8992 *** (25.4063)
Log(ECS)	0.0558 ** (2.1757)	0.0562 * (1.9259)	0.0559 * (1.9479)
Log(EP)	−0.0099 (−0.1343)	−0.0077 (−0.0920)	−0.0068 (−0.0778)
Log(OPEN)	−0.0061 (−0.4668)	−0.0060 (−0.4043)	−0.0236 (−1.6327)
W*Log(SI)	−0.7884 *** (−3.1937)	−0.7950 *** (−2.8319)	0.0076 (0.1216)
W*Log(IE)	−0.2033 ** (−2.3175)	−0.2072 ** (−2.0778)	−0.0896 (−0.9588)
W*Log(EI)	0.1854 * (1.8193)	0.1489 (1.3753)	0.1037 (0.9703)
W*Log(ECS)	−0.1113 ** (−1.9599)	−0.1145 * (−1.7750)	−0.1056 * (−1.6609)
W*Log(EP)	−0.1750 (−1.2573)	−0.1733 (−1.0945)	−0.2225 (−1.3693)
W*Log(OPEN)	0.0088 (0.5440)	0.0094 (0.5144)	0.0018 (0.0974)
teta	–	–	0.0778 *** (5.4872)
σ^2	0.0001	0.0002	0.0002
R^2	0.9982	0.9982	0.9974
Square correlation coefficient	0.8348	0.8350	0.9270
Log likelihood	465.4691	465.4691	363.4027
Wald test spatial Lag	30.3274 (0.000)	22.6643 (0.000)	10.4796 (0.1059)
LR test spatial Lag	28.0752 (0.000)	28.0752 (0.000)	NA
Wald test spatial error	30.7411 (0.000)	24.1089 (0.000)	11.0656 (0.0864)
LR test spatial error	27.7196 (0.000)	27.7196 (0.000)	NA
Hausman test	Statistics	df.	P-value
	27.5757	13	0.0104

Note: t or z-values are in the parentheses. P-values in the parentheses under the coefficients of the LM and Wald tests. * represents significance at 10%, ** 5%, and *** 1% respectively.

3.2.2. Scale of Investment

The significant and positive coefficients on *SI* (line 3, Table 3) suggest that industrial investment contributes considerably to China's carbon emissions. The reason is that the dramatically increased energy consumption of the fast-expanding industrial sector has led to rapid growth in carbon emissions. Table 4 shows that *SI* positively impacts carbon intensity, but the relationship is statistically insignificant, suggesting that an increase in *SI* does not boost China's industrial carbon emissions intensity. Given the emissions reduction target of the industrial sector, it should take full responsibility for saving energy and reducing emissions, as it consumes the most energy and resources while emitting the most pollutants. The provinces have set targets for energy consumption per unit of industrial value-added, which is intended to lead to reduced industrial carbon emissions intensity. The coefficient on the spatially lagged variable *W**Log (*SI*) significantly and positively impacts both the level and intensity of carbon emissions, suggesting that there are carbon emissions spillover effects from the level of investment. As provinces prioritize industrial carbon emissions reductions, investments flow to low-carbon industrial enterprises.

3.2.3. Industrial Economy

China is in an era of rapid industrialization, causing swift growth of both its industrial economy and its CO_2 emissions. The positive but insignificant impacts of industrial value-added per unit of labor (*IE*) on the level and intensity of carbon emissions, as found in this estimation, suggest that China's growth in industrial production is not the main driver of carbon emissions growth. Although industrial energy consumption and industrial carbon emissions are increasing, industrial carbon emissions intensity is decreasing. Thus, carbon emissions reductions are unlikely to be achieved by controlling industrial development but rather by adjusting the structure of industrial energy consumption, optimizing industrial structure, and following a low-carbon development path. The negative and significant coefficient on the spatially lagged variable *W**Log (*IE*) shows that the industrial economy in one region can affect carbon emissions of other regions through spillover effects. This is explained by the fact that the adjacent provinces compete to develop low-carbon industries. These resource-saving and environmentally friendly industries are aimed at creating a sustainable and low-carbon economy.

3.2.4. Energy Intensity

Decreased industrial energy intensity arises mainly from technological innovation, which affects both the scale and intensity of industrial carbon emissions. We found a significant and positive impact of energy intensity (*EI)* on carbon emissions levels and intensity, indicating that the industrial low-carbon technologies are advancing and energy consumption per unit of industrial value-added is decreasing. This is beneficial for lowering carbon emissions and aligns with expectations. Recently, with the progress of industrial technologies, optimization of industrial structures, and efforts at energy savings and emissions reductions, China's industrial sector energy intensity is decreasing. This lowering of energy intensity is an important strategy for reducing China's industrial carbon emissions. No spatial spillover effects of energy intensity on carbon emissions are found: the coefficient on *W**Log (*EI*) is positive but insignificant. This is because carbon emissions technology is hard to imitate across regions. The development of carbon emissions-reduction technologies is largely influenced by economic development and innovation capability of a given region. The large differences in research capabilities across regions block diffusion of these technologies.

3.2.5. Energy Consumption Structure

The rapid development of China's industrial economy has created need to optimize its industry consumption structure to slow long-term growth in carbon emissions. However, given resource constraints, a coal-intensive energy consumption structure is likely to remain in place for a long time. In this model estimation, energy consumption structure has a positive and significant effect on both the scale and intensity of carbon emissions, indicating that an increase in the coal intensity of total energy consumption has a negative effect on industrial carbon emissions reductions. Thus, increasing the ratio of non-fossil energies, such as wind, nuclear, and solar, to total energy consumption will support a reduction in the scale and intensity of carbon emissions. Furthermore, energy consumption structure has a spillover effect: the coefficient on *W**Log (ECS) has a negative and significant impact on the intensity of carbon emissions. This spillover is due to competition among adjacent regions to save energy and reduce emissions. All provinces are trying to optimize their industrial energy consumption structures and develop green technologies, such as non-fossil energies, so measures should be taken to let the availability of new and renewable energy types play an important role in adjusting the structure of industrial energy consumption.

3.2.6. Energy Price

In this paper, the energy price refers to the producer price index for manufactured products. In theory, factor prices are negatively correlated with factor demands. The energy price has a negative but insignificant effect on carbon emissions levels and intensity, indicating that the role of energy prices in shaping China's industrial carbon emissions is statistically insignificant. To meet the increasing industrial demand for energy (and accompanying growth in carbon emissions), China needs to scale upmarket supervision and inspection and enforce a price-forming mechanism for important energy types, such as electricity and gas. Under such a mechanism, energy prices can act as signals of energy demand and supply. When energy price in one region increases, producers in that region would be expected to purchase energy from adjacent regions, causing increases in total carbon emissions. However, we find no evidence that spillover effects exist for energy prices, probably because the release of regional energy price information is inefficient. When energy prices rise in the region, especially regions that rely on industrial development, it will purchase energy from the neighboring regions, resulting a slightly increase of industrial carbon emissions. Therefore, China should improve and perfect the price information release system and stabilize social expectations.

3.2.7. Openness

Accelerating urbanization has pushed up energy consumption and CO_2 emissions. In recent years, the need for a low-carbon economy has necessitated importing advanced technologies from abroad, and these foreign imports are also expected to be environmentally sustainable. Thus, openness should decrease carbon emissions. However, this hypothesis is rejected in the estimation results: the coefficient on *OPEN* is negative but insignificant, as is that on *W**Log (*OPEN*), suggesting that *OPEN* does not have spillover effects. It shows that the changes of opening level in China has a limited negative effect on China's industrial carbon emissions, but the rise in opening level can still help to reduce carbon emissions which make it of great significance. China buy low carbon technology and clean energy from abroad, and attract low carbon foreign investment. FDI technology spillover effect can reduce China's carbon emissions scale and intensity. These methods both play a positive role in reduce China's carbon emissions scale and intensity. In the Durbin Model which use natural logarithm Log *(CS)* (CS: carbon emission scale) and Log *(CI)* (CI: carbon emission intensity) as the explanatory variable, the opening level spatial lag estimated coefficient is negative, but it is not significant. This indicating that there is a limited spatial spillover effect of opening level. The reason is mainly reflected in the spillover effect of competition between neighboring regions, while they open the gate to the world. Regions have accelerated the pace of industrial opening, this laid a solid foundation for China to stick to the low-carbon economic development.

3.3. Spillover Effect of Regional Industrial Carbon Emissions

It is common to use point estimates from one or more spatial regressions to test the existence of spillover effects. Lesage and Pace [2] (pp. 19–44), however, argue that using point estimates from multiple spatial regressions will bias prediction of spillover effects. They further decompose spillover effects into direct and indirect effects.

The SDPDM model Equation (13) can be rearranged as Equations (14) and (15).

$$\begin{aligned} Y_{it} &= \delta \sum_{j=1}^{N} Y_{it} + \alpha + \sum_{i=1}^{m} \beta_i X_{it} + \sum_{J=1}^{N} w_{ij} X_{ijt}\theta + \mu_i + \lambda_t + \varepsilon_{it} \\ Y_{it} &= \delta W Y_{it} + X_{it}\beta + W X_{it}\theta + +\mu_i + \lambda_t + \varepsilon_{it} \end{aligned} \quad (14)$$

$$\begin{aligned} Y_{it} &= (I-\delta W)^{-1}(X_{it}\beta + W X_{it}\theta) + (I-\delta W)^{-1}\mu_i \\ &\quad + (I-\delta W)^{-1}\lambda_t + (I-\delta W)^{-1}\varepsilon_{it} \end{aligned} \quad (15)$$

where Y_{it} is the dependent variable for region i at time t, X_{it} is a vector of independent variables of region i at time t, α is the constant term, θ is similar to β, which is a $K \times 1$

vector of coefficients, μ_i is an individual-fixed effect, and λ_t is a time-fixed effect. Taking partial derivatives of the kth independent variable X on both sides, we obtain:

$$\begin{bmatrix} \frac{\partial Y}{\partial \chi_{ik}} & \frac{\partial Y}{\partial \chi_{Nk}} \end{bmatrix} = \begin{bmatrix} \frac{\partial y_1}{\partial \chi_{ik}} & \frac{\partial y_1}{\partial \chi_{ik}} \\ \frac{\partial y_N}{\partial \chi_{ik}} & \frac{\partial y_N}{\partial \chi_{ik}} \end{bmatrix} = (I - \delta W)^{-1} \begin{bmatrix} \beta_k & w_{12}\theta_k & w_{1N}\theta_k \\ w_{21}\theta_k & \beta_k & w_{2N}\theta_k \\ w_N\theta_k & w_{N2}\theta_k & \beta_k \end{bmatrix} \tag{16}$$

where W_{ij} is the (i, j) element of the matrix W. The direct effect is defined as the sum of the diagonal elements in the right matrix while the indirect effect is defined as the average of all the non-diagonal elements (Lesage and Pace, 2009). Identifying the direct and indirect effects via this method has drawbacks in that calculating $(I - \delta W)^{-1}$ is time-consuming. To solve this, Lesage and Pace [2] (pp. 19–44) propose another method, specified in Equation (17).

$$(I - \delta W)^{-1} = I + \delta W + \delta^2 W^2 + \delta^3 W^3 \tag{17}$$

The models estimating the direct and indirect effects in Equation (16) are denoted as Method 1 and Method 2 using Equation (17). Table 5 displays the direct and indirect effects estimated through Method 1, for Model 4 with Log (*CS*) as the dependent variable. The results differ slightly between Method 1 and Method 2, but all variables have direct and indirect effects. *SI*, *EI*, and *ECS* positively and significantly impact the scale of carbon emissions, while *IE*, *EP*, and *OPEN* insignificantly impact it, although their signs are as expected.

Table 6 displays the direct and indirect effect estimates for Model 4 with Log (*CI*) as the dependent variable. The results do not differ from the Model 1 results. All variables have direct and indirect effects, direct effects of *EI* and *ECS* on intensity of industrial carbon emissions are positive, and the direct effects of *SI*, *IE*, *EP*, and *OPEN* are all insignificant, although the signs are as expected. On the other hand, *SI*, *IE*, and *ECS* have significant indirect effects. We can thus argue that spillover effects do exist. Spillover effects imply that the independent variables affect the dependent variable via the spatially lagged variables. The differences indirect and indirect effects are substantial for all the control variables, indicating that failing to explicitly model the spatial correlation would lead to estimation bias.

Table 5. Direct, indirect and total effects of Spatial Durbin model (Method1, dependent variable: Log (*CS*)).

Variable	Direct Effect	Indirect Effect	Total Effect
Log(SI)	1.2012 *** (9.8339)	−0.7645 *** (−2.6558)	0.4367 (1.3678)
Log(IE)	0.0700 (1.3308)	−0.2086 ** (−2.1793)	−0.1386 (−1.4935)
Log(EI)	0.7803 *** (18.5641)	0.1058 (1.4353)	0.8861 *** (11.4843)
Log(ECS)	0.0576 ** (1.9643)	−0.1193 ** (−1.8808)	−0.0616 (−0.9326)
Log(EP)	−0.0106 (−0.1193)	−0.1709 (−1.1175)	−0.1815 (−1.0960)
Log(OPEN)	−0.0057 (−0.3914)	0.0094 (0.5229)	0.0037 (0.1572)

Note: * represents significance at 10%, ** 5%, and *** 1% respectively.

Table 6. Direct, indirect and total effects of Spatial Durbin model (Method1, dependent variable: Log (*CI*)).

Variable	Direct Effect	Indirect Effect	Total Effect
Log(SI)	0.1975 (1.6262)	−0.7795 *** (−2.9212)	−0.5820 ** (−1.9978)
Log(IE)	0.0717 (1.3607)	−0.2048 ** (−2.0105)	−0.1331 (−1.3790)
Log(EI)	0.7790 *** (18.7878)	0.1094 (1.4981)	0.8884 *** (11.7796)
Log(ECS)	0.0576 ** (1.9648)	−0.1138 * (−1.7426)	−0.0562 (−0.8170)
Log(EP)	−0.0033 (−0.0382)	−0.1658 (−1.0350)	−0.1692 (−1.0068)
Log(OPEN)	−0.0061 (−0.4183)	0.0098 (0.5407)	0.0037 (0.1646)

Note: * represents significance at 10%, ** 5%, and *** 1% respectively.

4. Conclusions

In this paper, an expanded Stochastic Impacts by Regression on Population, Affluence, and Technology (STIRPAT) model was conceptualized to investigate the factors driving regional industrial carbon emissions in China. Based on provincial panel data from 2006 to 2010, a spatial Durbin model was applied to test the impacts of these contributing factors on the scale and intensity of industrial carbon emissions, examine the spatial correlation of industrial carbon emissions among provinces, and consider the determinants' potential spillover effects. In addition, a geographically was applied to investigate the regional heterogeneity in the impacts of the driving factors on the scale and intensity of industrial carbon emissions. The main conclusions from these analyses are as follows.

First, most of the driving factors significantly impact the scale and intensity of carbon emissions, and their signs align with expectations. The coefficients on the spatially lagged dependent variables are estimated to be negative but insignificant, suggesting that industrial carbon emissions are highly correlated among regions.

As a result, the exemplar role of the low-carbon-emissions provinces is of great importance in fostering nationwide low-carbon economy.

Second, the driving factors impact carbon emission both directly and indirectly. Considering the indirect effects, all of the variables except for energy intensity, energy price, and openness have significant indirect effects. We can thus argue that spillover effects do exist: the independent variables affect the dependent variable via the spatially lagged variables. Using the transitional panel data model, which assumes the indirect effect to be zero, will bias the estimation. The differences in the direct and indirect effects of the control variables are substantial, indicating that ignoring spatial correlation would lead to estimation bias.

Third, the impacts of the influencing factors and their spillover effects vary across provinces. The results show substantial variation in the coefficients on the spatially lagged dependent variables of the control variables. The driving factors' influences on industrial carbon emissions and their spillover effects suggest that regional industrial spatial correlation, heterogeneity, and externalities must be taken into account during policy formulation. Considering the fact that industrial carbon emissions differ across regions, we propose the following strategies to reduce industrial carbon emissions: (1) a moderate increase in investment in low-carbon industries; (2) optimization of the structure of industrial energy consumption; (3) optimization of industrial structures; (4) innovation in industrial technology; and (5) effective coordination of planning related to regional industrial carbon emissions.

Acknowledgments: We thank the financial support provided by the National science foundation of China (71473242, 41461118), the National Basic Research 973 Program (2012CB955700), and the Strategic Priority Research Program "Climate Change: Carbon Budget and Related Issues" of the Chinese Academy of Sciences (CAS) (XDA05140300). This paper is also partly supported by IDE, JETRO—ICCS, Tsinghua Univ.—IGLCE, Nagoya Univ.—EAEF, Xian Jiaotong Univ.—CBS, Fudan Univ. international joint project "Tracing China's Greenhouse Gas and Pollution Sources in Global Value Chains (GVCs): Regional and Firm Heterogeneity", 2015–2016.

Author Contributions: Yu Liu performed research, analyzed the data and wrote the paper. Hongwei Xiao contributed to the conceptual framework of the methodology and interpreted the results, drafted and revised the manuscript. Ning Zhang performed the calculations and analyzed the data.

Conflicts of Interest: The authors declare no conflict of interest.

References

1. Shi, A. The impact of population pressure on global carbon dioxide emissions, 1975–1996: Evidence from pooled cross-country data. *Ecol. Econ.* **2003**, *44*, 29–42.
2. LeSage, J.P.; Pace, R.K. An introduction to spatial econometrics. *Revue D'économie Industrielle* **2009**, *123*, 19–44.
3. Huang, Y.; Wang, L.; Li, G.; Jiang, D. An empirical analysis for the environmental Kuznets curve in China based on spatial penal models. *South China J. Econ.* **2009**, *1*, 59–68.

4. Xu, H.; Deng, Y. Does FDI lead to environment pollution in China? A spatial econometric analysis based on provincial panel data. *Manag. World.* **2012**, *2*, 30–43.
5. Wang, H.; Teng, Y. Economic Development and Environmental Pollution Space Panel Data Analysis. *Technol. Econ. Manag. Res.* **2013**, *2*, 85–89.
6. Liu, Y.; Xiao, H.; Zikhali, P.; Lv, Y. Carbon Emissions in China: A Spatial Econometric Analysis at the Regional Level. *Sustainability* **2014**, *6*, 6005–6023.
7. Wang, J.; Li, L.; Zhang, F.; Xu, Q. Carbon Emissions Abatement Cost in China: Provincial Panel Data Analysis. *Sustainability* **2014**, *6*, 2584–2600.
8. Zhu, P.; Yuan, J.; Zen, W. Analysis of Chinese Industry Environmental Kuznets Curve—Empirical Study Based on Spatial Panel Model. *China Ind. Econ.* **2010**, *6*, 65–74.
9. Wang, L.; Guan, J.; Zhang, J. Environmental Pollution and Economic Growth in China: A Dynamic Spatial Panel Data Model. *Geogr. Res.* **2010**, *30*, 818–825.
10. Yao, Y.; Ni, Q. The impact of foreign direct investment on carbon intensity empirical study based on Chinese provincial dynamic panel data. *Econ. Geogr.* **2011**, *31*, 1432–1438.
11. Yu, Y. Innovation Cluster, Government Support and the Technological Innovation Efficiency: Based on Spatial Econometrics of Panel Data with Provincial Data. *China Econ. Rev.* **2011**, *2*, 93–101.
12. Wei, C.; Ni, J.; Shen, M. China's energy inefficiency: A cross-country comparison. *Soc. Sci. J.* **2011**, *48*, 478–488.
13. Wei, C.; Ni, J.; Du, L. Regional allocation of carbon dioxide abatement in China. *China Econ. Rev.* **2012**, *23*, 552–565.
14. Choi, Y.; Zhang, N.; Zhou, P. Efficiency and abatement costs of energy-related CO2 emissions in China: A slacks-based efficiency measure. *Appl. Energ.* **2012**, *98*, 198–208.
15. Ehrlich, P.R.; Holdren, J.P. Impact of population growth. *Science* **1971**, *171*, 1212–1217. PubMed]
16. Harrison, P. *Inside the Third World: The Anatomy of Poverty*; Penguin Books: New York, NY, USA, 1981.
17. Raskin, P.D. Methods for estimating the population contribution to environmental change. *Ecol. Econ.* **1995**, *15*, 225–233.
18. York, R.; Rosa, E.A.; Dietz, T. Bridging environmental science with environmental policy: Plasticity of population, affluence, and technology. *Soc. Sci. Q.* **2002**, *83*, 18–34.
19. Rosa, E.A.; York, R.; Dietz, T. Tracking the anthropogenic drivers of ecological impacts. *Ambio. J. Hum. Environ.* **2004**, *33*, 509–512.
20. Dietz, T.; Rosa, E.A. Rethinking the environmental impacts of population, affluence and technology. *Hum. Ecol. Rev.* **1994**, *1*, 277–300.
21. York, R.; Rosa, E.A.; Dietz, T. STIRPAT, IPAT and Impact: analytic tools for unpacking the driving forces of environmental impacts. *Ecol. Econ.* **2003**, *46*, 351–365.
22. CSY. *China Energy Statistical Yearbook*; China Statistics Press: Beijing, China, 2007–2011.
23. Lee, L.; Yu, J. Estimation of spatial autoregressive panel data models with fixed effects. *J. Econ.* **2010**, *154*, 165–185.

Regional Competition, Heterogeneous Factors and Pollution Intensity in China: A Spatial Econometric Analysis

Jianhuan Huang and Jiejin Xia

Abstract: Regional competition may play an important role in the balance of environmental protection and economic growth. However, it is a pending issue of whether the competition among Chinese local governments leads to a race to black development or green development. This paper aims to explore the strategic interactions in provincial development in terms of an environment-economic indicator, *i.e.*, the pollution intensity in China from 2000 to 2013. We divide four predominant industrial pollutants into two groups according to whether the pollutant is regulated, and then test the strategic interactions among regions based on the spatial lag term by employing the spatial Durbin model. The results show that the heterogeneous factors, such as various pollutants and regional difference, may give rise to diversified competition strategies. We find that the "race to black development" hypothesis is not supported at the national level, and the "race to green development" hypothesis is established in the developed eastern regions only in terms of the regulated industrial pollutants. We also detect how pollution intensity is influenced by the direct and spatial spillover effects of environmental regulation and find that environmental legislation has been effective in reducing regulated pollutants' pollution intensity, while the effects of environmental staff and investment are weak. Finally, some policy suggestions are discussed.

Reprinted from *Sustainability*. Cite as: Huang, J.; Xia, J. Regional Competition, Heterogeneous Factors and Pollution Intensity in China: A Spatial Econometric Analysis. *Sustainability* **2016**, *8*, 171.

1. Introduction

China, the largest developing country in the world, is confronted with a dilemma of economic growth and environmental protection right now [1]. China has achieved important economic development for more than 30 years (since the 1980s), with an annual economic growth rate of approximately 10%. However, its annual GDP (gross domestic product) growth decreased to nearly 7% in 2015 with the background of the global economic recession, which indicates the enormous pressure to maintain China's rapid economic growth. The pressure of environmental protection is also great for China, which has been among the top of all countries in terms of the scale of pollutant emission and has been struggling with increasingly severe environmental pollution problems, such as heavy haze, in recent years. Lowering the environmental costs of economic growth for achieving balance between environmental protection and economic growth is an urgent and crucial issue.

Interprovincial competition is believed to be one of the most essential driving forces for China's economic growth, as proposed by Li and Zhou [2], based on the promotion tournament model. However, on the other hand, the competition among provinces may lead to a serious pollution problem and high environment costs in the early developmental stage [3]. Pollution abatement in China is dominated by provincial-level implementation with enforcement standards created by the central government. If the development of a region and the assessment for provincial officials are economically oriented, local governments may ease environmental regulation stringency to attract more investment for economic growth in the competition, *i.e.*, developing in a "dirty" or "black" way. Fortunately, the Chinese central government has realized the importance of green development and proposed a scientific development concept for regional development in 2003, gradually replacing the economically oriented assessment system for local officials with an environment-economic–oriented one; this change may reverse the direction of regional competition. The environment-economic–oriented assessment system may induce the subnational regions to compete for economic growth with lower environmental costs and thus develop in a greener way, as indicated by some studies [4,5].

However, China is a country with many heterogeneous provinces, so the results of competition in different areas may depend on the regional development level. Moreover, there are many kinds of heterogeneous pollutants that are regulated with different degrees of stringency. The local governments may treat them differently and only compete in the abatement of the regulated pollutants, which implies a selective competition mechanism (see the explanation in Subsection 2.2). Using provincial panel data since 2000 and an environment-economic indicator, *i.e.*, pollution intensity, this paper examines the relationship between regional competition and environmental costs of economic growth with consideration of the heterogeneous factors in China, as well as the effects of environmental regulation. Based on the test procedures, we choose the spatial Durbin model and explore the direct effects and spatial spillover effects of related variables. This paper contributes to the literature in three ways. First, we study regional competition in a comprehensive way which simultaneously includes economic and environmental regulatory competition. By incorporating these two kinds of competition, the research may provide comprehensive and meaningful implications aimed at alleviating the conflict of economic growth and environmental protection in China. By extending the "race to the bottom" (RTB) and "race to the top" (RTT) theories about environmental regulation, we propose that regional governments may engage in a race to black development (*RBD*) or a race to green development (*RGD*). Therefore, we investigate the situation in China empirically in terms of variation of pollution intensity. Second, we find that the heterogeneous factors, such as various pollutants and regional difference, may lead to diverse results. We differentiate four industrial pollutants into two groups. The Chinese central government set up obligatory emission reduction targets only for the first group, which contains two kinds of pollutants, *i.e.*, SO_2 (sulfur dioxide) and COD (chemical oxygen demand), but did not establish obligatory targets for the others (waste water and waste gas). Thus, the local regulators and firms treat these pollutants differently, which leads to different environmental costs of economic growth and different results for regional competition. The empirical results are also different when we separate the samples into developed regions and developing ones. In the end, we find that the effectiveness of environmental regulation on the pollution intensity is also related to these heterogeneous factors. We identify regulation intensity variables from the perspectives of environmental legislation, staff and capital investment and then find some surprising results associated with the regulation status of the pollutants. Our research may help to better understand whether interactions between regions lead to *RBD* or *RGD* when considering regional and pollutant heterogeneities and provide insight into how environmental regulation could be fully explored to reduce pollution.

The remainder of this paper is organized as follows: In Section 2, we propose the theories of "*RBD*" and "*RGD*" about regional competition, then adding the heterogeneous factors of China into the theoretical analysis to get the hypotheses to be verified in this study. Section 3 describes the empirical methodology and data. Section 4 reports empirical results, including changes in provincial environmental costs and regulation intensity, the estimated spatial econometric models and a comparison of the eastern and western areas. In Section 5, we offer some conclusions and policy implications.

2. Theoretical Background

2.1. From RTB (RTT) to RBD (RGD)

The extant literature has proven that regional governments interact with each other in both developed and developing countries [2–23]. The policies and behavior of a local government are correlated with those of neighboring or benchmark jurisdictions, while interactions often take the form of competition in economic growth, environmental regulation, *etc.* Many empirical studies on the economic competition among governments have been presented [6–9]. Specifically in China, economic competition among provinces has been embedded in a centralized political system, and the role of political incentives for regional government officials has been emphasized in promoting local economic growth [2]. Evidence of Chinese provincial interaction can also be seen in other studies [10–13].

On the other hand, the environmental regulatory competition in local governments has been a fixture of studies for the past 20 years. At least four alternative theories have been proposed. Among them, the "race to the bottom" (RTB) theory suggests that local governments have incentives to ease environmental regulation to influence economic voting [14] or to meet business-oriented interest groups' demands [15]. However, the "race to the top" (RTT) theory argues that some regions may strive to reject those pollution-intensive industries by generating more stringent environmental standards [16] because some state-elected officials may view environmental protection as a more important factor in gaining the support of environmentally minded voters [17]. The third theory, proposed by Konisky [18,19], argues that neither RTB nor RTT entirely accounts for the complexity of governments' environmental regulatory behavior: Some states may race to the bottom, while others may race to the top in environmental regulatory competition. The above three theories assume the existence of environmental regulatory competition, whereas the fourth theory maintains that regulatory decisions are based solely on intrastate factors, namely local regulators not acting strategically [20]. The related empirical results are mixed so far. Using the US states as samples, Woods [21] provided empirical evidence for RTB, but Fredriksson and Milliment [22] and Konisky [18] did not find enough proof to support RTB. Li and Shen [23] and Yang *et al.* [3] showed evidence of RTB in China, while the RTT argument was also supported by some studies [4,5].

We realized that the extant studies about intergovernmental competition focus only on economic or environmental aspects. However, the environmental regulatory competition is strongly interrelated with economic competition. Relaxing environmental regulation is often stimulated by economic competition and is a potential instrument in regional economic competition in some cases [18]. In addition, environmental regulatory competition would influence the regional economic development as well. Nevertheless, economic growth is always a crucial subject for local governments, even if they execute stringent environmental regulation. Given the importance of balancing economic growth and environmental protection, providing comprehensive and meaningful implications appears to be imperative. Unfortunately, there have been few studies explicitly and simultaneously analyzing the economic and environmental regulatory competition among local governments via some ingenious devices.

A quick overview of developmental patterns may provide us new ideas for further research. With the increasing concern about global warming and environmental pollution, "green development" has become a new consensus around the world, which means lowering the environmental costs of economic development and maintaining balance between economic development and ecological environment [24]. In contrast, "black development", which has been conducted by many countries during industrialization, indicates that economic growth occurs with high environmental costs and huge resource consumption [25]. To achieve sustainable development, governments should transform the pattern of economic growth from black to green to realize a win-win solution for economic development and environmental protection [26]. Therefore, it is extremely necessary to investigate the economic and environmental regulatory competition among governments in a common framework, as environmental regulatory competition always appears to be the method for economic competition. Noting that both the RTB and RTT theories proposed by previous related literature emphasize competition only from the perspective of environmental regulation, we suggest extending the above theories into the "race to black development" (*RBD*) and "race to green development" (*RGD*) hypotheses, which combine economic and environmental factors in the research on intergovernmental competition. Further empirical research based on it may provide more comprehensive and fruitful results.

The *RBD* hypothesis suggests that interregional competition may give rise to increasingly higher environmental costs of economic development, which would imply a pattern of unsustainable ("black") development. The negative externality of environmental pollution provides governments the chance to gain economic growth while sharing transboundary pollution with their neighbors. Local governments may weaken environmental regulation stringency and try to attract or maintain "dirty" industries [18]. As a result, the environmental costs of economic growth will increase as time passes and be positively correlated with other competitive regions. In contrast, the *RGD* hypothesis suggests that environmental costs of economic development are lowered steadily under the influence of interregional competition, which would imply a pattern of sustainable ("green") development. Since regional regulators may endeavor to improve environmental performance to gain political promotion or votes [17,18], the corresponding competition among regions may cause more stringent environmental regulation, which will lead to development in a greener way. Consequently, environmental costs of economic development will decrease.

An interesting and important question is how the regional competition influences the environmental costs of economic growth and the developmental pattern in practice, *i.e.*, whether the *RBD* or *RGD* hypothesis can be supported by empirical tests. If either of them is confirmed, the related results will provide meaningful implications for the balance of economic growth and environmental protection. Specifically in China, pollution abatement is a combination of centralized standard-setting and provincial-level implementation. According to the promotion tournament model [27], China's local officials are engaged in competition for economic growth. In addition, the local environmental agencies are dominated by local governments. To ensure economic growth, reducing environmental protection efforts may be the choice for some local governments. Therefore, in the economic-oriented developmental stage, the regional competition may result in higher environmental costs and push regions toward black development as the *RBD* hypothesis suggests, while in the environment-economic–oriented developmental stage, local officials would engage in a race to lower the environmental costs of economic growth because their political promotion relies on the assessment of regional environment-economic performance. In this case, the inter-regional competition may lead to *RGD*, which is beneficial for the balance of economic growth and environmental protection. However, the *RBD* and *RGD* hypotheses are not fully explored in the existing literature.

2.2. Heterogeneous Factors and Hypotheses to Be Verified

When environmental regulation is dominated by local governments, relaxing environmental standards to absorb or maintain environmentally sensitive investment may be an applicable strategy due to the externality of transboundary pollution [21]. This may lead to the *RBD* generally if related governments behave in similar ways. However, heterogeneous factors may result in diverse regional competitive strategies, which would undermine the *RBD* or *RGD* hypothesis and give rise to diversified results. This paper focuses on two heterogeneous factors in China: regional difference and divergence of pollutants.

Beginning with regional difference, we presume that the *RBD* may occur in undeveloped regions, which are engaged in economic growth and do not have enough choices available for economic competition. Because the environmental pressure could be relatively low in the undeveloped regions, those local governments may tradeoff environmental protection for economic development and relax environmental standards, with the thought of treating after polluting. However, the developed regions have advantages in technological innovation, human capital, industrial agglomeration and market systems, which may offer them a number of competitive strategies. When facing environmental problems and economic competition, they may turn to alternative strategies, such as technological or capital incentive, instead of easing environmental protection efforts. To gain the support of environmentally sensitive voters or promotion from the central government, local officials in developed areas may even strengthen regulation, which may lead to the *RGB*. In sum, the logic of *RBD* or *RGD* relies on a homogeneous hypothesis.

Specifically in China, a large country with many heterogeneous provinces, neither the *RBD* nor the *RGD* might occur readily at the national level. As stated by Song *et al.* [28], the difference in geographical locations of distinct provinces might increase the disparity of abilities for economic development, which further affects their environmental performance and efficiency. For example, the average

GDP per capita of the Chinese eastern area (including 11 provincial-level regions) was approximately 3 (1.8) times as big as that of the western area (including 10 provincial-level regions) in 2000 (2013), and the average emission of industrial waste gas of the former was nearly 2(1.6) times as much as that of the latter in 2000 (2013). Provincial governments in the western regions may lower their environmental standards to absorb more investment because they lack enough capital and technology, whereas provincial governments in the eastern regions may diversify approaches to stimulating their economies because the strategy of relaxing environmental regulation is neither a unique nor a common choice. Therefore, it is difficult to prove the existence of the *RBD* at the national level. When environmental performance becomes one of the assessment indicators of local officials, governments may put more resources into pollutant abatement, which may cause the phenomenon known as the "California effect" [29], in which regulators tend to adopt the most stringent environmental standards and facilitate the regions to develop in a greener way. In a word, we propose the hypothesis that it is possible to find evidence of the *RBD* occurring in the undeveloped areas, while the *RGD* might occur in the developed areas in China.

Another heterogeneous factor is the divergence of regulation on the pollutants. Pollutants such as SO_2, COD and waste gas have different environmental regulation intensities. Because SO_2 and COD are under the stringent regulation of the central government, regional governments may pay more attention to controlling the pollution of related industries and firms. Then, the inter-regional competition may lead to lower environmental costs computed with the regulated pollutants, and provide evidence to support the *RGD* hypothesis. While other unregulated pollutants might receive scant attention from local governments, the *RBD* may be established when the environmental costs are calculated with the unregulated pollutants, because the local governments may select different competitive manners in terms of the regulation stringency of pollutants. We summarize it into the "selective competition mechanism", which indicates that the *RBG* and *RGD* are likely to be conditionally supported.

Regulation heterogeneity of pollutants has existed in China. China set incentive emission reduction targets of SO_2 (−10%) and COD (−10%) for the period of 2001–2005; however, the total emissions of industrial waste water and waste gas are not regulated [30]. Specifically, the obligatory emission reduction targets of SO_2 and COD were set for −10% during 2006–2010 [31]. As a result, we find that the emission volume of industrial waste gas increased approximately 380%, while the emission volume of SO_2 merely increased approximately 15% from 2000 to 2013. When considering the above heterogeneity, this paper infers that the establishment of *RBD* or *RGD* in China is associated with whether or not the pollutants (which are used to compute the environmental costs) are under the supervision of the central government. We examine the previously mentioned hypotheses in terms of pollution intensity, which is a simple and clear indicator used by previous studies [4].

3. Methodology, Variables and Data

3.1. Econometric Methodology

To test the interaction of regions empirically, previous studies such as Fredriksson and Millimet [22], Konisky [18] and Li *et al.* [5] focused on the spatial autoregressive coefficient ρ in Equation (1):

$$y_{it} = a + \rho \sum_{j=1, i \neq j}^{N} w_{ij} y_{jt} + \beta X_{it} + s_i + u_t + \varepsilon_{it} \quad (1)$$

where y_{it} is the dependent variable such as pollution intensity; a is the constant term; ρ is the spatial autoregressive coefficient and the absolute value of it is smaller than unity, *i.e.*, $|\rho| < 1$. A nonzero coefficient, *i.e.*, significant ρ, implies that a province's competitive strategy depends on the status of other provinces' strategies [18,22]. Specifically, a positive and significant ρ indicates similar strategies in the regional competition (RBD/RGD), while a negative or an insignificant positive ρ indicates diverse competitive strategies. w_{ij} denotes the weight describing the strategic interaction between the regions i and j; β is the coefficient vector of independent variables; s_i is regional fixed effects; and u_t is time fixed effects. X_{it} refers to independent variables and ε_{it} is the error term.

To obtain consistent and accurate estimation, some issues should be addressed here. The first is the endogeneity of $\sum_{j=1, i \neq j}^{N} w_{ij} y_{jt}$ in Equation (1). If the dependent variable (y) in region i is a function of y in region j, then y in region j should also be a function of y in region i [18]. OLS (ordinary least squares) estimates will be biased because of the above simultaneity problem. To address this issue, the instrumental variable approach is used by Konisky [18]. However, Zhang *et al.* [4] argued that it is very hard to find the proper instrumental variables, and the estimation results rely heavily on the choice of instrumental variables. Noting that the instrumental variable approach may lead to inaccurate estimation of ρ, they suggested the application of the spatial econometric method proposed by Anseilin [32] and used the maximum likelihood estimation to settle this endogeneity issue. Following LeSage and Pace [33,34] and Zhang *et al.* [4], this paper employs the ML (maximum likelihood) approach to estimate the spatial econometric model. It is worth noting that the ML estimation method requires an assumption about the normal distribution of the error term. If it is not satisfied, the estimated coefficients are still consistent in the ML estimation, whereas the estimated standard errors of coefficients may be biased. In such a case, the Quasi ML method should be applied to provide consistent and unbiased estimates. The Quasi ML and ML provide the same estimated coefficients, while their estimation of standard errors is different.

The second issue is spatial spillover effects of independent variables. The dependent variable may be influenced by the independent variables (X) of adjacent regions, which implies the spatial spillover effects and spatial Durbin model should be considered [33]. In the interaction models, the spatial weighted lag term of X should not be neglected, and the application of it may help to alleviate the auto-correlation of the error term [4]. This paper adds the spatial weighted lag term of X, *i.e.*, $\theta \sum_{j=1,i\neq j}^{N} w_{ij} X_{jt}$, to Equation (1) in the empirical study as follows:

$$y_{it} = a + \rho \sum\nolimits_{j=1,i\neq j}^{N} w_{ij} y_{jt} + \beta X_{it} + \theta \sum\nolimits_{j=1,i\neq j}^{N} w_{ij} X_{jt} + s_i + u_t + \varepsilon_{it} \quad (2)$$

where θ represents a parameter to be estimated. This paper uses a spatial adjacent matrix, *i.e.*, w_{ij}, if two provinces share a common border and zero otherwise. The spatial weights matrix W should be row-standardized, *i.e.*, row-stochastic, and the row-sums of W equal unity. The spatial and time-specific effects can be treated as fixed or random effects, which should be selected by the Hausman test. A spatial econometric model may provide a better estimation of the regulatory competition. However, there are several kinds of spatial models, such as the autoregressive model (SAR), the spatial error model (SEM) and the spatial Durbin model (SDM). SAR contains endogenous interaction effects (see Equation (1)), which are often used in the studies of regulatory competition [18,22], while the SEM contains interaction effects among the error terms. Both SAR and SEM do not include the weighted lag term of X mentioned above, while the SDM (see Equation (2)) does. Zhang *et al.* [4] used SDM to identify the interaction of Chinese provinces. A reasonable way to choose the proper model is based on econometric tests instead of subjective analyses. We select the proper panel model on the basis of the test procedures proposed by Elhorst [35] and find that SDM is the proper one (see the introduction of the specification test in Appendix).

Another endogeneity is related to the simultaneity of regulation intensity variables and the pollution intensity, which could be influenced by each other. To address this endogeneity, we take lags of regulation intensity variables, following the existing literature [36,37]. This method provides two advantages: (1) helping to ease the endogeneity to some extent and (2) considering the time-lag effects of environmental and economic policies.

The last issue is the choice of weights matrix. Some literature [38,39] used Bayesian MCMC (Markov chain Monte Carlo) methods to make the results more robust with respect to the choice of the weights matrix. Recently, LeSage and Pace [34] raised doubt about the view that estimates of spatial regression models are sensitive to the spatial weights, and pointed out that the view is of little theoretical basis. Following previous studies on the regional interaction [4,22], this paper uses the row-standardized spatial adjacent matrix for two reasons. The first one relies on the fact that regulatory competition originates from the externality of transboundary pollution, which is relatively obvious among adjacent regions. The second reason is that a region may be more inclined to compare itself with its neighbors than those not bordering it, which leads to competitive relationships between one region and its neighbors. The above analysis indicates an adjacent competition mechanism, which is an important theoretical basis of regional competition. We also use an inverse distance weights matrix in the robustness test and then get similar results.

3.2. *Variables and Data*

3.2.1. Dependent Variable and Strategic Interaction of Local Governments

To test the existence of strategic interactions between local governments, an intuitive method is to observe governments' behavior directly and use it as the dependent variable. For example, inspections and punitive actions [18], pollution abatement and control expenditures [22] have been regarded as proxy measures of environmental regulation. However, this direct approach is not reliable in this empirical study for two reasons: (1) the lack of data about environmental enforcement such as the annual number of inspections in China; and (2) the different measures implemented by different governments to relax environmental regulation for the sake of stimulating their economies. A local government may lower the frequency of inspection for pollution-intensive firms, while its neighbors may use other methods such as reducing the punishment stringency; therefore, the behavior and reactions of local governments will be diversified and asymmetric.

An alternative way is to observe the results of the governments' behavior, *i.e.*, the environmental and economic performance linked with their competition, such as the ratio of industrial added value over pollutant emission [4,40], and input-output efficiency [5]. If a region stimulates its economic growth by relaxing the environmental regulation, the emission volume of pollutants will increase at the same time. Assuming that the technological level and other determinants are constant, the pollutant emission may grow faster than the economic output when the region lowers the environmental standards to absorb "dirty" industries in the economic competition, which implies the increasing environmental costs of economic growth. If the neighboring regions adopt similar strategies and develop in a black way, their environmental costs will be elevated together and show positive correlations with each other. However, if the regions race to green development, the competition among them will lead to lower environmental costs. Following the above logic, we test the strategic interactions of local governments by focusing on the results of competitive behavior. Consistent with previous studies [4,40–43], we use the ratio of pollutant emission volume over economic output, *i.e.*, pollution intensity (PI), as the dependent variable (after a natural log transformation), which describes the environmental costs of per capita economic output as well as the coordination degree of the economy and environment. Note that a smaller PI is better.

As mentioned in Subsection 2.2, the empirical results may be diversified due to the heterogeneity of pollutant abatement. This paper observes two regulated industrial pollutants, SO_2 and COD, as well as two unregulated industrial pollutants, waste gas and waste water. Therefore, the PIs in this study include *SO_2_GDP*, *COD_GDP*, *Water_GDP* and *Gas_GDP*.

3.2.2. Independent Variables

Generally, measures of economic output, such as GDP, are functions of capital (K), labor (L) and other factors. Holdren and Ehrlich [44] suggested that the main determinants of environmental impact include the population, affluence and technology. According to Grossman and Krueger [45] and He [46], the pollutant emission is a function of the scale effect (SE), the technique effect (TE) and the composition effect (CE). The scale effect in China is paramount because achieving high economic growth has been the main political promotion criterion for local officials. The technique effect reflects the efforts from techniques that are likely to decrease the pollution intensity [46]. In addition, the composition effect denotes the association of a region's environmental performance and its industrial structure as well as other kinds of composition factors at a given production scale. Note that systematic factors such as environmental regulation (ER) are also important determinants of PI. Combining them, we have:

$$\mathrm{PI} = \mathrm{F}\left(\mathrm{ER}, \mathrm{SE}, \mathrm{TE}, \mathrm{CE}, \varnothing\right) \tag{3}$$

where ϕ refers to other determinants. Note that K and L can be included in Equation (3) (see additional explanation in Appendix).

A key point of this paper is to explore the relationship between environmental regulation (ER) and PI. We determine the main independent variables from each of the three programs in accordance with data availability and the previous literature. (1) According to Mu *et al.* [47], the legislation may be the legal systematic improvement to balance economic and environmental interest for gaining sustainable development. The number of local environmental bylaws and rules (*Sum_rule*) is a proxy for the developmental level of environmental legislation by a provincial unit. (2) The number of staff in the environmental protection departments per industrial firm above the designated size (*Regulator_perfirm*) represents the input on the human resources for environmental protection by a provincial unit. (3) The investment in the treatment of industrial pollution per capita industrial output measures the expenditure on the abatement by a provincial unit (*Inves_output*). As we explained in Section 2, a one-year lag of the above variables is used, considering the time-lag effects of environmental regulation. Moreover, the Pearson correlation test shows that the three variables are weakly correlated (see Appendix), so their multicollinearity should not be an issue.

Specifically, we consider the following control variables. (1) To proxy for informal environmental regulation, we employ the share of people with college degrees or above in the population (*S_college*), consistent with Yuan and Xie [48]. They noted that highly educated people may be more concerned with environmental protection and then exert more pressure on the government. (2) *L* is a measure of the scale effect, which is greatly correlated with the population. (3) Following Wang and Chen [49], we also measure the scale effect of economic activity by the GDP per square kilometer (*GDPSK*). (4) Many kinds of pollutants are generated during the process of consuming energy, and the energy intensity (energy consumption per capita GDP, *Engery_GDP*) represents the technical level of energy utilization [50], which is a proxy measure of the technique effect as well. (5) The industry structure will influence the environmental costs, especially for heavy industry which contributes considerably to the GDP while emitting lots of pollutants. We employ the share of heavy industrial output in the total industrial output (*S_HI*) to denote the composition effect. (6) The state-owned enterprises (SOEs) also play very important roles in China [51], which produced approximately 47% of the gross industrial output value from 2000 to 2013. Because the property right structure of these enterprises may influence environmental performance, we use the share of industrial output generated by SOEs (*S_SOE*) as a measure of the composition effect. (7) *K_L* is the capital-labor ratio representing both capital intensity and endowment structure. Previous studies have found a clear positive correlation between *K_L* and environmental performance [52,53]. (8) Moreover, we employ the energy composition, *i.e.*, the share of coal in the regional energy consumption (*S_coal*), to control its potential impacts on the environment because the extensive exploitation of coal, the largest energy source in China at present, has caused serious environmental pollution problems [54]. (9) Lastly, the ratio of total export-import volumes to GDP (*Open*) is used to measure the influence of regional openness on the local environment according to previous studies [55,56]. We anticipate that a higher degree of openness may cause better environmental performance by facilitating the regional technological innovation. Following some literature [10,13], we use the natural log transformation for most variables during the empirical analysis, except those ratio indicators such as *S_coal*, *S_SOE* and *S_HI*, *etc.*

3.2.3. Samples and Data Source

Our samples consist of a panel of 30 provincial regions covering the period of 2000–2013. These 30 provinces are divided into three groups. The eastern provinces include Beijing, Tianjin, Hebei, Liaoning, Shanghai, Jiangsu, Zhejiang, Fujian, Shandong, Guangdong, and Hainan. The western ones include Inner Mongolia,

Qinghai, Ningxia, Shaanxi, Sichuan, Chongqing, Guizhou, Yunnan, Guangxi, Gansu and Xinjiang. The central ones include Jilin, Heilongjiang, Shanxi, Henan, Anhui, Jiangxi, Hubei and Hunan. The firms referred to in our study are only those industrial enterprises above the designated size (the annual sales are not less than 5 million yuan before 2010, and not less than 20 million yuan since 2011. Tibet, Hong Kong, Macau and Taiwan are excluded due to data constraints. We collect the data from a number of official sources, including a few Chinese national yearbooks [57–61] and statistical yearbooks of each province. Table A1 in Appendix describes the data sources of each variable in detail, and all acronyms of variables are defined in the above paragraphs. Table 1 reports the descriptive statistics for all variables.

Table 1. Descriptive statistics.

Variable	Observations	Mean	Std. Dev.	Min	Max	Unit
SO_2_GDP	420	136.316	121.881	4.394	640.461	Tons per 10^8 yuan
COD_GDP	420	37.884	51.673	0.515	477.328	Tons per 10^8 yuan
Water_GDP	420	12.508	8.638	0.801	52.611	Tons per 10^4 yuan
Gas_GDP	420	2.062	1.566	0.297	18.129	m^3 per yuan
Sum_rule	420	43.333	31.046	6.000	191.000	Pieces
Regulator_perfirm	420	1.055	0.717	0.102	4.413	Persons per firm
Inves_output	420	0.186	0.155	0.008	1.083	%
S_college	420	7.533	5.285	1.64	39.46	%
L	420	2444.124	1630.484	262.100	6580.400	Persons
Engery_GDP	420	1.703	0.916	0.621	5.176	Tons per 10^4 yuan
S_HI	420	0.721	0.113	0.370	0.954	
S_SOE	420	0.467	0.206	0.107	0.891	
GDPSK	420	0.127	0.334	0.000	2.823	10^8 yuan per square kilometer
K_L	420	10.973	9.088	1.491	49.161	10^4 yuan per person
S_coal	420	0.555	0.142	0.135	0.825	
Open	420	0.339	0.430	0.040	1.840	

Notes: If prices are involved, the data is at 2000 constant price. For structure variables, the current year's price is adopted. Referring to Zhang *et al.* [62] and Huang *et al.* [63], we compute gross fixed capital by the perpetual inventory method.

4. Empirical Results

In this section, we report the results of the provincial interactive behavior, namely the environmental costs of economic growth, as well as the estimated econometric models. We also compare the cases of the eastern and western areas to examine the hypotheses proposed in Subsection 2.2.

4.1. PI and Environmental Regulation

To make it easy to compare, we set the pollution emission scale in 2000 equivalent to the value of 100 and calculate the emission index of each industrial pollutant from 2000 to 2013 in China, which shows that there exists large difference in the variation of the emissions of the four industrial pollutants during this period.

Figure 1a shows that the emission index of a kind of regulated pollutant (COD) was reduced substantially to approximately 50 in 2013. However, the emission index of SO_2 increased quickly to approximately 140 before 2006 and then was reduced gradually to 110 in 2013. The emission change of waste water is similar to that of SO_2. The emission volume of waste gas has continued increasing since 2000, and it grew faster than the GDP (see Figure 1b). The fast-growing emission of waste gas and its emission accumulation may be the main source of air pollution, which helps to explain why many cities in China have experienced severe haze in recent years. Although much progress has been achieved in the emission abatement of COD, the emissions of unregulated pollutants such as waste gas are still rising, which should be paid more attention. All of these findings indicate that the severity of different pollutants covers a wide range, and then the empirical results based on them may be different.

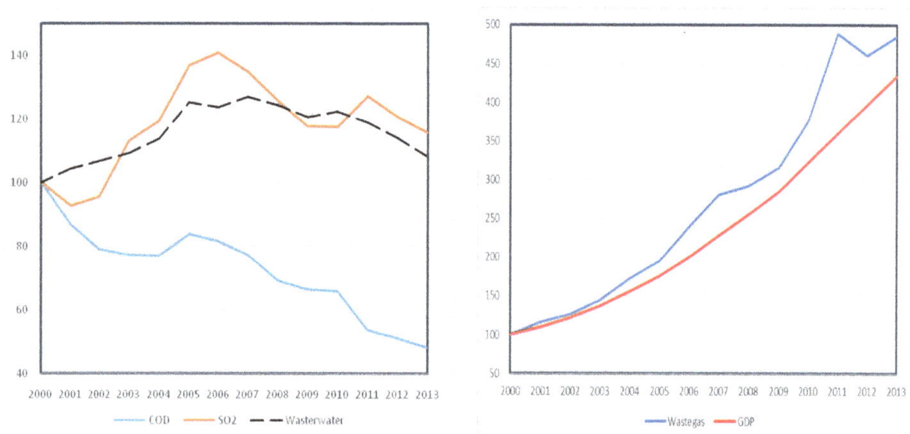

(a) (b)

Figure 1. Change in emission volume of pollutants. Notes: We set the emission scale in 2000 equivalent to 100.

Table 2 reports the average PI computed with different pollutants respectively each year. For the sake of comparing and analyzing conveniently, we also set the PI in 2000 equivalent to 100. Figure 2 sketches the change in PI calculated by using the four pollutants respectively in the period of 2000–2013. The results indicate that most PIs have been reduced substantially except that of waste gas (*Gas_GDP*). Specifically, the PI of *Gas_GDP* has increased to approximately 130, which indicates that the environmental costs of economic growth have increased from this perspective. However, others have been reduced to less than 40 in 2013, which provides indications of the lower environmental costs of economic growth from these pollutants' perspective. If one merely considers a single kind of pollutant, a full-scale observation may not be reached, which implies related heterogeneity should be taken into account.

Table 2 reports the average PI in the eastern and western areas, which enables a comparison between them. The PI of the western area is far worse than that of eastern one in all of the indicators in this study. For example, the average emission of SO_2 per capita GDP (SO_2_GDP) in the western area was 324.881 (116.007) in 2000 (2013), while it was only 105.906 (22.939) in the eastern area in 2000 (2013). This result implies that the western area overpaid the environmental costs about three times (five times) the size of the eastern area in 2000 (2013). Moreover, the gap between the two areas is enlarged, although their average emission intensities are both reduced. Similar evidence could be found in the cases of *COD_GDP* and *Gas_GDP*, *etc.*

Table 2. PI (pollution intensity) from 2000 to 2013.

Year	SO_2_GDP			COD_GDP			Water_GDP			Gas_GDP		
	National	East	West	National	East	West	National	East	West	National	East	West
2000	205.965	105.906	324.881	86.955	41.900	137.694	21.030	14.901	25.366	1.724	1.114	2.314
2001	176.111	87.607	276.981	70.276	30.292	114.546	19.350	15.217	22.763	1.763	1.250	2.259
2002	164.217	81.530	262.825	53.352	25.252	82.483	17.761	14.067	20.745	1.765	1.215	2.296
2003	181.461	81.663	303.940	47.532	21.302	75.264	16.381	12.663	19.509	1.803	1.211	2.323
2004	168.836	73.032	280.749	40.767	18.857	63.213	14.838	11.885	17.340	1.929	1.217	2.701
2005	174.110	71.401	289.384	44.612	18.417	73.961	15.098	11.814	19.139	1.898	1.288	2.481
2006	165.996	62.154	287.451	39.096	14.822	65.729	12.985	10.146	15.995	2.131	1.393	2.959
2007	140.117	52.095	242.229	33.142	12.079	56.469	11.783	8.796	15.114	2.220	1.389	3.209
2008	116.590	42.411	201.679	27.014	9.155	46.522	10.419	7.727	13.683	2.071	1.219	2.861
2009	97.367	34.994	167.184	23.233	8.014	39.898	9.126	6.901	11.767	2.031	1.235	2.908
2010	87.283	31.373	149.518	20.106	7.059	34.563	8.052	6.048	10.107	2.510	1.261	4.059
2011	88.275	30.572	153.645	16.959	5.828	31.561	6.971	5.685	8.172	2.567	1.422	3.731
2012	75.696	26.499	132.025	14.592	5.221	27.230	6.052	4.723	7.011	2.248	1.215	3.373
2013	66.398	22.939	116.007	12.746	4.495	24.047	5.268	4.079	6.112	2.203	1.318	3.253

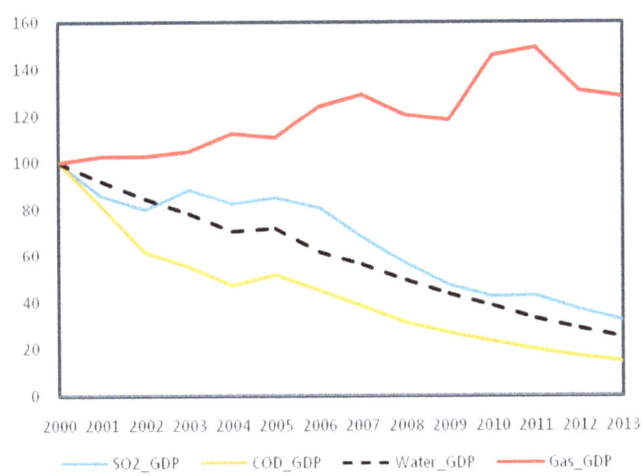

Figure 2. PI from 2000 to 2013. Notes: We set the PI in 2000 equivalent to 100.

Table 3 displays the variation of the environmental regulation variables. The average number of local environmental bylaws and rules (*Sum_rule*) continued to rise from 2000 to 2013, and increased from approximately 23 in 2000 to 61 in 2013. However, the number of staff in the environmental protection departments per industrial firm above the designated size (*Regulator_perfirm*) remained relatively stable in this period, of which the average value was about one in most years. The average investment in the treatment of industrial pollution per capita industrial output (*Inves_output*) went down from 0.39% in 2000 to 0.13% in 2013, although, in fact, the scale of investment increased during this period. The above results indicate that local governments have made progress in legislation for environmental protection; however, the invested capital aimed at protecting the environment did not keep up with the GDP growth. Compared with the western area, the eastern area shows higher scale in the variable of *Sum_rule*, which indicates more stringent regulation in these regions. However, the cases of the other two variables are complicated to some extent. The result indicates that the western area has more environmental human resources per firm than the eastern area, which could lead to more stringent regulation in the west. However, we should realize the reason behind it is that the number of firms is relatively small in the west. Similarly, although the *Inves_output* of the western area is relatively high, its abstract amount of investment in the treatment of industrial pollution is far smaller than that of the eastern area.

Table 3. Three kinds of environmental regulation from 2000 to 2013.

Year	Sum_rule			Regulator_perfirm			Inves_output		
	Whole Nation	East	West	Whole Nation	East	West	Whole Nation	East	West
2000	23.333	22.000	19.364	1.081	0.632	1.252	0.390	0.238	0.512
2001	26.167	25.000	20.818	1.169	0.621	1.403	0.243	0.138	0.323
2002	30.267	27.500	25.636	1.213	0.617	1.469	0.232	0.145	0.284
2003	33.733	31.000	28.273	1.198	0.569	1.454	0.187	0.143	0.234
2004	36.900	35.200	30.273	1.005	0.489	1.273	0.219	0.118	0.305
2005	39.600	37.400	32.091	1.064	0.519	1.331	0.223	0.175	0.261
2006	43.033	40.800	33.364	1.004	0.518	1.284	0.226	0.137	0.300
2007	45.633	44.400	35.364	1.002	0.566	1.289	0.187	0.096	0.258
2008	47.833	46.800	37.091	0.881	0.506	1.164	0.163	0.077	0.252
2009	49.500	49.100	38.455	0.891	0.568	1.145	0.138	0.056	0.221
2010	54.267	50.100	49.000	0.877	0.579	1.118	0.091	0.041	0.149
2011	56.033	52.400	50.364	1.199	0.864	1.561	0.087	0.053	0.141
2012	58.533	55.600	52.455	1.107	0.827	1.407	0.092	0.069	0.145
2013	61.833	60.000	55.909	1.079	0.813	1.367	0.130	0.073	0.195

Notes: The unit of *Inves_output* is %. The indicators of central area are not reported due to space limitation, the data of which is available upon request.

4.2. *Econometric Analysis*

4.2.1. Results of Full Samples

4.2.1.1. Proper Econometric Models

Firstly we investigate the spatial interdependence of PI with the two standard spatial auto-correlation tests of Moran's *I* [64] and Geary's *C* [65]. In the interest of brevity, Table 4 summarizes the results in 2013. Moran's *I* statistic shows that the variables (SO_2_GDP and COD_GDP) are spatially auto-correlated at the 10% statistical significance level. Geary's *C* shows similar results, indicating the necessity of using the spatial econometric models for the empirical analysis. To get robust results, we carry out further test procedures for all dependents.

Table 4. Spatial auto-correlation of dependents (2013).

Variable	Moran's *I* Test		Geary's *C* Test	
	Statistic	*p*-Value	Statistic	*p*-Value
SO_2_GDP	0.236	0.007	0.547	0.048
COD_GDP	0.142	0.036	0.538	0.138
Water_GDP	−0.007	0.796	0.728	0.179
Gas_GDP	0.094	0.235	0.749	0.165

Notes: The two-tail test is used to compute *p*-values. The null hypothesis is that the distribution of variables in different provinces is considered to be spatially independent.

We perform the Lagrange multiplier (LM) test and the robust LM test for the null hypothesis of no spatially lagged dependent variable and no spatially correlated error term. For example, when using the natural log transformation of SO_2_GDP as the dependent variable, the LM test results reject the null hypothesis of no spatial lag and no error dependence, indicating that the spatial Durbin model (SDM) specification should be tested. However, the SDM is a general spatial model, which can be interpreted as a spatial autoregressive model (SAR) or a spatial error model (SEM), and we further perform the Wald test to determine the appropriate model (see details in Appendix). The Wald test of the spatial lag and spatial error indicates that both SAR and SEM are rejected, implying that the general SDM is the most proper model to describe the data [66]. Moreover, we execute the Hausman test to make a choice between the fixed and the random effects model and find that random effects should be considered in most cases of our study.

4.2.1.2. Strategic Interaction: RBD or RGD

Note that if the *RBD* argument is the case, it can be inferred that ρ should be positive and the PI should increase, indicating higher environmental costs. However, if the *RGD* is the case, ρ should also be positive, while the PI should decrease. Otherwise, if the regions perform different strategies and lack competition, ρ should be negative or statistically insignificant. The above logic enables us to observe the regional interactions and examine the proposed hypotheses. The primary interest of this paper is the spatial autoregressive coefficient ρ, which reveals the strategic interaction of samples. As shown in Table 5, the coefficients (ρ) in all models are significantly negative, which implies that the samples perform differentiation strategies. In sum, the *RBD* hypothesis is not supported at the national level, which is in line with the predictions from our theoretical analysis. Moreover, the *RGD* hypothesis is also not supported in the observation period of 2000–2013 at the national level.

4.2.1.3. Direct Effects of Environmental Regulation and Control Variables

The empirical studies relying on the simple spatial regression estimation (which is called point estimation by LeSage and Pace [33]) may lead to biased conclusions because they do not represent the marginal effect of a certain change in an independent variable [33,64]. LeSage and Pace suggested that a more appropriate way should be taken through direct, indirect, and total effects. The direct effects gauge the impact that a certain variation in an independent variable has on the dependent variable of a sample. The indirect effects examine the impact of a change in the independent variables of other provinces on the dependent variable of a certain province, and they are also known as spatial spillover effects. The total effects are the combination of direct effects and indirect effects [67]. Previous literature [67,68] provided the corresponding statistical test. The statistical significance of the direct and indirect effects is obtained by simulating the distribution using the variance-covariance matrix implied by the maximum likelihood estimated coefficients.

Table 5. Direct, indirect and total effects of full samples.

	(1)	(2)	(3)	(4)
	ln_SO_2_GDP	**ln_COD_GDP**	**ln_Water_GDP**	**ln_Gas_GDP**
Panel A: Direct Effects				
ln_Sum_rule	−0.535 *** (−3.997)	0.325 (1.338)	−0.488 *** (−3.421)	0.162 (1.178)
ln_Regulator_perfirm	0.230 *** (6.271)	0.542 *** (7.992)	0.088 ** (2.123)	0.023 (0.618)
Inves_output	0.061 (0.735)	0.371 ** (2.330)	0.181 * (1.865)	0.048 (0.503)
S_college	−0.006 (−0.817)	0.005 (0.394)	−0.024 *** (−3.198)	−0.027 *** (−3.950)
ln_L	−0.120 (−1.160)	−0.466 ** (−2.624)	0.165 (1.527)	−0.200 ** (−2.461)
ln_GDPSK	0.033 (0.458)	0.142 (1.052)	−0.038 (−0.411)	−0.057 (−1.036)
ln_Engery_GDP	0.832 *** (8.255)	0.652 *** (3.783)	0.784 *** (7.735)	0.679 *** (7.057)
S_HI	−1.205 *** (−4.181)	−2.031 *** (−4.195)	−0.719 ** (−2.380)	0.335 (1.266)
S_SOE	−0.359 ** (−2.422)	−0.353 (−1.299)	0.135 (0.828)	−0.503 *** (−3.481)
ln_K_L	−0.289 *** (−3.604)	−0.323 ** (−2.186)	−0.050 (−0.555)	−0.065 (−0.873)
S_coal	1.091 *** (4.035)	1.412 *** (2.952)	0.450 (1.476)	0.555 ** (2.084)
Open	0.017 (0.202)	0.021 (0.135)	−0.055 (−0.552)	0.064 (0.747)

Table 5. *Cont.*

	(1)	(2)	(3)	(4)
	ln_SO_2_GDP	ln_COD_GDP	ln_Water_GDP	ln_Gas_GDP
Panel B: Indirect Effects				
ln_Sum_rule	−1.686 (−1.576)	5.632 *** (3.171)	1.711 * (1.895)	2.361 ** (2.377)
ln_Regulator_perfirm	0.294 (1.499)	0.044 (0.138)	−0.394 ** (−2.304)	0.429 ** (2.163)
Inves_output	−0.639 (−1.425)	−1.863 ** (−2.421)	−0.734 * (−1.779)	−0.920 * (−1.923)
S_college	0.027 (1.090)	−0.042 (−1.160)	0.008 (0.398)	0.036 (1.583)
ln_L	−1.697 ** (−2.121)	−4.311 *** (−3.487)	0.678 (1.016)	−0.782 (−1.170)
ln_GDPSK	0.766 (1.671)	1.011 (1.464)	0.702 * (1.761)	0.458 (1.234)
ln_Engery_GDP	2.758 *** (3.474)	0.523 (0.456)	1.161 * (1.837)	0.521 (0.709)
S_HI	−17.277 *** (−6.535)	−6.905 * (−1.979)	−3.548 * (−1.969)	−0.616 (−0.305)
S_SOE	4.494 *** (4.364)	1.099 (0.718)	2.153 *** (2.685)	−1.522 (−1.613)
ln_K_L	−0.736 (−1.037)	−4.458 *** (−3.888)	−1.023 * (−1.731)	−1.060 (−1.660)
S_coal	4.268 ** (2.428)	5.597 * (2.040)	2.717 * (1.822)	0.114 (0.065)
Open	0.904 ** (2.311)	2.094 *** (3.127)	0.279 (0.821)	0.170 (0.431)
Panel C: Total Effects				
ln_Sum_rule	−2.220 * (−1.958)	5.958 *** (3.177)	1.224 (1.286)	2.523 ** (2.393)
ln_Regulator_perfirm	0.523 ** (2.729)	0.586 * (1.923)	−0.306 * (−1.904)	0.453 ** (2.347)
Inves_output	−0.578 (−1.250)	−1.492 * (−1.874)	−0.553 (−1.330)	−0.872 * (−1.768)
S_college	0.021 (0.797)	−0.038 (−0.952)	−0.016 (−0.767)	0.009 (0.366)
ln_L	−1.816 ** (−2.129)	−4.777 *** (−3.620)	0.843 (1.202)	−0.982 (−1.368)
ln_GDPSK	0.800 * (1.766)	1.153 * (1.733)	0.664 * (1.793)	0.401 (1.085)
ln_Engery_GDP	3.590 *** (4.278)	1.175 (0.973)	1.945 *** (2.936)	1.200 (1.549)
S_HI	−18.482 *** (−6.811)	−8.937 ** (−2.616)	−4.268 ** (−2.451)	−0.281 (−0.139)
S_SOE	4.135 *** (3.854)	0.746 (0.476)	2.288 *** (2.794)	−2.025 ** (−2.070)
ln_K_L	−1.025 (−1.400)	−4.781 *** (−4.064)	−1.073 * (−1.793)	−1.125 * (−1.707)
S_coal	5.359 *** (2.909)	7.009 ** (2.448)	3.167 ** (2.055)	0.669 (0.364)
Open	0.921 ** (2.288)	2.115 *** (3.139)	0.224 (0.682)	0.234 (0.592)
ρ	−0.250 * (−1.867)	−0.443 *** (−2.942)	−0.628 *** (−4.071)	−0.304 ** (−2.002)
Hausman test & *p*-value	26.024 [0.406]	18.493 [0.821]	28.045 [0.306]	22.324 [0.617]

Notes: 1. ***, ** or * denotes significance at the level of 1%, 5% or 10%, respectively. The *t*-values of coefficients are reported in the parentheses; 2. Considering the endogeneity between regulation intensity variables and the pollution intensity, as well as the time-lag effects of environmental regulation, we use a one-year lag of all the regulation intensity variables; 3. "ln_" indicates the form of the variables after a natural log transformation; 4. According to the Hausman test, the reported models (1)–(4) are random effects models, so the year and provincial fixed effects are not reported.

Table 5 summarizes the estimation results of the direct, indirect and total effects of regulation intensity variables and other control variables. We report the direct effects in a simple way based on Panel A. It shows that *Sum_rule* (in natural logarithm transformation with a one-year lag) is significantly and negatively associated with the PIs of SO_2, suggesting that the higher number of local environmental bylaws and rules may help to decrease the PIs of SO_2. However, it is not the case for the model using *Gas_GDP* (Model (4) in Table 5) as the dependent variable. As a whole, the above results indicate that environmental legislation is effective in reducing PI, *i.e.*, curtailing the environmental costs, especially in the cases of regulated pollutants.

Regarding other environmental regulation variables, *i.e.*, *Regulator_perfirm* and *Inves_output*, we have a surprising finding: No evidence supports the effectiveness of the environmental staff and the invested capital for the reduction of environmental costs. Among all models, the coefficients of *Regulator_perfirm* (in natural log form with a one-year lag) in most cases are significantly positive, except for *Gas_GDP* (see Table 5). We also perform the estimation only using *Regulator_perfirm* as the regulation variable and find that the results are robust. The comparison between western and eastern areas may help us understand the above counterintuitive results. The PI of the western area is higher (see Table 2) than that of the eastern area, and *Regulator_perfirm* in the western area is also relatively high (see Tables 2 and 3), which leads to a positive correlation between the two variables. However, this result implies that the higher input of environmental human resources does not lead to a lower PI, although it is intuitive that higher *Regulator_perfirm* may help to reduce pollutant emissions. The result is meaningful for policymakers with the implication that increasing the number of environmental staff blindly may not be an effective strategy. Further measures should be taken to avoid overstaffing the environmental protection departments in the western area. We have similar findings for the variable of *Inves_output* (with a one-year lag), the coefficients of which are insignificant in most models, such as the model with SO_2*_GDP*, although the coefficient is significant in the case of *COD_GDP*. The results are also understandable based on the fact that the *Inves_output* and PI of the western area are higher than those of the eastern area. Higher environmental capital investment does not lead to a lower PI, implying that environmental capital investment is inefficient in the western area of China.

The observation of the other variables may provide some evidence on the driving forces of PI. The *S_college* is significantly and negatively associated with PI in the models with *Water_GDP* or *Gas_GDP*, suggesting that informal regulation plays a positive role in environmental protection. The coefficients of *L* are significantly negative in the models with *COD_GDP* or *Gas_GDP*, indicating that the scale effect works for only two pollutants, *i.e.*, COD and waste gas. The *GDPSK* is insignificantly associated with PI in all models, implying a weak connection between the intensity of economic activity and PI.

In regard to the technique effect, we find that the coefficients of *Engery_GDP* are significantly positive in all models. Note that a lower *Engery_GDP* is better, so the results indicate that higher energy efficiency will lead to better environmental performance.

While the results of the composition effects are mixed, the coefficients of *S_HI* are significantly negative in some models, such as those with *COD_GDP* or SO_2*_GDP*, and they are insignificant in the other models. This result seems somewhat surprising and counterintuitive because the results indicate that a high share of heavy industrial output in total industrial output may be associated with a lower PI in terms of COD and SO_2. However, this behavior is understandable for two reasons: The first one is related to the environment-economic indicator, *i.e.*, the PI we used as the dependent variable. Although heavy industry causes a lot of pollution, it is also currently a major source of GDP in China, which means that both the numerator and denominator of the PI may be high. Therefore, the PI may not be high, which results in the outcome that a large share of heavy industrial output may not lead to a high PI. The second reason is that those pollutants mainly generated by heavy industry have obligatory emission reduction targets, which means that governments have to exert stringent environmental standards on them. The combination of these two reasons may cause the result of PI being negative with the variable (*S_HI*).

The coefficients of *S_SOE* are also negative in some models such as those with SO_2*_GDP* or *Gas_GDP*, hinting that SOEs are positive with pollutant abatement. Because SOEs are nationally owned in China, the country will pay for the costs of environmental governance. The managers of SOEs will gain political records by better performance of environmental governance, and they do not pay for it. Therefore, managers are motivated to invest in the environment and execute stringent environmental standards to reduce pollution. The SOEs may have less negative environmental impact than private firms or others, which causes the phenomenon that the higher the proportion of SOE in a region, the lower the PI may be.

However, it is interesting that the role of *K_L* is two-sided and depends on whether the pollutants are constrained by emission reduction targets. The coefficients are significantly negative in the models using regulated pollutant variables (SO_2*_GDP* and *COD_GDP*) as the dependent variables separately, while they are insignificantly negative in the models with unregulated ones (*Water_GDP* and *Gas_GDP*).

Moreover, *S_coal* is positively associated with the PI in most models except the model using *Water_GDP* as the dependent variable, which supports the consensus that coal consumption is one of the most predominant sources of severe pollution problems in China. In the end, the coefficients of *Open* are insignificant, indicating that the regional openness has little influence on the PI.

In short, we find strong evidence that technical progress in energy utilization and endowment structure are the important driving forces to lower environmental costs in terms of most pollutants. We also find that the informal environmental regulation, scale effects, economic agglomeration and property right structure have positive effects on the decrease of the PI of some pollutants. In addition, the effectiveness of environmental regulation variables relies on the variety of the pollutants, which implies that the heterogeneity related to pollutant emission regulation does matter to some extent.

4.2.1.4. Spillover Effects and Total Effects of Environmental Regulation

Panel B of Table 5 reports the estimation results of the spillover effects. We focus on the spillover effects of the regulation intensity variables. The coefficients of *Sum_rule* are significantly positive in the models using *Water_GDP* or *Gas_GDP* as the dependent, which indicates that the local PI is negatively influenced by environmental legislation of adjacent regions, especially for those related to regulated pollutants. The coefficients of *Regulator_perfirm* are significantly negative only in the model with *Water_GDP* and significantly positive in the model with *Gas_GDP*. It implies that the large scale of input on the environmental human resources in the contiguous regions may lead to a low PI of waste water in the local region, while it is the opposite for the PI of waste gas. In addition, the coefficients of *Inves_output* are significantly negative in most models except the model with SO_2*_GDP*, which indicates that environmental investment in the adjacent regions may help to lower the PIs of COD, waste water and waste gas in the local region. The above results reveal that the input of neighboring regions on the environmental capital investment shows effective spillover effects in reducing the PI of the local region. In all, the spillover effects also rely on the types of environmental intensity variables and the pollutants observed in the models.

With respect to the total effects, we have mixed findings (see Panel C of Table 5). The coefficient of *Sum_rule* is significantly negative in the model with SO_2*_GDP*, while it is significantly positive in the model using *Gas_GDP* as the dependent. The results indicate that the environmental legislation exerts a positive influence on the PI of regulated pollutants, such as SO_2. The coefficients of *Regulator_perfirm* are significantly positive in most models, such as those with the SO_2*_GDP* or *COD_GDP*, indicating the overstaffing of environmental human resources. In addition, the total effects of *Inves_output* are significantly negative in the models with *COD_GDP* or *Gas_GDP*, which shows that it helps to lower the PI in terms of total effects. The above results indicate that the environmental protection institutions and the capital invested for environmental protection are more effective in reducing PI, while the input in environmental human resources is inefficient.

4.2.2. Results of Subsamples: Different Areas

We divide the samples into three groups according to the geographic location. The first one lies in the west of China and includes 11 provincial units, while the second lies in the east and includes 11 provincial units. Another one with only 8 provinces refers to the central area of China. The western regions are undeveloped, while the eastern regions are relatively developed. In addition, the economic development level of the central ones is between those of the former two. The western and eastern groups provide us with good samples to test for the existence of *RBD* or *RGD*. We carry out test procedures and find that the SDM is the most proper one. Table 6 reports the spatial autoregressive coefficient ρ in the western area and the eastern area (see details of the models in Appendix). For the western area, ρ is significantly negative in most models except the model using *Gas_GDP* as the dependent (the ρ of which is insignificantly negative), indicating that the samples perform differentiation strategies. This finding does not support the conjecture that undeveloped regions are inclined to the *RBD*.

Table 6. The ρ in different areas.

	(1)	(2)	(3)	(4)
	ln_SO_2_GDP	**ln_COD_GDP**	**ln_Water_GDP**	**ln_Gas_GDP**
National	−0.250 * (−1.867)	−0.443 *** (−2.942)	−0.628 *** (−4.071)	−0.304 ** (−2.002)
Western	−0.491 *** (−4.839)	−0.683 *** (−6.638)	−0.599 *** (−5.729)	−0.145 (−1.338)
Central	−0.121 (−1.321)	0.119 (1.414)	0.075 (0.875)	−0.198 ** (−2.342)
Eastern	0.212 *** (2.830)	0.157 ** (1.984)	0.077 (0.965)	−0.062 (−0.745)

Notes: ***, ** or * denotes significance at the level of 1%, 5% or 10%, respectively. "ln_" indicates the form of the variables after a natural log transformation.

For the eastern area, ρ is significantly positive in the models with *SO_2_GDP* or *COD_GDP*, indicating that the samples perform similar strategies in regional competition. Combined with the decreasing PIs of the three pollutants shown in Table 2, there is some evidence to support the *RGD* hypothesis in the east of China. However, China's central government set up targets on the abatement of the SO_2 and COD, *i.e.*, respectively 10% decrease of those two pollutants in 2010 compared with the scale in 2005. The targets were assigned to provincial units, which could bring top-down pressure to the samples. On the one hand, the competition among regions and the desire for environmental protection will bring bottom-up pressure to the local governments, which should be a source of *RGD*. However, on the other hand, if the decrease in emissions is dominated by top-down pressure, *i.e.*, mandatory orders from the central government, similar strategies taken by the eastern provinces would no longer show enough evidence to support *RGD*. To figure out this issue, we compute the growth rate of the emission volumes of the three pollutants (see Table 7) since 2005. The average annual growth rates of SO_2 and COD in the eastern area are −2.0% and −4.7%, respectively, since 2005, and their cumulative growth rates are −21.8% and −26.2% during the period of 2006–2010. The cumulative decrement rates of emissions are far more than the assigned targets (10%), which implies that there is an abatement competition on some related pollutants in the eastern area. This finding also provides stronger evidence to support the *RGD* argument in the developed regions, as proposed in the theoretical analysis.

Table 7. Growth of some industrial pollutants since 2005.

	All Samples				Eastern Samples			
	Yearly Growth		Cumulative Growth since 2005		Yearly Growth		Cumulative Growth since 2005	
	SO_2	COD	SO_2	COD	SO_2	COD	SO_2	COD
2005	14.6%	8.8%	-	-	10.6%	10.0%	-	-
2006	3.0%	−2.6%	3.0%	−2.6%	−1.3%	−6.8%	−1.3%	−6.8%
2007	−4.1%	−5.4%	−1.3%	−7.9%	−5.2%	−7.1%	−6.4%	−13.4%
2008	−6.9%	−10.5%	−8.2%	−17.5%	−8.5%	−16.0%	−14.3%	−27.2%
2009	−6.3%	−3.9%	−13.9%	−20.7%	−7.7%	−1.9%	−21.0%	−28.6%
2010	−0.1%	−1.1%	−14.0%	−21.6%	−1.1%	3.3%	−21.8%	−26.2%
2011	8.2%	−18.4%	−7.0%	−36.0%	6.7%	−15.3%	−16.5%	−37.5%
2012	−5.2%	−4.6%	−11.8%	−39.0%	−5.7%	−2.3%	−21.3%	−39.0%
2013	−4.0%	−5.6%	−15.4%	−42.4%	−5.6%	−6.3%	−25.7%	−42.8%
Average	−0.1%	−4.8%			−2.0%	−4.7%		

However, it is worth noting that ρ is insignificant in the models with unregulated pollutants, such as waste water or waste gas, implying that the *RGD* argument is not supported in those cases. Therefore, we may conclude that the achievement in the emission abatement of the eastern provinces is associated with both the top-down pressure from the central government and the competition pressure from adjacent regions. Therefore, the adjacent competition mechanism has proven to be effective in such cases.

To check the robustness of our findings, this paper performs two tests as follows. First, we use a two-year lag of environmental regulation variables in robustness tests instead of a one-year lag. We find that the regression results are essentially in agreement with those that have been reported in previous sections (the results of the robustness tests are not reported due to space limitations but are available upon request). Second, we replace the adjacent weights matrix with the inverse distance weights matrix. Basically, it shows that the results in the previous sections are robust. Our result is consistent with LeSage and Pace [34] and indicates the specification of the spatial weights matrix should have no effect on the effects estimates.

5. Conclusions

In contrast to previous studies that focused on either environmental or economic enforcement in regional competition, this paper focuses on the results of competition in a comprehensive way. Using an environment-economic indicator, we study how the interaction of regions influences the balance of economic growth and environmental protection in China based on the proposed *RBD* and *RGD* hypotheses. Noticing that the empirical results of arguments regarding the *RBD* or *RGD* in China may rely on the heterogeneity of pollutants and regions, we differentiate two types of industrial pollutants and test the hypotheses. Our results contribute to the literature in three ways. First, the *RBD* is not the case at the national level, and provincial governments do perform different strategies for environmental protection in China. We suggest that this result should be due to the heterogeneity among regions in the largest developing country. Second, the *RGD* hypothesis is supported in the eastern area; however, this result is only the case for pollutants with obligatory emission reduction targets, such as SO_2 and COD. The results imply that the selective competition mechanism has been established in China. In the end, we cast new light on the driving forces of environmental costs. Although, on average, the national PI has decreased and environmental performance has improved, we find that some methods of environmental regulation have contributed little to these achievements. In particular, the current environmental staff and capital investment for environmental protection do not lead to better environmental performance. Only increasing environmental legislation has been effective in reducing PI, especially for pollutants with obligatory reduction targets.

The previously mentioned empirical results offer some implications for policymakers. The existence of the *RGD* may lead to better balance of environment and economy. However, in China, it is the case only for pollutants with obligatory emission reduction targets in the developed eastern regions, which implies that top-down pressure from the central government is necessary right now. To achieve quick improvements in environmental protection, a combination of top-down pressure from the central government and bottom-up pressure from adjacent competitors may be a more effective approach. On the other hand, more attention should be paid to the pollutants whose emission volumes have been growing fast in recent years, such as waste gas. The control of toxic components such as SO_2 does not guarantee that the emissions of all toxic components will be reduced. Therefore, establishing obligatory abatement targets for pollutants such as waste gas and waste water should be considered. The effectiveness of environmental regulation should also be reexamined. We find that Chinese regional environmental staff and capital investment have not exerted powerful effects in environmental protection thus far, which implies a low efficiency of human resources and capital investment in China. However, this also implies that the potential of these two inputs could be more fully explored for pollutant abatement.

As stated in our analysis, the concepts of green and black development incorporate economic and environmental elements. Therefore, future research with a more systematic analysis of *RBD* or *RGD* is warranted as it might be inadequate just from the perspective of pollution intensity in this paper. Furthermore, a data-driven approach proposed by Piribauer [69] to treat structural heterogeneity in spatial autoregressive models could be considered as a future extension of the research. In the end, to provide more robust and fruitful results, city-level data and other environmental regulation variables should be examined in further studies.

Acknowledgments: The authors sincerely appreciate the helpful comments on the early drafts of this paper offered by the two anonymous referees. All remaining errors are the authors'. The research in this paper is jointly funded by the National Natural Science Foundation of China (41571524; 41461118), National Social Science Foundation of China (14ZDB144) and CICSU Collaborative Innovation Center for the Development of Modern Services and New Urbanization in Hunan Province.

Author Contributions: All authors conceived and designed the research. Jiejin Xia collected and processed the data. Jianhuan Huang performed the theoretical analysis, analyzed the data, and wrote the manuscript.

Conflicts of Interest: The authors declare no conflict of interest.

Appendix

A1. The Determinants of Environmental Costs of Economic Growth

We suppose that the economic production takes the Cob-Douglas production function as follows:

$$Y = AK^{\alpha}L^{\delta} \tag{A1}$$

where Y is the economic output of a region; K and L represent capital stock and labor employed, respectively; A is the total factor productivity; α and δ are related coefficients.

As analyzed in Subsection 3.2.2, the pollutant emission is a function of the scale effect (SE), the technique effect (TE), the composition effect (CE) and environmental regulation (ER). Therefore, we have:

$$\mathrm{E} = \mathrm{e}\,(\mathrm{ER}, \mathrm{SE}, \mathrm{TE}, \mathrm{CE}, \varnothing) \tag{A2}$$

where ϕ refers to other determinants. Combining Equations (A1) and (A2), we have:

$$PI = \frac{E}{Y} = \frac{e\,(ER, SE, CE, TE, \phi)}{AK^{\alpha}L^{\delta}} \tag{A3}$$

Supposing the function of E takes the form of the log-linear model, we have:

$$\begin{aligned} \ln(PI) &= \beta_1\ln(ER) + \beta_2\ln(SE) + \beta_3\ln(TE) + \beta_4\ln(CE) - \ln(A) - \alpha\ln(K/L) - (\alpha+\delta)\ln(L) + \varepsilon \\ &= \beta_0 + \beta_1\ln(ER) + \beta_2\ln(SE) + \beta_3\ln(TE) + \beta_4\ln(CE) + \beta_5\ln(L) + \beta_6\ln(K/L) + \varepsilon \end{aligned} \tag{A4}$$

where β_k (k = 1, 2, ... , 6) are the coefficients, and $\beta_0 = -\ln(A)$, $\beta_5 = -(\alpha+\delta)$, $\beta_6 = -\alpha$. We regard L (labor) as a measure of the scale effect, while K/L (capital-labor ratio) is a measure of the composition effect. Then the variables of the determinants can be written as follows:

$$\beta X_{it} = \beta_1 \ln(ER_{it}) + \beta_2 \ln(SE_{it}) + \beta_3 \ln(TE_{it}) + \beta_4 \ln(CE_{it}) \quad \text{(A5)}$$

A2. Spatial Econometric Models and the Specification Test

To observe the spatial interaction effects of the economic phenomenon, many methods were suggested in previous literature. Among them, Moran's I [64] and Geary's C [65] are widely used. If the key variable is spatially auto-correlated significantly, further tests should be taken for choosing proper econometric models. To a space-time model with a panel of N observations over T time periods, the panel form of the general nesting spatial model (GNS) with spatial specific and time-period specific effects is as follows [32,33]:

$$y_{it} = a + \rho \sum\nolimits_{j=1, i \neq j}^{N} w_{ij} y_{jt} + \beta X_{it} + \theta \sum\nolimits_{j=1, i \neq j}^{N} w_{ij} X_{jt} + s_i + u_t + \varepsilon_{it} \quad \text{(A6)}$$

$$\varepsilon_{it} = \lambda \sum\nolimits_{j=1, i \neq j}^{N} w_{ij} \varepsilon_{jt} + \xi_{it} \quad \text{(A7)}$$

where α is the constant term and ρ is the spatial autoregressive coefficient, while θ, β and λ represent fixed but unknown parameters; λ is the coefficient of the spatial error term; w_{ij} denotes the weight describing the spatial arrangement of the samples. This paper uses the spatial adjacent matrix, *i.e.*, $w_{ij} = 1$, if two provinces share a common border and zero otherwise. Then the spatial weights matrix W is row-stochastic and the row-sums of W equal unity. S_i is regional fixed effects and u_t is time fixed effects, while ε_{it} is the error term.

The most frequently used panel models are the special cases of GNS. When $\lambda = 0$ and $\rho, \theta \neq 0$, GNS turns to the spatial Durbin model (SDM). When $\rho = 0$ and $\lambda, \theta \neq 0$, the model turns to the spatial Durbin error model (SDEM). When $\lambda, \theta = 0$ and $\rho \neq 0$, the model takes the form of the spatial autoregressive model (SAR). When $\lambda = 0$, $\rho \neq 0$ and $\theta = -\rho\beta$, the model turns to the spatial error model (SEM). When $\lambda, \rho = 0$ and $\theta \neq 0$, it turns to the spatial lag of X model (SLX). In the end, when $\lambda, \rho, \theta = 0$, the model turns to the general panel model without spatial interaction effects as follows:

$$y_{it} = a + \beta X_{it} + s_i + u_t + \varepsilon_{it} \quad \text{(A8)}$$

The final selected form of model is determined by a series of statistical tests using the data of the empirical study as follows.

According to Florax *et al.* [70], Mur and Angula [71] and Elhorst [66], we adopt the following test procedure to figure out the most proper model to describe the quantitative relationship among data.

(i) Estimate the panel model without spatial interaction effects (Equation (A8)). We use the classic LM test proposed by Anselin [32], as well as the robust LM test proposed by Anselin *et al.* [72], to verify the existence of spatial error effect (LM_E) or spatial lag effect (LM_L), and then turn to step (ii) if the results reject the hypothesis H_0 that assumes the absence of spatial interaction effects; otherwise, turn to step (iii).

(ii) Estimate the panel SDM model (Equation (A6)). We test hypotheses $H_0^1 : \theta = 0$ and $H_0^2 : \theta = -\rho\beta$. If hypothesis H_0 in step (i) is rejected and both these two hypotheses are also rejected, then the SDM is the most suitable method to characterize the data. If hypothesis H_0 of LM_L and hypothesis $H_0^1 : \theta = 0$ cannot be rejected, then the spatial autoregressive model is the most proper method to describe the data. If hypothesis H_0 of LM_E and hypothesis $H_0^2 : \theta = -\rho\beta$ cannot be rejected, then the spatial error model is the most appropriate. In addition to the above three kinds of possibilities, SDM should be adopted to get robust results.

(iii) Estimate the SLX model, *i.e.*, add the term $\theta \sum_{j=1,i\neq j}^{N} w_{ij}X_{jt}$ to Equation (A8), and test hypothesis $H_0 : \theta = 0$. If the hypothesis cannot be rejected, then the model without spatial interaction effects is the best. Otherwise, the SLX is the proper one.

A3. Tables

Table A1. Variables and their measures.

	Variable	Measure	Abbr.
Dependent variable	Pollution intensity	The ratio of pollutant emission volumes over economic output (GDP)	PI
Key independent variable	Local environmental bylaws and rules	The number of local environmental bylaws and rules	Sum_rule
	Staff of environmental protection departments per firm	The number of employees in the environmental protection departments per industrial firm above the designated size	Regulator_perfirm
	Investment in treating industrial pollution	The investment in the treatment of industrial pollution per capita industrial output	Inves_output
Control Variable	Informal regulation	The share of people with the college degree or above in the population	S_college
	Capital intensity	The ratio of gross fixed capital per worker	K_L
	Labor	Total number of employees	L
	GDP per square kilometer	Gross domestic product per square kilometer of urban lands	GDPSK
	Energy consumption intensity	Energy consumption per capita GDP	Engery_GDP

Table A1. *Cont.*

	Variable	Measure	Abbr.
	Property structure	Share of industrial output generated by state-owned enterprises (SOEs)	S_SOE
	Industrial structure	Share of heavy industrial output in the whole industrial output	S_HI
	Energy structure	Share of coal in energy consumption	S_coal
Control Variable	Openness	The ratio of total export-import volumes over GDP	Open

Table A2. Variables and data sources.

Variable	Definition/Source
PI	Pollution intensity, the ratio of pollutant emission volumes to GDP.
GDP	At 2000 constant price. The data source is the *China Statistical Yearbook for Regional Economy*, 2001–2014.
Engery_GDP	Energy consumption per capita GDP. The data of energy consumption is from the *China Energy Statistical Yearbook*, 2001–2014.
L	Labor. The data is from the *China Statistical Yearbook*, 2001–2014.
K_L	The ratio of capital to labor. The amount of capital is computed with fixed capital using the perpetual inventory method. The data comes from the *China Statistical Yearbook*, 2001–2014 and the *China Compendium of Statistics* (1949–2008). At 2000 constant price.
GDPSK	GDP per square kilometer of urban lands. The data comes from the *China Statistical Yearbook for Regional Economy*, 2001–2014.
S_HI	Share of heavy industrial output in GDP. The data is from the *China Industrial Economy Statistical Yearbook*, 2001–2013.
S_coal	Share of coal in the total energy consumption. The data is from the *China Industrial Economy Statistical Yearbook*, 2001–2014.
S_SOE	Share of industrial output generated by SOEs. The data is from the *China Industrial Economy Statistical Yearbook*, 2001–2014.
Open	The ratio of total export-import volumes over GDP. The source is the *China Statistical Yearbook for Regional Economy*, 2001–2014.

Table A3. Correlation of dependent variables.

	ln_SO_2_GDP	ln_COD_GDP	ln_Water_GDP	ln_Gas_GDP
ln_SO_2_GDP	1			
ln_COD_GDP	0.806	1		
ln_Water_GDP	0.659	0.843	1	
ln_Gas_GDP	0.715	0.493	0.267	1

Notes: It shows that there are high correlations among most of the dependent variables. "ln_" indicates the form of the variables after a natural log transformation.

Table A4. Correlation of independent variables.

	ln_Sum_rule	ln_Regulator_perfirm	Inves_output	S_college	ln_L	ln_K_L	ln_Engery_GDP	S_SOE	ln_GDPSK	S_HI	S_coal
ln_Regulator_perfirm	−0.168	1									
Inves_output	−0.207	0.377	1								
S_college	0.236	−0.411	−0.337	1							
ln_L	0.373	−0.229	−0.22	−0.267	1						
ln_K_L	0.315	−0.384	−0.45	0.778	−0.311	1					
ln_Engery_GDP	−0.141	0.633	0.552	−0.346	−0.37	−0.319	1				
S_SOE	−0.415	0.634	0.48	−0.164	−0.441	−0.412	0.593	1			
ln_GDPSK	0.356	−0.73	−0.441	0.563	0.302	0.516	−0.706	−0.676	1		
S_HI	0.0294	0.386	0.204	0.314	−0.454	0.272	0.548	0.443	−0.259	1	
S_coal	0.129	0.347	0.329	−0.468	0.271	−0.425	0.541	0.123	−0.314	0.027	1
Open	0.134	−0.694	−0.292	0.622	−0.0677	0.492	−0.553	−0.369	0.683	−0.139	−0.595

Notes: We use the natural log transformation for most variables in the estimation except those structural/proportional indicators such as *S_coal*, *S_SOE* and *S_HI*, *etc*. It shows that there are low correlations among the independent variables. For example, the maximum correlation coefficient is 0.778 between *ln_K_L* and *S_college*, which is smaller than 0.8. Therefore, we use all the independent variables at the same time in the empirical estimation.

Table A5. Estimated results for total regions (point estimation).

	(1)	(2)	(3)	(4)
	ln_SO_2_GDP	ln_COD_GDP	ln_Water_GDP	ln_Gas_GDP
ln_Sum_rule	−0.565 *** (−3.935)	0.502 * (1.909)	−0.405 ** (−2.558)	0.209 (1.481)
ln_Regulator_perfirm	0.237 *** (6.543)	0.544 *** (8.170)	0.067 * (1.687)	0.031 (0.852)
Inves_output	0.053 (0.604)	0.307 * (1.887)	0.147 (1.538)	0.026 (0.274)
S_college	−0.005 (−0.708)	0.004 (0.293)	−0.024 *** (−3.196)	−0.026 *** (−3.862)
ln_L	−0.142 (−1.330)	−0.609 *** (−3.206)	0.198 (1.599)	−0.219 ** (−2.525)
ln_GDPSK	0.043 (0.581)	0.171 (1.338)	−0.008 (−0.089)	−0.045 (−0.873)
ln_Engery_GDP	0.878 *** (8.706)	0.670 *** (3.608)	0.833 *** (7.519)	0.694 *** (6.834)
S_HI	−1.483 *** (−5.685)	−2.264 *** (−4.737)	−0.878 *** (−3.077)	0.328 (1.251)
S_SOE	−0.300 ** (−2.004)	−0.318 (−1.156)	0.233 (1.433)	−0.537 *** (−3.456)
ln_K_L	−0.296 *** (−3.597)	−0.470 *** (−3.144)	−0.098 (−1.059)	−0.089 (−1.185)
S_coal	1.161 *** (4.141)	1.595 *** (3.118)	0.590 * (1.895)	0.554 ** (2.078)

Table A5. *Cont.*

	(1)	(2)	(3)	(4)
	ln_SO$_2$_GDP	ln_COD_GDP	ln_Water_GDP	ln_Gas_GDP
Open	0.033 (0.394)	0.089 (0.571)	−0.039 (−0.419)	0.066 (0.785)
W×ln_Sum_rule	−2.224 * (−1.749)	7.869 *** (3.393)	2.441 * (1.740)	2.993 ** (2.404)
W×ln_Regulator_perfirm	0.420 * (1.762)	0.316 (0.716)	−0.554 ** (−2.155)	0.551 ** (2.263)
W×Inves_output	−0.747 (−1.355)	−2.420 ** (−2.370)	−1.045 * (−1.746)	−1.141 * (−1.924)
W×S_college	0.031 (1.094)	−0.058 (−1.109)	−0.003 (−0.106)	0.037 (1.288)
W×ln_L	−2.060 ** (−2.218)	−6.250 *** (−3.719)	1.193 (1.134)	−1.102 (−1.319)
W×ln_GDPSK	0.929 * (1.793)	1.498 (1.601)	1.069 * (1.809)	0.592 (1.294)
W×ln_Engery_GDP	3.527 *** (3.984)	1.072 (0.663)	2.287 ** (2.347)	0.901 (1.033)
W×S_HI	−21.333 *** (−8.550)	−10.672 ** (−2.344)	−5.943 ** (−2.165)	−0.733 (−0.302)
W×S_SOE	5.358 *** (4.635)	1.321 (0.619)	3.452 *** (2.733)	−2.051 * (−1.690)
W×ln_K_L	−0.952 (−1.160)	−6.414 *** (−4.265)	−1.626 * (−1.789)	−1.413 * (−1.801)
W×S_coal	5.514 ** (2.510)	8.402 ** (2.098)	4.545 * (1.919)	0.422 (0.195)
W×Open	1.123 ** (2.420)	2.946 *** (3.443)	0.406 (0.802)	0.244 (0.502)
ρ	−0.250 * (−1.867)	−0.443 *** (−2.942)	−0.628 *** (−4.071)	−0.304 ** (−2.002)
R-squared	0.965	0.909	0.907	0.920
Corr-squared	0.808	0.586	0.356	0.812
Log-likelihood	129.809	128.432	87.211	114.140
LM test for no spatial lag & p-value	46.798 *** [0.000]	1.195 [0.274]	8.332 *** [0.004]	0.503 [0.478]
Robust LM test for no spatial lag & p-value	53.505 *** [0.000]	49.706 *** [0.000]	32.429 *** [0.000]	4.001 ** [0.045]
LM test for no spatial error & p-value	4.962 ** [0.026]	11.139 *** [0.001]	0.415 [0.519]	9.074 *** [0.003]
Robust LM test for no spatial error & p-value	11.670 *** [0.001]	59.650 *** [0.000]	24.512 *** [0.000]	12.572 *** [0.000]
Wald spatial lag test & p-value	127.731 *** [0.000]	71.717 *** [0.000]	53.554 *** [0.000]	26.613 *** [0.009]
Wald spatial error test & p-value	128.370 *** [0.000]	65.948 *** [0.000]	48.717 *** [0.000]	28.833 *** [0.004]
Hausman test & p-value	26.024 [0.406]	18.493 [0.821]	28.045 [0.306]	22.324 [0.617]

Notes: 1. ***, ** or * denotes significance at the level of 1%, 5% or 10%, respectively. The t-values of coefficients are reported in the parentheses; 2. Considering the endogeneity between regulation intensity variables and the pollution intensity as well as the time-lag effects of environmental regulation, we use a one-year lag of all the regulation intensity variables; 3. "ln_" indicates the form of the variables after a natural log transformation; 4. According to the Hausman test, models (1)–(4) are random effects models, so the year and provincial fixed effects are not reported. Meanwhile the constant term is not reported.

References

1. Zheng, X.Y.; Yu, Y.H.; Wang, J.; Deng, H.H. Identifying the determinants and spatial nexus of provincial carbon intensity in China: A dynamic spatial panel approach. *Reg. Environ. Chang.* **2014**, *4*, 1651–1661.
2. Li, H.B.; Zhou, L.A. Political turnover and economic performance: The incentive role of personnel control in China. *J. Public Econ.* **2005**, *89*, 1743–1762.
3. Yang, H.S.; Chen, S.L.; Zhou, Y.Z. Local Government Competition and Environmental Policy—-Empirical Evidence from Province's Governments in China. *South China J. Econ.* **2008**, *6*, 15–30. (In Chinese).
4. Zhang, W.B.; Zhang, L.F.; Zhang, K.Y. The competition form and evolution of China' provincial environmental regulation intensity—-Based on the analysis of a two-regime spatial Durbin model with fixed effects. *Manag. World* **2010**, *12*, 34–44. (In Chinese).
5. Li, S.L.; Chu, S.B.; Shen, C. The competition of local governments, the environmental regulation and regional ecological efficiency. *World Econ.* **2014**, *4*, 88–110. (In Chinese).
6. Devereux, M.P.; Lockood, B.; Redoano, M. Horizontal and vertical indirect tax competition: Theory and some evidence from the USA. *J. Public Econ.* **2007**, *91*, 451–479.
7. Albert, S.-O. Expenditure spillovers and fiscal interactions: Empirical evidence from local governments in Spain. *J. Urban Econ.* **2006**, *59*, 32–53.
8. Agrawal, D.R. The tax gradient: Spatial aspects of fiscal competition. *Am. Econ. J. Econ. Policy* **2015**, *2*, 1–30.
9. Emilie, C.; Martial, F.; Gregoire, R. Decentralization in Africa and the nature of local governments' competition: Evidence from Benin. *Int. Tax Public Financ.* **2015**, *22*, 1048–1076.
10. Wang, W.J.; Zhang, J.T.; Qin, C.L. Fiscal Decentralization, Competition between Local Governments, and the Economic Growth of FDI. *Manag. World* **2007**, *3*, 13–23. (In Chinese).
11. Emilie, C. Yardstick competition in a federation: Theory and evidence from China. *China Econ. Rev.* **2012**, *23*, 878–897.
12. Yu, Y.H.; Zhang, L.; Li, F.H.; Zheng, X.Y. On the determinants of public infrastructure spending in Chinese cities: A spatial econometric perspective. *Soc. Sci. J.* **2011**, *3*, 458–467.
13. He, L.; Bao, X. Fiscal Expenditure Competition of China's Local Governments: The Characteristics and Its Effects on Capital Allocation. *Int. J. u- e-Serv. Sci. Technol.* **2015**, *2*, 91–100.
14. Ebeid, M.; Rodden, J. Economic Geography and Economic Voting: Evidence from the US States. *Br. J. Political Sci.* **2006**, *36*, 527–547.
15. Engel, K.H. State Environmental Standard-Setting: Is There a "Race" and Is It "To the Bottom"? *Hastings Law J.* **1997**, *48*, 271–398.
16. Vogel, D. *Trading up: Consumer and Environmental Regulation in a Global Economy*; Harvard University Press: Cambridge, MA, USA, 1995.
17. Meyer, S.M. The Economic Impact of Environmental Regulation. *J. Environ. Law Pract.* **1995**, *3*, 4–15.

18. Konisky, D.M. Regulatory Competition and Environmental Enforcement: Is There a Race to the Bottom? *Am. J. Political Sci.* **2007**, *51*, 853–871.
19. Konisky, D.M. Inequities in Enforcement? Environmental Justice and Government Performance. *J. Policy Anal. Manag.* **2009**, *28*, 102–121.
20. Wilson, J.D. *Capital Mobility and Environmental Standards: Is There a Theoretical Basis for a Race to the Bottom?*; MIT Press: Cambridge, MA, USA, 1996; pp. 393–427.
21. Woods, N.D. Interstate Competition and Environmental Regulation: A Test of the Race-to-the-Bottom Thesis. *Soc. Sci. Q.* **2006**, *87*, 174–189.
22. Fredriksson, P.G.; Millimet, D.L. Strategic Interaction and the Determination of Environmental Policy across U.S. States. *J. Urban Econ.* **2002**, *51*, 101–122.
23. Li, Y.Y.; Shen, K.R. The reduction effect of pollution control policy in China—Based on the empirical analysis of provincial industrial pollution data. *Manag. World* **2008**, *7*, 7–17. (In Chinese).
24. Hu, A.G. *Innovating Green Development in China*; China Renmin University Press: Beijing, China, 2012. (In Chinese)
25. Xu, G.Y. The Paradigm Transition from Black Development to Green Development. *West Forum* **2014**, *1*, 53–60. (In Chinese).
26. Du, M.Z.; Wang, B.; Wu, Y.R. Sources of China's economic growth: An empirical analysis based on the BML index with green growth accounting. *Sustainability* **2014**, *6*, 5983–6004.
27. Zhou, L.A.; Tao, J. Government Size, Market Development and Corruption. *Econ. Res. J.* **2009**, *1*, 57–69. (In Chinese).
28. Song, M.L.; Song, Y.Q.; An, Q.X.; Yu, H.Y. Review of environmental efficiency and its influencing factors in China: 1998–2009. *Renew. Sustain. Energy Rev.* **2013**, *20*, 8–14.
29. Vogel, D. Environmental Regulation and Economic Integration. *J. Int. Econ. Law* **2000**, *3*, 265–279.
30. The Tenth Five-year Plan Outline of National Economic and Social Development of the People's Republic of China. Available online: http://www.npc.gov.cn/wxzl/gongbao/2001-03/19/content_5134505.htm (accessed on 15 March 2001).
31. The Eleventh Five-year Plan Outline of National Economic and Social Development of the People's Republic of China. Available online: http://www.gov.cn/gongbao/content/2006/content_268766.htm (accessed on 14 March 2006).
32. Anselin, L. Lagrange multiplier test diagnostics for spatial dependence and spatial heterogeneity. *Geogr. Anal.* **1988**, *20*, 1–17.
33. LeSage, J.; Pace, R.K. *Introduction to Spatial Econometrics*; Chapman & Hall/CRC: Boca Raton, FL, USA, 2009.
34. LeSage, J.P.; Pace, R.K. The biggest myth in spatial econometrics. *Econometrics* **2014**, *2*, 217–249.
35. Elhorst, J.P. Spatial Panel Data Models. In *Handbook of Applied Spatial Analysis*; Fischer, M.M., Getis, A., Eds.; Springer: New York, NY, USA, 2010; pp. 377–407.
36. Stiebale, J. Do financial constraints matter for foreign market entry? A firm-level examination. *World Econ.* **2011**, *34*, 123–153.

37. Vergara, R. Taxation and private investment: Evidence for Chile. *Appl. Econ.* **2010**, *42*, 717–725.
38. Jesús, C.C.; Martin, F. Spatial filtering, model uncertainty and the speed of income convergence in Europe. *J. Appl. Econom.* **2013**, *28*, 720–741.
39. Piribauer, P.; Manfred, M.F. Model uncertainty in matrix exponential spatial growth regression models. *Geogr. Anal.* **2015**, *47*, 240–261.
40. Cole, M.A.; Elliott, R.J.R. Determining the trade—-Environment composition effect: The role of capital, labor and environmental regulations. *J. Environ. Econ. Manag.* **2003**, *46*, 363–383.
41. Michael, T.R. Pollution Intensity of GDP and Trade Policy: Can the World Bank Be Wrong? *World Dev.* **1996**, *24*, 471–479.
42. Maria, L. Economic structure and pollution intensity within the environmental input-output framework. *Energy Policy* **2007**, *35*, 3410–3417.
43. Wursthorn, S.; Poganietz, W.-R.; Schebek, L. Economic-environmental monitoring indicators for European countries: A disaggregated sector-based approach for monitoring eco-efficiency. *Ecol. Econ.* **2011**, *70*, 487–496.
44. Holdren, J.P.; Ehrlich, P.R. Human Population and the Global Environment: Population growth, rising per capita material consumption and disruptive technologies have made civilization a global ecological force. *Am. Sci.* **1974**, *62*, 282–292.
45. Grossman, G.M.; Krueger, A.B. Economic growth and the environment. *Q. J. Econ.* **1995**, *110*, 353–377.
46. He, J. Pollution haven hypothesis and environmental impacts of foreign direct investment: The case of industrial emission of sulfur dioxide (SO_2) in Chinese provinces. *Ecol. Econ.* **2006**, *60*, 228–245.
47. Mu, Z.L.; Bu, S.C.; Xue, B. Environmental legislation in China: Achievements, challenges and trends. *Sustainability* **2014**, *6*, 8967–8979.
48. Yuan, Y.J.; Xie, R.H. Research on the effect of environmental regulation to industrial restructuring—-Empirical test based on provincial panel data of China. *China Ind. Econ.* **2014**, *8*, 57–69. (In Chinese).
49. Wang, D.T.; Chen, W. Foreign direct investment, institutional development and environmental externalities: Evidence from China. *J. Environ. Manag.* **2014**, *135*, 81–90.
50. Luo, N.S.; Li, J.J.; Luo, F.Z. Empirical Analysis on the Relationship between the China Urbanization and Regional Eco-efficiency. *China Popul. Resour. Environ.* **2013**, *23*, 53–60. (In Chinese).
51. Wang, B.; Wu, Y.R.; Yan, P.F. Environmental Efficiency and Environmental Total Factor Productivity Growth in China's Regional Economies. *Econ. Res. J.* **2010**, *5*, 95–109. (In Chinese)
52. Cole, M.A.; Elliott, R.J.R. FDI and the Capital Intensity of "Dirty" Sectors: A Missing Piece of the Pollution Haven Puzzle. *Rev. Dev. Econ.* **2005**, *9*, 530–548.
53. Chen, S.Y. The evaluation indicator of ecological development transition in China's regional economy. *Ecol. Indic.* **2015**, *51*, 42–52.

54. Cole, M.A.; Elliott, R.J.R.; Zhang, J. Growth, foreign direct investment, and the environment: Evidence from Chinese cities. *J. Reg. Sci.* **2011**, *51*, 121–138.
55. Chintrakarn, P.; Millimet, D.L. The Environmental Consequences of Trade: Evidence from Subnational Trade Flows. *J. Environ. Econ. Manag.* **2006**, *52*, 430–453.
56. McAusland, C.; Millimet, D.L. Do National Borders Matter? Intranational Trade, International Trade, and the Environment. *J. Environ. Econ. Manag.* **2013**, *65*, 411–437.
57. National Bureau of Statistics of China. *China Statistical Yearbook*; China Statistics Press: Beijing, China, 2001–2014.
58. The Editorial Board of China Environment Yearbook. *China Environment Yearbook*; The Press of China Environment Yearbook: Beijing, China, 2001–2014.
59. National Bureau of Statistics of China. *China Statistical Yearbook for Regional Economy*; China Statistics Press: Beijing, China, 2001–2014.
60. National Bureau of Statistics of China. *China Energy Statistical Yearbook*; China Statistics Press: Beijing, China, 2001–2014.
61. National Bureau of Statistics of China. *China Compendium of Statistics (1949–2008)*; China Statistics Press: Beijing, China, 2010.
62. Zhang, J.; Wu, G.Y.; Zhang, J.P. The estimation of China's provincial capital stock: 1952–2000. *Econ. Res. J.* **2004**, *10*, 35–44. (In Chinese).
63. Huang, J.H.; Yang, X.G.; Cheng, G.; Wang, S. A comprehensive eco-efficiency model and dynamics of regional eco-efficiency in China. *J. Clean. Prod.* **2014**, *67*, 228–238.
64. Moran, P.A. Notes on continuous stochastic phenomena. *Biometrika* **1950**, *37*, 17–23.
65. Geary, R.C. The contiguity ratio and statistical mapping. *Inc. Stat.* **1954**, *5*, 115–145.
66. Elhorst, J.P. *Spatial Econometrics: From Cross-Sectional Data to Spatial Panels*; Springer Briefs in Regional Science; Springer-Verlag: Berlin, Germany; Heidelberg, Germany, 2014.
67. Elhorst, J.P. Applied spatial econometrics: Raising the bar. *Spat. Econ. Anal.* **2010**, *5*, 9–28.
68. Elhorst, J.P.; Fréret, S. Evidence of political yardstick competition in France using a two-regime spatial Durbin model with fixed effects. *J. Reg. Sci.* **2009**, *49*, 931–951.
69. Piribauer, P. Heterogeneity in spatial growth clusters. *Empir. Econ.* **2015**, *10*, 1–22.
70. Florax, R.J.G.M.; Folmer, H.; Rey, S.J. Specification searches in spatial econometrics: The relevance of Hendry's methodology. *Reg. Sci. Urban Econ.* **2003**, *33*, 557–579.
71. Mur, J.; Angulo, A. Model selection strategies in a spatial setting: Some additional results. *Reg. Sci. Urban Econ.* **2009**, *39*, 200–213.
72. Anselin, L.; Bera, A.K.; Florax, R.; Yoon, M.J. Simple diagnostic tests for spatial dependence. *Reg. Sci. Urban Econ.* **1996**, *26*, 77–104.

Commercially Available Materials Selection in Sustainable Design: An Integrated Multi-Attribute Decision Making Approach

Rui Zhao, Han Su, Xiaolang Chen and Yanni Yu

Abstract: This paper presents an integrated multi-attribute decision-making (MADM) approach to aid selection of commercially available materials in the context of sustainable design. The MADM couples grey relational analysis (GRA) with an analytic hierarchy process (AHP) to rank alternative materials in terms of their economic, environmental, and social performance. AHP is used to determine the corresponding weighting values for the selected indicators. In addition, a case example is used to verify the proposed MADM method and demonstrate its practical application. Three alternative polymer materials, *i.e.*, poly(vinyl chloride) (PVC), polypropylene (PP), and polyethylene (PE), are examined to determine their sustainability for plastic pipe design. The associated MADM result and the limitations of the approach are discussed to lay the foundation for further improvement.

Reprinted from *Sustainability*. Cite as: Zhao, R.; Su, H.; Chen, X.; Yu, Y. Commercially Available Materials Selection in Sustainable Design: An Integrated Multi-Attribute Decision Making Approach. *Sustainability* **2016**, *8*, 79.

1. Introduction

In recent years, low-carbon development has been gradually entering product life-cycle management to combat climate change under umbrella terms such as "green design", "eco-design", "environmentally conscious design" or "design for environment" [1–3]. Green design, also called sustainable design and environmental design is a philosophy which seeks to improve design performance by incorporating environmental, health, and safety attributes into product design decisions, thus reducing the entire life-cycle impact [4–8].

Materials selection plays a crucial role in product design [9]. A number of studies have given emphasis on material selection by "design-oriented", *i.e.*, by functional demand of a product . For example, Chauhan and Vaish (2012) applied the VlseKriterijumska Optimizacija I Kompromisno Resenje (VIKOR) and the technique for order preference by similarity to ideal solution (TOPSIS) methods to soft and hard magnetic materials classification and selection [10]. Khare *et al.* (2012) proposed a multi-objective optimization model to identify the most appropriate phase change materials (PCMs) for latent heat storage design [11]. Similarly, Mansor *et al.* (2013)

employed the multi-objective optimization methodology for materials selection in design of an automotive brake lever [12]. O'Connor *et al.* (2016) proposed materials selection criteria for a new skin-wearable organic solar cell design based on its mechanical compliance [13].

Some studies have focused on materials selection by market demand. For instance, Prasad and Chakraborty (2013) used a quality function deployment (QFD)-based approach to select the suitable materials by integrating the voice of the customers with product functionality [14]. Tamani *et al.* (2015) designed a multi-criteria decision support system for fresh food packaging, in which consumers' preferences were considered [15]. Kwong *et al.* (2016) employed an artificial intelligence (AI) based methodology for materials selection in design of electric iron, by which the factors of affective design, engineering, and marketing issues were taken into account [16].

Materials selection is also oriented by cost consideration. Rahman *et al.* (2012) developed a decision support system for roofing materials selection, in which the conceptual cost for the roof element was assessed in the early stage of building design [17]. Kumar *et al.* (2014) used the entropy with TOPSIS method to evaluate the optimum material for exhaust manifold design, in which cost was taken as an important criterion [18]. Ilg *et al.* (2016) calculated life cycle cost of several high-performance materials (HPM) in infrastructure. Discount rate, expected service life and external costs were identified as the major factors in affecting the performances of materials [19].

However, most of them have not taken environmental management as a key aspect of materials selection in product design. With raw materials being extracted at an increasing rate, product design is affected by a growing public attention to sustainable development [20–23]. In this context, selection of materials should follow the principles of green design to mitigate adverse impact on both environmental and human health, to enhance product quality and to minimize toxicity of materials [23,24].

In order to provide better, safer, and more sustainable products for publics, it is important to predefine the material in the initial period of product design. Materials for further industrial manufacture that are more environmentally friendly, e.g., minimizing environmental impact and maximizing economic and social benefits, should be identified. A number of studies have taken the life cycle assessment into materials selection as an environmental impact assessment [25–29]. For instance, Ribeiro *et al.* (2013) assessed the life-cycle-based performance of injection-molded samples made of four commercial biodegradable polymers [30]. Hosseinijou *et al.* (2014) proposed social-life cycle assessment (S-LCA) for building materials selection, in which three stages as pre-building, building and post-building were taken into account [31]. Van der Velden *et al.* (2015) employed LCA approach for materials selection to promote the sustainability of smart textile garment, in which eco-cost of resource depletion, carbon footprint, human health were taken as the typical indicators [32]. However, life cycle assessment (LCA) is difficult for engineering designers to implement in real applications, due to the predicament of required data acquistion, data quality control, product system boundary division etc. [33,34]. In this context, there is a need for an approach that is simple to use.

This study contributes to the literature by developing a computational method for the use of engineers or designers to aid in the selection of commercially available materials by taking some typical indicators into account, e.g., economic, environmental, and social performance, thus promoting green design and improving product quality.

The selection of materials may be influenced by many factors, such as product function, cost, environment, or consumer satisfaction [35,36]. Because of conflicting or competing objectives, such as economic cost and environmental restrictions, the selection of materials for engineering design may be treated as a multi-criteria decision-making problem (MCDM) [37–39]. MCDM is often divided into two categories, the multiple-objective decision-making (MODM) and multiple-attribute decision-making (MADM) methods [40]. The former is an optimization technique based on prioritized functional relationships such as the maximization of sales profit or the minimization of cost. By contrast, the latter one mainly focuses on the comparison and, subsequently the ranking of alternatives based on attribute weighting or user defined data [10,41,42].

In real process of materials selection, a number of variables should be taken into account. However, only several variables can be involved when using the MODM techniques for materials selection [43,44]. MADM methods such as analytic hierarchy process (AHP), TOPSIS, VIKOR, and Grey Relational Analysis (GRA) are able to rank material alternatives and help designers determine the best material, defined as the highest degree of satisfaction [45–49].

The MADM methodology for materials selection in our study provides an important decision-making tool when confronted with a number of closely differing materials for diversified needs. This study further integrates two common MADM methodologies: AHP, used for weighting of selected indicators, and GRA for further comparison. GRA is based on the measurement of the degrees of similarity or difference among several numerical sequences, which is used to identify the relative influences of the sequences for comparison with a predetermined reference sequence [50,51]. The GRA gives the preferential order of various investigated objects in terms of a certain performance index. However, the relative importance of the various indices is difficult to determine, e.g., some of the indicators may be conflicting [52]. In comparison, the analytic hierarchy process (AHP) is a classic method to fill this gap, as it is used to assess the relative importance among indices. Generally, the purpose of AHP is to decompose a complex decision issue into a variety of hierarchies and factors, which is a typical subjective weighting method based upon the judgment and synthesis of decision makers [53]. An ordered hierarchical decision is thus composed of a dominance relationship in which the weights among different indicators are determined through pairwise comparison [54]. Therefore, integration of the two methods can give rise to a maximization of their advantages, which also facilitates the multi-attributive decision-making of a complex system.

This study further gives preferences of selection between similar materials (with similar physical and chemical properties) to examine their performance. A case example is used to verify the proposed MADM method and demonstrate its practical application. Three alternative polymer materials, *i.e.*, poly(vinyl chloride) (PVC), polypropylene (PP), and polyethylene (PE), are examined to determine their sustainability for plastic pipe design. It is further expected that this integrated approach may help manufacturers to implement sustainable design, and thus to enhance product sustainability.

2. Integrated MADM Approach for Materials Selection

As environmental resources are increasingly depleted, products are gradually being developed with less impact to the environment, thus moving towards "sustainability." In this context, this study has identified a number of typical indicators to represent product sustainability as a basis for materials selection, based upon the basic principles and guidelines of sustainable design to construct a decision matrix, shown in Table 1 [2,24,55,56].

Table 1. Selected indicators for material selection.

Attributes	Indicators
Economic	Direct cost Manufacturing cost Manufacture defects Sales profit
Environmental	Energy consumption Health risk Waste-gas production Wastewater production Solid waste production
Social	Product satisfaction Material recyclability Material reusability Use of recycled Materials

For any manufacturer, there are triple bottom lines (TBL) to run a sustainable business from bottom to top, *i.e.*, as basic market demand, alternative materials and energy for production, and pursuing societal value for environmental protection [57–62]. The manufacturer is primarily driven by profit, as economic issues play a dominant role in business operations [61,63–65]. Thus, four economic indicators are taken into consideration including direct cost, manufacturing cost, manufacture defects and sales profit. With the conflicts between economic growth and environmental protection being paid more and more attention, the most preferable strategy for encompassing sustainable development is to implement cleaner production [66,67].The mitigation of adverse impact of products on environment and human health can be achieved partly by the careful selection of materials. Thus, five indicators are proposed to represent the product environmental performances: energy consumption, health risk, waste-gas production, wastewater production, and solid waste production [68,69].

The social attributes proposed in this study is from a perspective of corporate social responsibility (CSR), in order to extend manufacturer's responsibility through the improvement of product sustainability, such as improving product life cycle (including reusability and recyclability), reforming the quality of products, *etc.*, to provide better, safer and sustainable products for public consumption [70–72]. Proposed by McDonough and Brauggart (2002), all materials can be classified into two categories in a closed-loop social system: "technical" or "biological" nutrients [73]. The technical nutrients are those non-toxic and non-harmful materials, which can be reused directly in continuous cycles without losing their integrity or quality, and parts of which can be recycled for their quality and functionality, e.g., the cola bottle [74]. In this context, "material reusability", "material recyclability" and "use of recycled materials" are selected to reflect the manufacturer's social responsibility for product waste prevention.

The use of these evaluation indicators, by taking economic, environmental and social dimensions into account, is expected to help improve and monitor process of product design [75,76]. The proposed MADM approach combines GRA with AHP to rank alternative materials by their economic, environmental, and social performance. AHP is used to determine the weights of the selected indicators. GRA is then used to develop an evaluation approach to aid material selection. The detailed computational approach is presented below.

2.1. Weighting by AHP

Generally, the weighting approach can be divided into two categories: objective and subjective weighting determination [77]. The former obtains the weights derived from the collected data corresponding to each indicator. However, the different indicators may result in weighting variation, and the actual importance of indicators could be conflicted [78]. The latter is mainly dependent upon the decision maker's preferences regarding the attributes, e.g., AHP is a subjective decision-making tool, which is widely used for weighting determination [79]. Generally speaking, there are three steps to the AHP calculation [80]:

(1) decompose an issue into a hierarchy system: the above attributes are classified into three categories: economic, environmental, and social,

(2) establish a pairwise comparison matrix to compare a set of n attributes in relative importance:

$$A = \begin{bmatrix} a_{11} & \cdots & a_{1j} & \cdots & a_{1n} \\ \vdots & & \vdots & & \vdots \\ a_{i1} & \cdots & a_{ij} & \cdots & a_{in} \\ \vdots & & \vdots & & \vdots \\ a_{n1} & \cdots & a_{nj} & \cdots & a_{nn} \end{bmatrix} \quad (1)$$

where $a_1, a_2, ..., a_n$, denote various attributes, $a_{ij} = 1/a_{ji}$ and $a_{ij} = a_{ik}/a_{jk}$ indicate how much more important objective i is than objective j to the decision maker. The relative importance is measured by using a scale from 1 to 9, where 1 represents that objective i and j are equally important, 3 represents that objective i is just slightly more important than j, 5 represents objective i is somewhat more important than j, 7 represents that objective i is much more important than j, and 9 represents that objective i is far more important than j [53,79]. We measure the relative importance by using the scale 1 to 5 in this study.

(3) determine the relative weights; as a_{ij} approximates to w_i/w_j, i.e., $a_{ij} \cong w_i/w_j$, matrix A can be transformed as follows:

$$W = \begin{array}{c} \\ w_1 \\ \vdots \\ w_i \\ \vdots \\ w_n \end{array} \begin{array}{c} \begin{array}{ccccc} w_1 & \cdots & w_j & \cdots & w_n \end{array} \\ \begin{bmatrix} w_1/w_1 & \cdots & w_1/w_j & \cdots & w_1/w_n \\ \vdots & & \vdots & & \vdots \\ w_i/w_1 & \cdots & w_i/w_j & \cdots & w_i/w_n \\ \vdots & & \vdots & & \vdots \\ w_n/w_1 & \cdots & w_n/w_j & \cdots & w_n/w_n \end{bmatrix} \end{array} \quad (2)$$

The relative weights are derived by addressing the eigenvector w with the largest eigenvalue $\lambda_{\max}$ of matrix A, which should satisfy the equation $Aw = \lambda_{\max}w$. In addition, the consistency index *C.I.* and the consistency ratio *C.R.*, as two examination indicators, should be calculated to ensure the accuracy of the relative weights and the consistency of the subjective decision. These are:

$$C.I. = (\lambda_{\max} - n)/(n - 1) \quad (3)$$

$$C.R. = C.I./R.I. \tag{4}$$

where *R.I.* is a random consistency index which varies with the size of the matrices, e.g., 1.11 for a 5×5 matrix. *C.I.* should be less than 0.1 in order for a suitable result [53,80].

2.2. Grey Relational Analysis for Materials Evaluation

GRA or grey incidence analysis, proposed by Professor Deng Julong in 1984, has been developed as a quantitative approach to aid decision-making by examining the degree of similarity or difference between two sequences based upon the grade of relation [50,81–83]. The more similar the two sequences are, the greater the grey relational correlation is, and *vice versa* [50,84].

To use GRA, it is necessary to normalize the data related to the sequence of selected criteria to a common scale for comparison. Basically, there are three approaches for normalization [50,51]:

(1) If the ideal value of sequence of selected criteria is preferred to be larger, such as indicators of sales profit, materials recyclability, or reusability, the consequence can be normalized as follows:

$$\beta_i(k) = \frac{\beta_i(k) - \min\beta_i(k)}{\max\beta_i(k) - \min\beta_i(k)} \tag{5}$$

where $\beta_i(k)$ is the comparable sequence of selected evaluation criteria.

(2) If the ideal value of sequence of selected criteria is preferred to be smaller, such as direct cost, manufacturing cost, or energy consumption, the consequence can be normalized as follows:

$$\beta_i(k) = \frac{\max\beta_i(k) - \beta_i(k)}{\max\beta_i(k) - \min\beta_i(k)} \tag{6}$$

where $\beta_i(k)$ is the comparable sequence of selected evaluation criteria.

(3) If the ideal value is determined within a numerical value range, the sequence can be normalized as follows:

$$\beta_i(k) = 1 - \frac{|\beta_i(k) - \beta_i|}{\max\beta_i(k) - \beta_i} \tag{7}$$

where $\beta_i(k)$ is the comparable sequence of selected evaluation criteria, and β_i is the determined ideal value.

Following the above normalization, a grey relational coefficient is calculated. It demonstrates the relationship between the non-dimensional expected optimal sequence and the original sequence [24,85]. The grey relational coefficient is:

$$\delta_i(k) = \frac{\min_i \min_j ||\beta_0(k) - \beta_j(k)|| + \varepsilon \max_i \max_j ||\beta_0(k) - \beta_j(k)||}{||\beta_0(k) - \beta_j(k)|| + \varepsilon \max_i \max_j ||\beta_0(k) - \beta_j(k)||} \tag{8}$$

where ε is defined as the distinguishing coefficient and $\varepsilon \in (0,1)$. Generally, ε is set to 0.5 for further calculation [84].

Thus, the degree of grey incidence can be expressed as:

$$\gamma_i = \sum_{k=1}^{n} w_k \cdot \delta_i(k) \tag{9}$$

where w_k is the weighting value of the selected indicator k derived from the AHP approach.

The degree of grey incidence indicates the relationship between the comparability sequence and the reference sequence [83,86]. The reference sequence represents the best performance that can be achieved by any among the comparability sequences. When γ_i grows much bigger, the comparability sequence approaches the reference sequence. When γ_i reaches the maximum value, a comparability sequence of the corresponding material obtains the highest degree of grey incidence, which means the alternative material is better than its counterparts. Moreover, the performance of all the alternative materials can be ranked in a sequence, from most optimal to least.

3. A Case Example

A case example is presented in this section to demonstrate the application of the proposed methodology of materials selection for plastic pipe design. The plastic pipes are widely used in the water supply system of Sichuan Province, China, especially for indoor hot and cold water supply. They are manufactured out of three alternative materials: poly(vinyl chloride) (PVC), polypropylene (PP) and polyethylene (PE). Here, the PVC is specified as un-plasticized PVC (uPVC), with a synthesis of 15% chlorinated polyethylene (CPE) and 75% PVC, and PP as a polypropylene random (PPR), with 5% PE and 95% PP.

The case example employs a set of weight values determined by the principal engineer of the investigated manufacturer and adjusted by the retrieval results of the selected indicators from the online Scirus scientific database in order to avoid subjective judgment in comparing their importance. Scirus.com is a comprehensive academic search engine with over 440 million scientific items, including published scientific papers, patents, web sources, courseware, and current research progress [87]. Based on the retrieval results of each indicator using "hits" as the measure of importance, the relative importance is ranked from 1 to 13, as shown in Figure 1.

The grey relational result will be compared with the actual material selection in the investigated manufacturer of plastic pipe to verify the proposed approach. Table 2 shows how the selected indicators for the three alternative materials are measured.

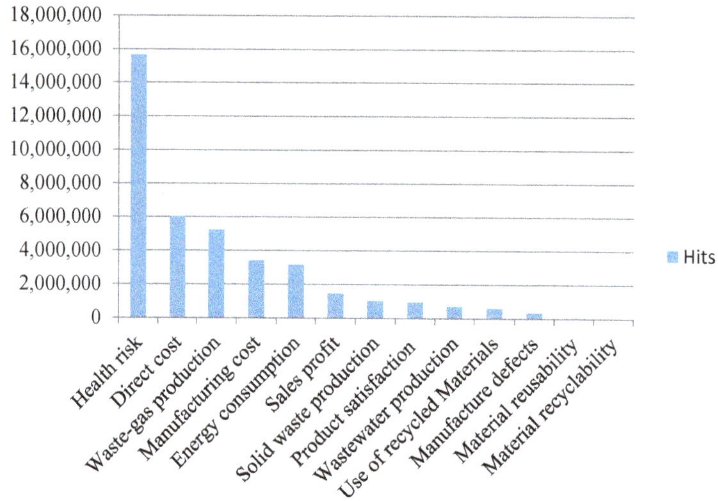

Figure 1. Rank of selected indicators as determined from Scirus.com retrieval.

The data for the selected indicators are derived from a feasibility study report of a plastic pipe manufacturer, located in Leshan City, Sichuan Province, China. According to the measurements, the data have been processed dimensionlessly by percentage for further application of GRA, as shown in Table 3.

Table 2. Measurements of the selected indicators.

Indicators	Measurements
Direct cost	$p_i = \sum_{i=1}^{n} c_i/C$ where p_i is the ratio of direct cost of the investigated pipe manufacturer, c_i is the material cost corresponding to the *i*th activity, e.g., $i = 1$ denotes the purchasing cost of a specific material (PVC, PP, PE), $i = 2$ denotes transportation cost of a specific material (PVC, PP, PE), $i = 3$ denotes labor cost of a specific material (PVC, PP, PE), and C is the total investment the investigated pipe manufacturer.
Manufacturing cost	$p_m = c_m/C$ where p_m is the ratio of the investigated plastic pipe manufacturing cost, c_m is the plastic pipe manufacturing cost of a specific material (PVC, PP, or PE), and C is the total investment of the pipe manufacturer.
Manufacture defects	$p_d = M_d/Q_0$ where p_d is the ratio of the investigated plastic pipe manufacture defects, M_d is the annual quantity of the manufacture defects of the investigated plastic pipe manufacturer, and Q_0 is the annual production of plastic pipes.
Sales profit	$p_p = c_p/c_s$ where p_p is the ratio of sales profit of the investigated plastic pipe manufacturer, c_p is the expected sales profit from using a specific material for plastic pipe production (PVC, PP, or PE), and c_s is the sales revenue of the investigated plastic pipe manufacturer.
Energy consumption	$p_e = e_i/\sum_{i=1}^{n} e_i$ where p_e is the ratio of energy consumption of the investigated plastic pipe manufacturer, e_i is the energy consumption from using the *i*th fuel, e.g., $i = 1$ denotes the natural gas consumption, $i = 2$ denotes electricity consumption, and $i = 3$ denotes coal consumption.
Health risk	Health risk is defined as the potential harm to human health in this study, especially the toxic and poisonous components posed to the workers in the material processing and manufacturing [88,89]. $p_h = r_i/\sum_{i=1}^{n} r_i$ where p_h is the ratio of health risk, r_i is the toxic risk from using the *i*th material for plastic pipe production, *i.e.*, $i = 1$ denotes PVC, $i = 2$ denotes PP, and $i = 3$ PE. The toxic risk r_i resulted from using the *i*th material (PP, PE, PVC) is given as follows [90]: $r_i = P_i C_i$ where P_i is the probability of personal injury resulting from the *i*th material process or manufacture, C_i the consequence of the personal injury, such as the number of workers suffered by allergic rhinitis, dermatosis *etc.*

Table 2. *Cont.*

Indicators	Measurements
Waste-gas production	$p_g = Q_g/Q_o$ where p_g is the ratio of waste-gas production of the investigated plastic pipe manufacturer, Q_g is the annual emission of waste gas from using a specific material for production (PVC, PP, or PE), and Q_0 is the annual production of plastic pipes.
Wastewater production	$p_w = Q_w/Q_o$ where p_w is the ratio of waste water production of the investigated plastic pipe manufacturer, Q_w is the annual emission of waste water from using a specific material for production (PVC, PP, or PE), and Q_0 is the annual production of plastic pipes.
Solid waste production	$P_s = Q_s/Q_o$ where p_s is the ratio of solid waste production of the investigated plastic pipe manufacturer, Q_w is the annual emission of solid waste from using a specific material for production (PVC, PP, or PE), and Q_0 is the annual production of plastic pipes.
Product satisfaction	p_z defined as the market share of plastic pipes while using a specific material (PVC, PP, or PE), which is based upon the market of the downstream distribution centers.
Material recyclability	Material recyclability is the extent to which waste materials can be converted into reusable material, which are used to make new items, given as follows [91]: $p_r = Q_r/Q$ where Q_r is the theoretical quantity of recycled materials (PVC, PP, or PE), and Q is the total quantity of waste materials produced of the investigated plastic pipe manufacturer.
Material reusability	Material reusability is defined as the extent to which the material can be reused directly in continuous cycles without losing their integrity or quality, which is given as follows [92]: $p_R = Q_R/Q$ where Q_R is the theoretical quantity of reused materials (PVC, PP, or PE), and Q is the total quantity of waste materials produced of the investigated plastic pipe manufacturer.
Use of recycled Materials	$p_t = Q'/Q_r$ where $Q\prime$ is the actual recycled materials (PVC, PP, or PE), and Q_r is the theoretical quantity of recycled materials of the investigated plastic pipe manufacturer.

Table 3. Processed data of the selected indicators.

Indicators	Alternative Materials		
	PE	PP	PVC
Direct cost	95.5	90.3	90.0
Manufacturing cost	1.43	1.40	10.00
Manufacture defects	1.5	2.0	2.0
Sales profit	29.4	20.0	38.2
Energy consumption	92.5	99.6	39.7
Health risk	0	0	0.00084
Waste-gas production	0.006	1.130	0.230
Wastewater production	11.0	150.0	45.5
Solid waste production	0.09	50.00	0.54
Product satisfaction	25	10	55
Material recyclability	94.50	0.28	65.90
Material reusability	30.7	2.5	16.5
Use of recycled Materials	100	100	98

4. Results and Discussion

Based upon the relative importance ranking list, the weighting value is calculated by the AHP method, as shown in Table 4. In order to simplify the computation, both AHP and GRA have been embodied into a compiled mathematical program implemented by MATLAB 2010. The degree of grey incidence γ_i of the PVC, PP, and PE is calculated as 0.800457709, 0.634945104, and 0.604756842, respectively. Different evaluated alternatives can be ranked in a sequence from the highest to the lowest grade. For instance, the ranking of degrees of grey incidence is deemed as the order of superiority for each material in this study. The assessed alternative material corresponding to the maximum grey correlative degree is deemed as the optimal material, and vice versa. Thus, $\gamma_{PVC} > \gamma_{PP} > \gamma_{PE}$ is obtained, which shows that the PVC is the optimum alternative material and PE is the least optimal choice.

From Table 4, it is clear that the indicator "health risk" accounts for about 40% of the weightings. This is because the manufacturer's employees are likely exposed to toxic substances when PVC materials are processed [93]. However, PVC has the highest sales profit and market share (product satisfaction), along with the lowest proportion of direct cost and energy consumption. That is why PVC exhibits the best overall performance among the three alternative materials, even if its inherent health risk is higher than PP or PE. Similarly, PP's degree of grey incidence is higher than PE, although PE material is superior to PP based on most of the data for the selected indicators. This is because PP shows better performance in both direct cost and manufacturing cost, which accounts for a large proportion of the relative importance. This shows that the weighting value may play a critical role in the sensitivity of the grey relational model in this case study.

Table 4. Weighting value of the selected indicators based on the Scirus.com retrieval results.

Indicators	Weighting Value
Direct cost	0.155623917
Manufacturing cost	0.088494826
Manufacture defects	0.008907485
Sales profit	0.038003123
Energy consumption	0.082127819
Health risk	0.404081921
Waste-gas production	0.135173217
Wastewater production	0.018335576
Solid waste production	0.027011282
Product satisfaction	0.024883922
Material recyclability	0.000545566
Material reusability	0.000975515
Use of recycled Materials	0.015835831

In the real world, a manufacturer may define profits as the main criterion of business success, while placing little emphasis on life-cycle environmental impact of its products [94,95]. Accordingly, the selection of materials is strongly influenced by market demand, cost-benefit analysis, materials properties, and fashion and popularity, as the fundamental drivers of design [24]. For instance, the investigated manufacturer in this study was originally cost-oriented in material selection, and PVC was used for plastic pipe manufacturing. This verifies the evaluation results of the integrated MADM approach. However, a sustainable business should focus on both economic prosperity, environmental and social performances [96,97]. In green design, the weighting assignment for materials selection should pay closer attention to the environmental and social perspectives. This could be expressed by increased weighting values for materials recyclability and reusability to enhance the sustainability of products.

This integrated MADM approach is flexible, which can provide a comprehensive matrix to judge a range of alternative materials, including index of environmental impact, economic perspective, *etc.* For instance, a similar study can be found by Zhao *et al.* (2012), who has employed "binary-dominance matrix" with GRA to aid decision-making on materials selection while taking environmental evaluation into account. Five PVC materials with close properties have been subjected to environmental evaluation to determine their environmentally friendliness for handbag manufacture [24]. Furthermore, Kumar *et al.* (2014) have offered a MCDM approach in which criteria importance was computed by entropy method, and the alternative materials were ranked by using the TOPSIS method [18]. Wang *et al.* (2015) have employed AHP combined with GRA to select a biomass briquette fuel-system scheme based on a variety of evaluation criteria, such as economy, cleanness and environment protection, production capacity, product quality and production stability [52]. Yang *et al.* (2015) have focused on the materials selection from remanufacturing perspective and developed a MCDM model by using the Fuzzy TOPSIS approach to evaluate the performance of the alternative automotive materials [98]. These similar studies have reflected the extension and validity of the proposed method.

However, the weightings play a critical role to affect the sensitivity of the GRA, which is subject to the preference of the indicators. AHP is a subjective weighting approach, depending upon the judgment and synthesis of decision makers [99,100]. It may be replaced by the objective weighting methods, such as the Entropy method, TOPSIS, *etc.*, in which the weights are derived from the collected data corresponding to each indicator. Thus, the influence of weighting variation on materials selection can be investigated to identify the sensitivity of the integrated approach.

5. Conclusions

This paper introduces a computational approach for materials selection in the context of green design by using GRA, integrated with AHP. A set of selected indicators is determined using the principles of green design, while the AHP method is used to examine the relative importance and assign the corresponding weighting values for each indicator. Moreover, GRA is applied to rank the alternative materials based upon the multiple attributes of their performances. In order to demonstrate how to use the computational approach, a case example is provided. Three materials, PVC, PP, and PE, are investigated to determine their best economic, environmental, and social performances for plastic pipe design. With the weighting value for the selected criteria being determined by the principal engineer of the investigated manufacturer and adjusted by keyword retrieval from the Scirus database, PVC is determined as the optimal material, and PE the least optimal.

However, limitations of this computational approach still remain. Although the selected criteria are generic and may be easily extended to various applications, whether these indicators can be used to fully cover the three dimensions of sustainability, namely, the economic, environmental, and social dimensions, still requires further investigation. Second, the weighting determination should be re-adjusted by taking the engineering designer's preferences into account, while giving full priority to environmental and social considerations. This approach is expected to be applied to more case scenarios in order to examine and refine its validity and sensitivity in the future.

Acknowledgments: This study is sponsored by National Natural Science Foundation of China (No. 41301639; No. 41571520; No. 41461118), Sichuan Provincial Key Technology Support (No. 2014GZ0168), and Innovative Research Team of Ministry of Education of China (No. IRT13092).

Author Contributions: Rui Zhao was involved in conceptualizing the whole study, proposing the integrated MADM model for material selection and writing the whole paper. Han Su implemented the calculation of the MADM model. Xiaolang Chen conducted the case study to collect the data and verify the proposed model. Yanni Yu improved the whole structure of the paper.

Conflicts of Interest: The authors declare no conflict of interest.

References

1. Brezet, H.; van Hemel, C. *Eco-Design: A Promising Approach to Sustainable Production and Consumption*; UNEP and TU Delft: Delft, The Netherlands, 1997.
2. Fiksel, J. *Design for Environment: A Guide to Sustainable Product Development*; McGraw-Hill: New York, NY, USA, 2009.
3. Bevilacqua, M.; Ciarapica, F.E.; Giacchetta, G. *Eco-Design: A Promising Approach to Sustainable Production and Consumption*; Springer: London, UK, 2011.
4. Ray, D.L.; Guzzo, L. *Environmental Overkill: Whatever Happened to Common Sense*; Harper Collins: New York, NY, USA, 1993.
5. McLennan, J.F. *The Philosophy of Sustainable Design*; Ecotone LLC: Kansas City, MO, USA, 2004.
6. Aoe, T. Eco-efficiency and eco design in electrical and electronic products. *J. Clean. Prod.* **2007**, *15*, 1406–1414.
7. Clark, G.; Kosoris, J.; Hong, L.N.; Crul, M. Design for Sustainability: Current Trends in Sustainable Product Design and Development. *Sustainability* **2009**, *1*, 409–424.
8. He, B.; Huang, S.; Wang, J. Product Low-Carbon Design Using Dynamic Programming Algorithm. *Int. J. Precis. Eng. Manuf.-Green Technol.* **2015**, *2*, 37–42.
9. Ashby, M.F.; Johnson, K. *Materials and Design: The Art and Science of Material Selection in Product Design*; Butterworth-Heinemann: Oxford, UK, 2013.
10. Chauhan, A.; Vaish, R. Magnetic material selection using multiple attribute decision-making approach. *Mater. Des.* **2012**, *36*, 1–5.
11. Khare, S.; Dell'Amico, M.; Knight, C.; McGarry, S. Selection of materials for high temperature latent heat energy storage. *Sol. Energy Mater. Sol. Cells* **2012**, *107*, 20–27.
12. Mansor, M.R.; Sapuan, S.M.; Zainudin, E.S.; Nuraini, A.A.; Hambali, A. Hybrid natural and glass fibers reinforced polymer composites material selection using Analytical Hierarchy Process for automotive brake lever design. *Mater. Des.* **2013**, *51*, 484–492.
13. O'Connor, T.F.; Zaretski, A.V.; Savagatrup, S.; Printz, A.D.; Wilkes, C.D.; Diaz, M.I.; Lipomi, D.J. Wearable organic solar cells with high cyclic bending stability: Materials selection criteria. *Sol. Energy Mater. Sol. Cells* **2016**, *144*, 438–444.
14. Prasad, K.; Chakraborty, S. A quality function deployment-based model for material selection. *Mater. Des.* **2013**, *49*, 525–535.

15. Tamani, N.; Mosse, P.; Croitoru, M.; Buche, P.; Guillard, V.; Guillaume, C.; Gontard, N. An argumentation system for eco-efficient packaging material selection. *Comput. Electron. Agric.* **2015**, *113*, 174–192.
16. Kwong, C.K.; Jiang, H.; Luo, X.G. AI-based methodology of integrating affective design, engineering, and marketing for defining design specifications of new products. *Eng. Appl. Artif. Intell.* **2016**, *47*, 49–60.
17. Rahman, S.; Odeyinka, H.; Perera, S.; Bi, Y. Product-cost modelling approach for the development of a decision support system for optimal roofing material selection. *Expert Syst. Appl.* **2012**, *39*, 6857–6871.
18. Kumar, R.; Jagadish; Ray, A. Selection of Material for Optimal Design Using Multi-Criteria Decision Making. *Procedia Mater. Sci.* **2014**, *6*, 590–596.
19. IIg, P.; Hoehne, C.; Guenther, E. High-performance materials in infrastructure: A review of applied life cycle costing and its drivers—The case of fiber-reinforced composites. *J. Clean. Prod.* **2016**, *112*, 926–945.
20. Tseng, K.C.; Lin, B.; Han, C.M. An Intelligent System for Sustainable Product Design at the Concept Development Stage. *Comput. Aided Des. Appl.* **2012**, *9*, 397–408.
21. Riberto, I.; Peças, P.; Henriques, E. A Life Cycle Framework to Support Materials Selection for Eco Design: A Case Study on Biodegradable Polymers. *Mater. Des.* **2013**, *51*, 300–308.
22. Gerstlberger, W.; Knudsen, M.P.; Stampe, I. Sustainable Development Strategies for Product Innovation and Energy Efficiency. *Bus. Strategy Environ.* **2014**, *23*, 131–144.
23. Beng, L.G.; Omar, B. Integrating Axiomatic Design Principles into Sustainable Product Development. *Int. J. Precis. Eng. Manuf.-Green Technol.* **2014**, *1*, 107–117.
24. Zhao, R.; Neighbour, G.; Deutz, P.; McGuire, M. Materials Selection for Cleaner Production: An Environmental Evaluation Approach. *Mater. Des.* **2012**, *37*, 429–434.
25. Umezawa, O.; Shinohara, Y.; Halada, K. New Aspects of Ecomaterials from the Viewpoints of the Consumer and Regional Communities. *Mater. Trans.* **2014**, *55*, 745–749.
26. Giudice, F.; la Rosa, G.; Risitano, A. Materials selection in the life-cycle design process: A method to integrate mechanical and environmental performances in optimal choice. *Mater. Des.* **2005**, *26*, 9–20.
27. Bribián, I.Z.; Capilla, A.V.; Us6n, A.A. Life cycle assessment of building materials: Comparative analysis of energy and environmental impacts and evaluation of the eco-efficiency improvement potential. *Build. Environ.* **2011**, *46*, 1133–1140.
28. Chan, H.K.; Wang, X.; Raffoni, A. An integrated approach for green design: Life-cycle, fuzzy AHP and environmental management accounting. *Br. Account. Rev.* **2014**, *46*, 344–360.
29. Pereira, L.; Mafalda, R.; Marconcini, J.M.; Mantovani, G.L. The Use of Sugarcane Bagasse-Based Green Materials for Sustainable Packaging Design. In *ICoRD'15—Research into Design Across Boundaries*; Chakrabarti, Ed.; Springer: London, UK, 2014; Volume 2, pp. 113–123.
30. Ribeiro, I.; Peças, P.; Henriques, E. A life cycle framework to support material selection for Ecodesign: A case study on biodegradable polymers. *Mater. Des.* **2013**, *51*, 300–308.

31. Hosseinijou, S.A.; Mansour, S.; Shirazi, M.A. Social life cycle assessment for material selection: A case study of building materials. *Int. J. Life Cycle Assess.* **2014**, *19*, 620–645.
32. Van der Velden, N.M.; Kuusk, K.; Köhler, A.R. Life cycle assessment and eco-design of smart textiles: The importance of material selection demonstrated through e-textile product redesign. *Mater. Des.* **2015**, *84*, 313–324.
33. Caputo, J.; Balogh, S.B.; Volk, T.A.; Johnson, L.; Puettmann, M.; Lippke, B.; Oneil, E. Incorporating uncertainty into a life cycle assessment (LCA) model of short-rotation willow biomass (*Salix* spp.) crops. *BioEnergy Res.* **2014**, *7*, 48–59.
34. Chen, X.; Corson, M.S. Influence of emission-factor uncertainty and farm-characteristic variability in LCA estimates of environmental impacts of French dairy farms. *J. Clean. Prod.* **2014**, *81*, 150–157.
35. Van Kesteren, I.E.H. Product Designers' Information Needs in Materials Selection. *Mater. Des.* **2008**, *29*, 133–145.
36. Ashby, M.F.; Shercliff, H.; Cebon, D. *Materials: Engineering, Science, Processing and Design*; Butterworth-Heinemann: Oxford, UK, 2013.
37. Rao, R.V.; Patel, B.K. A subjective and objective integrated multiple attribute decision-making method for material selection. *Mater. Des.* **2010**, *31*, 4738–4747.
38. Huang, H.H.; Zhang, L.; Liu, Z.F.; Sutherland, J.W. Multi-Criteria Decision Making and Uncertainty Analysis for Materials Selection in Environmentally Conscious Design. *Int. J. Adv. Manuf. Technol.* **2011**, *52*, 421–432.
39. Athawale, V.M.; Chakraborty, S. Material Selection Using Multi-Criteria Decision-Making Methods: A Comparative Study. *Proc. Insist. Mech. L* **2012**, *226*, 266–285.
40. Jahan, A.; Edwards, K.L. *Multi-Criteria Decision Analysis for Supporting the Selection of Engineering Materials in Product Design*; Butterworth-Heinemann: Oxford, UK, 2013.
41. Zhou, C.C.; Yin, G.F.; Hu, X.B. Multi-Objective Optimization of Material Selection for Sustainable Products: Artificial Neural Networks and Genetic Algorithm Approach. *Mater. Des.* **2009**, *30*, 1209–1215.
42. Chauhan, A.; Vaish, R. Hard Coating Material Selection Using Multi-Criteria Decision Making. *Mater. Des.* **2013**, *44*, 240–245.
43. Karande, P.; Chakraborty, S. Application of Multi-Objective Optimization on the Basis of Ratio Analysis (MOORA) Method for Materials Selection. *Mater. Des.* **2012**, *37*, 317–324.
44. Zander, J.; Sandström, R. Materials Selection with Several Sizing Variables Taking Environmental Impact into Account. *Mater. Des.* **2012**, *37*, 243–250.
45. Jahan, A.; Mustapha, F.; Ismail, M.Y.; Sapuan, S.M.; Bahraminasab, M. A Comprehensive VIKOR Method for Materials Selection. *Mater. Des.* **2011**, *32*, 1215–1221.
46. Girubha, R.J.; Vinodh, S. Application of Fuzzy VIKOR and Environmental Impact Analysis for Materials Selection of an Automobile Component. *Mater. Des.* **2012**, *37*, 478–486.
47. Jahan, A.; Barhraminasab, M.; Edwards, K.L. A Target-Based Normalization Technique for Materials Selection. *Mater. Des.* **2012**, *35*, 647–654.

48. Caliskan, H.; Kursuncu, B.; Kurbanoglu, C.; Guven, S.Y. Material Selection for the Tool Holder Working under Hard Milling Conditions Using Different Multi Criteria Decision Making Methods. *Mater. Des.* **2013**, *45*, 473–479.
49. Chan, J.W.K.; Tong, T.K.L. Multi-Criteria Material Selections and End-of-Life Product Strategy: Grey Relational Analysis Approach. *Mater. Des.* **2007**, *28*, 1539–1546.
50. Deng, J.L. *The Essential Methods of Grey Systems*; Huazhong University of Science and Technology Press: Wuhan, China, 1992. (In Chinese)
51. Wei, G. Grey relational analysis model for dynamic hybrid multiple attribute decision-making. *Knowl.-Based Syst.* **2011**, *24*, 672–679.
52. Wang, Z.; Lei, T.; Chang, X.; Shi, X.; Xiao, J.; Li, Z.; Yang, S. Optimization of a biomass briquette fuel system based on grey relational analysis and analytic hierarchy process: A study using cornstalks in China. *Appl. Energy* **2015**.
53. Saaty, T.L. Analytic hierarchy process. In *Encyclopedia of Operations Research and Management Science*; Harris, G., Ed.; Springer: London, UK, 2001; pp. 19–28.
54. Xu, G.; Yang, Y.P.; Lu, S.Y.; Li, L.; Song, X. Comprehensive evaluation of coal-fired power plants based on grey relational analysis and analytic hierarchy process. *Energy Policy* **2011**, *39*, 2343–2351.
55. Vezzoli, C.; Manzini, E. *Design for Environmental Sustainability*; Springer: London, UK, 2008.
56. Dornfeld, D.A. Moving towards Green and Sustainable Manufacturing. *Int. J. Precis. Eng. Manuf.-Green Technol.* **2014**, *1*, 63–66.
57. Stewart, W.H., Jr.; Carland, J.C.; Carland, J.W.; Watson, W.W.; Sweo, R. Entrepreneurial dispositions and goal orientations: A comparative exploration of United States and Russian entrepreneurs. *J. Small Bus. Manag.* **2003**, *41*, 27–46.
58. Smith, P. The Relationship between Management Accounting, Profitability and Operations in an Uncertain World: Evidence from Literature and Practice. Ph.D. Thesis, Cranfield University, Cranfield, UK, 2007.
59. Kane, G. *The Three Secrets of Green Business: Unlocking Competitive Advantage in a Low Carbon Economy*; Earthscan: London, UK, 2010.
60. Wang, H.F.; Gupta, S.M. *Green Supply Chain Management: Product Life Cycle Approach*; McGraw-Hill: New York, NY, USA, 2011.
61. Amah, E. Corporate Culture and Organizational Effectiveness: A Study of the Nigerian Banking Industry. *Eur. J. Bus. Manag.* **2012**, *4*, 212–229.
62. Zhao, R.; Peng, D.; Li, Y. An Interaction between Government and Manufacturer in Implementation of Cleaner Production: A Multi-Stage Game Theoretical Analysis. *Int. J. Environ. Res.* **2015**, *9*, 1069–1078.
63. Mezher, T.; Ajam, M. Integrating Quality, Environmental and Supply Chain Management Systems into the Learning Organisation. In *Greening the Supply Chain*; Sarkis, Ed.; Springer: London, UK, 2006; pp. 67–85.
64. Gummesson, E. Commentary on "The role of innovation in driving the economy: Lessons from the global financial crisis". *J. Bus. Res.* **2014**, *67*, 2743–2750.

65. Dubey, R.; Gunasekaran, A.; Ali, S.S. Exploring the relationship between leadership, operational practices, institutional pressures and environmental performance: A framework for green supply chain. *Int. J. Prod. Econ.* **2015**, *160*, 120–132.
66. Testa, F.; Iraldo, F. Shadows and lights of GSCM (green supply chain management): Determinants and effects of these practices based on a multinational study. *J. Clean. Prod.* **2010**, *18*, 953–962.
67. Madanhire, I.; Mugwindiri, K.; Mbohwa, C. Enhancing cleaner production application in fertilizer manufacturing: Case study. *Clean. Technol. Environ.* **2015**, *17*, 667–679.
68. Ljungberg, L.Y. Materials selection and design for development of sustainable products. *Mater. Des.* **2007**, *28*, 466–479.
69. Serafini, M.; Russo, D.; Rizzi, C. Multi Criteria Material Selection for Eco-Design. *Comput. Aided Des. Appl.* **2015**, *12*, 526–536.
70. Binhussain, M.A.; Marangoni, M.; Bernardo, E.; Colombo, P. Sintered and glazed glass-ceramics from natural and waste raw materials. *Ceram. Int.* **2014**, *40*, 3543–3551.
71. Akadiri, P.O.; Olomolaiye, P.O.; Chinyio, E.A. Multi-criteria evaluation model for the selection of sustainable materials for building projects. *Autom. Constr.* **2013**, *30*, 113–125.
72. Piccinno, F.; Hischier, R.; Saba, A.; Mitrano, D.; Seeger, S.; Som, C. Multi-perspective application selection: A method to identify sustainable applications for new materials using the example of cellulose nanofiber reinforced composites. *J. Clean. Prod.* **2016**, *112*, 1199–1210.
73. William, M.; McDonough, W. *Cradle to Cradle: Remaking the Way We Make Things*; North Point Press: New York, NY, USA, 2002.
74. Bjørn, A.; Hauschild, M.Z. Absolute *versus* relative environmental sustainability. *J. Ind. Ecol.* **2013**, *17*, 321–332.
75. Figge, F.; Hahn, T. Sustainable value added—Measuring corporate contributions to sustainability beyond eco-efficiency. *Ecol. Econ.* **2004**, *48*, 173–187.
76. Roscoe, S.; Cousins, P.D.; Lamming, R.C. Developing eco-innovations: A three-stage typology of supply networks. *J. Clean. Prod.* **2016**, *122*, 1948–1959. Available online: http://doi:10.1016/j.jclepro.2015.06.125 (accessed on 11 January 2016).
77. Lo, T.P.; Guo, S.J. Effective weighting model based on the maximum deviation with uncertain information. *Expert Syst. Appl.* **2010**, *37*, 8445–8449.
78. Alemi-Ardakani, M.; Milani, A.S.; Yannacopoulos, S.; Shokouhi, G. On the effect of subjective, objective and combinative weighting in multiple criteria decision-making: A case study on impact optimization of composites. *Expert Syst. Appl.* **2016**, *46*, 426–438.
79. Winston, W.L.; Albright, S.C. *Practical Management Science*; South-Western Cengage Learning: Boston, MA, USA, 2008.
80. Tzeng, G.H.; Huang, J.J. *Multiple Attribute Decision Making: Methods and Applications*; CRC Press: Boca Raton, FL, USA, 2011.
81. Deng, J.L. Introduction of Grey System. *J. Grey Syst.* **1989**, *1*, 1–24.
82. Liu, S.F.; Lin, Y. *Grey Information: Theory and Practical Application*; Springer: London, UK, 2006.

83. Tzeng, M.L. Using Linguistic Preferences and Grey Relational Analysis to Evaluate the Environmental Knowledge Management Capacity. *Expert Syst. Appl.* **2010**, *37*, 70–81.
84. Liu, S.F.; Lin, Y. *Grey Systems: Theory and Application*; Springer: London, UK, 2010.
85. Khan, Z.A.; Kamaruddin, S.; Siddiquee, A.N. Feasibility Study of Use of Recycled High Density Polyethylene and Multi Response Optimization of Injection Moulding Parameters using Combined Grey Relational and Principal Component Analyses. *Mater. Des.* **2010**, *31*, 2925–2931.
86. Lee, W.S.; Lin, Y.C. Evaluating and Ranking Energy Performance of Office Buildings Using Grey Relational Analysis. *Energy* **2011**, *36*, 2551–2556.
87. Zhao, R. Environmental Risk Management System Design for Hazardous Waste Materials. Ph.D. Thesis, University of Hull, Hull, UK, 2012.
88. Li, Z.; Ma, Z.; van der Kuijp, T.J.; Yuan, Z.; Huang, L. A review of soil heavy metal pollution from mines in China: Pollution and health risk assessment. *Sci. Total Environ.* **2014**, *468*, 843–853.
89. Wang, Y.; Hu, J.; Lin, W.; Wang, N.; Li, C.; Luo, P.; Liu, Y. Health risk assessment of migrant workers' exposure to polychlorinated biphenyls in air and dust in an e-waste recycling area in China: Indication for a new wealth gap in environmental rights. *Environ. Int.* **2016**, *87*, 33–41.
90. Sundblad, E.L.; Biel, A.; Gärling, T. Intention to change activities that reduce carbon dioxide emissions related to worry about global climate change consequences. *Eur. Rev. Appl. Psychol.* **2014**, *64*, 13–17.
91. Cimpan, C.; Rothmann, M.; Hamelin, L.; Wenzel, H. Towards increased recycling of household waste: Documenting cascading effects and material efficiency of commingled recyclables and biowaste collection. *J. Environ. Manag.* **2015**, *157*, 69–83.
92. Su, B.; Heshmati, A.; Geng, Y.; Yu, X. A review of the circular economy in China: Moving from rhetoric to implementation. *J. Clean. Prod.* **2013**, *42*, 215–227.
93. Fong, J.P.; Lee, F.J.; Lu, I.S.; Uang, S.N.; Lee, C.C. Estimating the Contribution of Inhalation Exposure to Di-2-Ethylhexyl Phthalate (DEHP) for PVC Production Workers, Using Personal Air Sampling and Urinary Metabolite Monitoring. *Int. J. Hyg. Environ. Health* **2014**, *217*, 102–109.
94. Deutz, P.; Neighbour, G.; McGuire, M. Integrating Sustainable Waste Management into Product Design: Sustainability as a Functional Requirement. *Sustain. Dev.* **2010**, *18*, 229–239.
95. Zhao, R.; Neighbour, G.; McGuire, M.; Deutz, P. A Software Based Simulation for Cleaner Production: A Game between Manufacturers and Government. *J. Loss Prev. Process Ind.* **2013**, *26*, 59–67.
96. Savitz, A.W.; Weber, K. *The Triple Bottom Line: How Today's Best-Run Companies Are Achieving Economic, Social and Environmental Success and How You Can too*; Jossey-Bass: San Francisco, FL, USA, 2006.
97. Pardo, R.J.H.; Bhamra, T.; Bhamra, R. Sustainable Product Service Systems in Small and Medium Enterprises (SMEs): Opportunities in the Leather Manufacturing Industry. *Sustainability* **2012**, *4*, 175–192.

98. Yang, S.S.; Nasr, N.; Ong, S.K.; Nee, A.Y.C. Designing automotive products for remanufacturing from material selection perspective. *J. Clean. Prod.* **2015**.
99. Tjader, Y.; May, J.H.; Shang, J.; Vargas, L.G.; Gao, N. Firm-level outsourcing decision-making: A balanced scorecard-based analytic network process model. *Int. J. Prod. Econ.* **2014**, *147*, 614–623.
100. Arce, M.E.; Saavedra, Á.; Míguez, J.L.; Granada, E. The use of grey-based methods in multi-criteria decision analysis for the evaluation of sustainable energy systems: A review. *Renew. Sustain. Energy Rev.* **2015**, *47*, 924–932.

Optimal Path for Controlling Sectoral CO_2 Emissions Among China's Regions: A Centralized DEA Approach

Zuoren Sun, Rundong Luo and Dequn Zhou

Abstract: This paper proposes a centralized data envelopment analysis (DEA) model for industrial optimization based on several different production technologies among several regions. We developed this model based on improved Kuosmanen environmental DEA technology, which avoids positive shadow price on undesirable outputs. We also designed a dual model for our centralized DEA model, and used it to analyze shadow prices on CO_2 emissions. We further employed the proposed model to determine the optimal path for controlling CO_2 emissions at the sector level for each province in China. At sectoral level, manufacturing showed the highest potential emissions reduction, and transportation was the largest accepter of emission quotas. At regional level, western and northeastern areas faced the largest adjustments in allowable emissions, while central and eastern areas required the least amount of adjustment. Because our model represents increase or decrease in emissions bidirectionally in terms of shadow price analysis, this setting makes the shadow price on CO_2 emissions lower than strong regulation (decreasing CO_2 emissions along with increasing value added) used by directional distance function (DDF).

Reprinted from *Sustainability*. Cite as: Sun, Z.; Luo, R.; Zhou, D. Optimal Path for Controlling Sectoral CO_2 Emissions Among China's Regions: A Centralized DEA Approach. *Sustainability* **2016**, *8*, 28.

1. Introduction

Climate change caused by greenhouse gas (GHG) emitted by human activity is one of the most urgent global issues of our time. According to Stern [1], if no action is taken to reduce GHG emissions, the overall cost of climate change will be equivalent to the loss of at least 5% of the world's GDP per year. Numerous countries have taken action to reduce their greenhouse gas emissions by enforcing market-oriented or administration-oriented regulations, including energy/carbon taxation, cap and trade systems, energy efficiency standards, and subsidies for new and renewable energy. China, the world's largest developing country and possessor currently of the most dynamic economy, has seen a dramatic increase in energy consumption and CO_2 emissions over the past three decades. According to International Energy Agency (IEA) report, Mainland China (excluding Hong Kong) emitted 8205.9 million tons of CO_2 in 2012, making it the world's largest CO_2 emitter at 25.9% of the world's total.

China's CO_2 emissions sources were mainly fossil fuel combustion (90%) and cement production (10%); coal consumption accounted for about 70% of CO_2 emissions from fossil fuel combustion. As for sector shares of China's CO_2 emissions, manufacturing and power generation are the largest contributors, accounting for more than 70% of China's total CO_2 emissions. Transportation and household sectors contributed only about 10% [2]. These phenomena are particularly notable when comparing China to developed countries, where CO_2 emissions come mostly from transportation and household sectors. Controlling CO_2 emissions in the production sector and enforcing dramatic, but proven-feasible, adjustments to industrial structure is highly necessary to ensure rational distribution of CO_2 emissions among transport, consumption, production, and processing industries, and to decrease China's overall carbon footprint.

In order to control the country's excessive energy consumption and resulting CO_2 emission problem, China's central government has established a series of energy conservation and emissions reduction regulations. Specifically, China has committed to reducing its CO_2 emissions by 40%–45% from 2005 levels per unit of GDP by 2020. Considering the sizeable differences in natural environment, resources, and economic development across different provinces, the government decomposed the national emissions reduction target to the province level, as outlined in the 11th and 12th Five Year Plans (FYPs) (in the 12th FYP, China outlined policies to decrease energy consumption and cut CO_2 emissions 16% and 17%, respectively, in 2015 compared to the end of the 11th FYP). In an effort to most effectively share the burden of carbon emissions, emissions reduction plans at both provincial and sectoral levels require further research.

Previous studies have examined CO_2 allocation according to well-established environmental production technology [3]. For example, Färe *et al.* [4] developed a generalized DEA framework for determining the optimal inter-temporal CO_2 allocation for 28 OECD countries. They found that OECD countries should be with weak regulations during 1991–1998 and strong regulations during 2000–2006 on CO_2 emissions, which is different to immediate CO_2 emissions reduction regulations advocated by Stern. The same research team further extended this model to multi-undesirable outputs (e.g., CO_2, NO_x, and SO_2) and proposed methods of allocating undesirable outputs for U.S. electric power plants under command-and-control and tradable permit regulations [5]. The results indicate that tradable permit regulation yields more potential gains than command-and-control regulation and in most cases good output increases more when three or more bad outputs are traded. Zhou *et al.* [6] transformed the CO_2 allocation problem into an optimal path for controlling CO_2 emissions under specific spatial, temporal, and spatial-temporal allocation strategies in China's 30 provinces. It is found that the spatial-temporal allocation strategy can achieve optimal CO_2 emissions distribution.

We established the model proposed here according to our own research and a careful review of the literature. First, we transformed the existing model from a constant returns to scale (CRS) to variable returns to scale (VRS) assumption, in terms of desirable and undesirable output joint-production framework, which better represents the real-world production in industry. Next, we improved the method of representing "environmental production technology", avoiding the emergence of positive shadow price on undesirable outputs. Of course, the traditional approach to manage weak disposability in the DEA model is to set undesirable output constraints as equal, resulting in positive shadow price, which does not reflect real-world production conditions. We set undesirable constraints as less than or equal to each other (Färe (2014) introduced an inequality for undesirable outputs to avoid any condition when less desirable output can be yielded by producing more undesirable output; he also indicated that the inequality can yield unbounded output sets, which can be avoided by setting the right-hand side of output sets equal to a bound such as the largest observed undesirable output value), which avoids positive shadow price but still meets weak disposability and null-jointness assumptions [5,7,8]. Finally, we conducted an empirical study from the provincial level to the sector level in different provinces and regions, allowing us to examine production technology homogeneously in sectors among provinces rather than heterogeneously in provinces across the entire nation, exploring China's optimal CO_2 emissions path with new and valuable information.

The remainder of this paper is organized as follows. Section 2 provides a literature review focusing on allocation methods based on DEA techniques. In Section 3, we propose new environmental DEA technology and space-temporal centralized DEA model, plus strategies for controlling emissions at various levels (sectoral and regional) in China. Section 4 presents the data and variable descriptions for China, and the results obtained using the proposed model. Section 5 makes conclusions and policy recommendations based on our findings.

2. Literature Review

CO_2 emission is closely related to industrial production, and especially the combustion of fuel necessary for production processes. For the sake of analysis, it is an "undesirable output" that is produced simultaneously with "desirable outputs". If we consider the environmental capacity for CO_2 emissions to be limited, CO_2 emissions are an exhaustible resource. The optimal utilization of CO_2 emission is both temporal and spatial, and achieves the maximum economic outputs with limited impact on the environment.

DEA is a specific method of using non-parametric mathematical programming to approximate production technology [9]. The principle of DEA is to allow data to speak for itself, rather than forcing any rigid, arbitrarily specified functional form [10].

This principle was originated by activity analysis [11–15] and non-parametric production analysis [16–20]. According to established DEA methodology, it can be implemented to solve resource allocation among peer decision making units (DMUs). Some researchers assume that there is a powerful, centralized decision-making environment for allocating resources to DMUs so as to maximize the summarized benefits or minimize the summarized costs. Golany and Tamir [21], for example, integrated efficiency, cost-benefit, and equality into an aggregated non-radial oriented DEA allocation model. Lozano and Villa [22] proposed a centralized reallocation DEA model, implementing step-by-step optimization according to the productive efficiency of a group of DMUs [23]; this model set unique shadow prices of inputs and outputs, as opposed to Golany and Tamir. Asmild *et al.* [24] extended the centralized reallocation DEA model to the BCC model, which only adjusts any previously inefficient DMUs. Mar-Molinero *et al.* [25] simplified the BCC centralized model using common intensity variables, which made measuring the productive efficiency of DMU groups as easy as measuring the single productive efficiency of average inputs and outputs of the DMUs. Fang [26] explored several centralized reallocation DEA models and proposed a generalized model that incorporates several advantages.

In addition to the series of Lozano models, researchers have used other optimization models based on DEA for resource allocation. Korhonen and Syrjänen [27] developed a multiple-objective linear programming model based on production technology for allocating resources. Hadi-Vencheh *et al.* [28] proposed a strongly efficient resource allocation model based on inverse DEA which maintains DMU efficiency levels after allocation. Amirteimoori and Tabar [29] provided a DEA-based fixed-resource allocation model that considers lower and upper goals achieved by individual DMUs. Athanassopoulos [30] developed the goal programming and data envelopment analysis (GoDEA) model, which combines the centralized target-based planning model and data envelopment analysis.

A wealth of previous researchers have used the DEA model to investigate CO_2 allocation, specifically. Gomes [31] developed the zero-sum game DEA (ZSG-DEA) model, taking CO_2 as the input variable allocated among Kyoto Protocol Contracting Parties—ZSG-DEA was proven to realize technical efficiency for all DMUs after optimization by multiplying an allocation coefficient. Other researchers applied ZSG-DEA, including Serro [32], Wang *et al.* [33], Pang *et al.* [34], and Chiu *et al.* [35], and provided valuable findings; Lozano *et al.* [36] used the centralized reallocation DEA model for emissions allocation. Notable studies on the centralized DEA model were conducted by Wu *et al.* [37], Singh and Majumdar [38], Sun *et al.* [39], Homayounfar *et al.* [40], and Feng *et al.* [41].

As reviewed above, prior research on allocation models and methods based on DEA has mostly concentrated on weight restrictions with user-specified preference structure [42]. Decision maker preference information can also be incorporated into DEA by employing the multiple criteria decision-making concept [43,44]. A well-known drawback of weight restrictions is the fact that their use in the multiplier model implicitly changes the model of production technology in the envelopment form [45], and that they inappropriately represent the demands of real-world productive activities [46]; in other words, their use makes the traditional meanings of "technical efficiency" and "technologically feasible improvement" become unsubstantiated [47,48]. Comparing to the other research up to now, this paper considers the weights of the outputs and inputs will only be restricted by the constraints used to form environmental production technology, without any user-specified preference restriction on weights. In that way, we can ensure the optimized activities will be projected on the facet of environmental production technology, then allocates CO_2 emissions in effort to maximize the aggregated desirable outputs.

3. Methodology

3.1. Environmental DEA Technology

The joint production of desirable and undesirable outputs can be described as "environmental production technology". It depicts the process of converting input vectors ($\mathbf{x}$) to desirable output vectors ($\mathbf{y}$) and undesirable output vectors ($\mathbf{b}$) [3]:

$$\mathbf{T} = \{(\mathbf{x}, \mathbf{y}, \mathbf{b}) : \mathbf{x} \text{ can produce } (\mathbf{y}, \mathbf{b})\} \tag{1}$$

The production technology set ($\mathbf{T}$) is assumed to satisfy the following assumptions:

(1) If $(\mathbf{x}, \mathbf{y}, \mathbf{b}) \in \mathbf{T}$ and $0 \leqslant \theta \leqslant 1$, then $(\mathbf{x}, \theta\mathbf{y}, \theta\mathbf{b}) \in \mathbf{T}$ (weak disposability for desirable and undesirable outputs).

(2) If $(\mathbf{x}, \mathbf{y}, \mathbf{b}) \in \mathbf{T}$ and $\mathbf{b} = \mathbf{0}$, then $\mathbf{y} = \mathbf{0}$ (null-jointness of desirable and undesirable outputs).

The conceptual definition of environmental production technology can be approximated by piecewise linear combinations of the observed data, which is similar theoretically to DEA form; as such, it can also be called "environmental DEA technology" [49,50]. Under CRS assumption, environmental production technology **T** can be approximately formulated as follows:

$$\hat{\mathbf{T}}_W = \{(\mathbf{x},\mathbf{y},\mathbf{b}) : \begin{array}{l} \sum_{k=1}^{K} z_k x_{mk} \leqslant x_m, \quad m = 1,2,\cdots,M \\ \sum_{k=1}^{K} z_k y_{nk} \geqslant y_n, \quad n = 1,2,\cdots,N \\ \sum_{k=1}^{K} z_k b_{jk} = b_j, \quad j = 1,2,\cdots,J \\ z_k \geqslant 0, \quad k = 1,2,\cdots,K\} \end{array} \tag{2}$$

Under VRS assumption, environmental production technology **T** can be approximately characterized as Shephard environmental DEA technology [51] and Kuosmanen environmental DEA technology [52].

The formulation of Shephard environmental DEA technology is as follows:

$$\hat{\mathbf{T}}_S = \{(\mathbf{x},\mathbf{y},\mathbf{b}) : \begin{array}{l} \sum_{k=1}^{K} z_k x_{nk} \leqslant x_n, \ n = 1,2,\cdots,N \\ \theta \sum_{k=1}^{K} z_k y_{mk} \geqslant y_m, \ m = 1,2,\cdots,M \\ \theta \sum_{k=1}^{K} z_k b_{jk} = b_j, \ j = 1,2,\cdots,J \\ 0 \leqslant \theta \leqslant 1, \ \sum_{k=1}^{K} z_k = 1 \\ z_k \geqslant 0, \ k = 1,2,\cdots,K \} \end{array} \tag{3}$$

where θ is the abatement factor, which ensures that $\hat{\mathbf{T}}_S$ satisfies the two assumptions above. The second and third constraints in Model (3) are non-linear, and must be converted to linear form. We defined a new variable, $\tilde{z}_k$, then let $\tilde{z}_k = \theta z_k$ replace the old variable. The converted model can be expressed as follows:

$$\hat{\mathbf{T}}_S = \{(\mathbf{x},\mathbf{y},\mathbf{b}) : \begin{array}{l} \sum_{k=1}^{K} \tilde{z}_k x_{nk} \leqslant \theta x_n, \ n = 1,2,\cdots,N \\ \sum_{k=1}^{K} \tilde{z}_k y_{mk} \geqslant y_m, \ m = 1,2,\cdots,M \\ \sum_{k=1}^{K} \tilde{z}_k b_{jk} = b_j, \ j = 1,2,\cdots,J \\ 0 \leqslant \theta \leqslant 1, \ \sum_{k=1}^{K} \tilde{z}_k = \theta \\ \tilde{z}_k \geqslant 0, \ k = 1,2,\cdots,K \\ x_n \geqslant \min\{x_{nk}\} \quad \} \end{array} \tag{4}$$

It is worth noting that we added constraint $x_n \geqslant \min\{x_{nk}\}$, which prevents the condition $\theta = 0$ causing $\widetilde{z}_k = 0$ so that any activity $(\mathbf{x}, \mathbf{0}, \mathbf{0})$ will be feasible. This is because according to the first constraint in Model (3), input vectors $\mathbf{x}$ cannot be smaller than the observed practical value.

An improvement upon Shephard environmental DEA technology, Kuosmanen environmental DEA technology can be expressed as follows:

$$\hat{\mathbf{T}}_K = \{(\mathbf{x}, \mathbf{y}, \mathbf{b}) : \begin{array}{l} \sum\limits_{k=1}^{K} z_k x_{nk} \leqslant x_n,\ n = 1, 2, \cdots, N \\ \sum\limits_{k=1}^{K} \theta_k z_k y_{mk} \geqslant y_m,\ m = 1, 2, \cdots, M \\ \sum\limits_{k=1}^{K} \theta_k z_k b_{jk} = b_j,\ j = 1, 2, \cdots, J \\ 0 \leqslant \theta_k \leqslant 1,\ \sum\limits_{k=1}^{K} z_k = 1 \\ z_k \geqslant 0,\ k = 1, 2, \cdots, K\ \} \end{array} \tag{5}$$

The difference between Models (3) and (5) is that Model (5) takes multiple abatement factors with desirable and undesirable outputs [53]. The strength of Model (5) is that it allows for each observed activity to abate by a different abatement factor before the convex combination is formed, opening the possibility that a target DMU with less abatement cost reduces more. Model (5) also includes non-linear constraints, so it must be converted to solvable form with linear constraints. The converted model is as follows:

$$\hat{\mathbf{T}}_K = \{(\mathbf{x}, \mathbf{y}, \mathbf{b}) : \begin{array}{l} \sum\limits_{k=1}^{K} (z_k + \lambda_k) x_{nk} \leqslant x_n,\ n = 1, 2, \cdots, N \\ \sum\limits_{k=1}^{K} z_k y_{mk} \geqslant y_m,\ m = 1, 2, \cdots, M \\ \sum\limits_{k=1}^{K} z_k b_{jk} = b_j,\ j = 1, 2, \cdots, J \\ \sum\limits_{k=1}^{K} (z_k + \lambda_k) = 1 \\ z_k \geqslant 0,\ \lambda_k \geqslant 0,\ k = 1, 2, \cdots, K\ \} \end{array} \tag{6}$$

Because Kuosmanen environmental DEA technology sets equal undesirable output constraints, positive shadow price may emerge on undesirable outputs (as also discussed by Kuosmanen and Kazemi Matin [54]). To avoid this, we adopted a suggestion by Leleu [55] and set an $\leqslant$ on undesirable output constraints (the same work done under CRS assumptions can be found in a study by Färe *et al.* (2014) [5]). Our proposed environmental DEA technology can be expressed as follows:

$$\hat{\mathbf{T}}_{\text{New}} = \{(\mathbf{x}, \mathbf{y}, \mathbf{b}) : \quad \begin{aligned} &\sum_{k=1}^{K} z_k x_{nk} \leqslant x_n, \; n = 1, 2, \cdots, N \\ &\sum_{k=1}^{K} \theta_k z_k y_{mk} \geqslant y_m, \; m = 1, 2, \cdots, M \\ &\sum_{k=1}^{K} \theta_k z_k b_{jk} \leqslant b_j, \; j = 1, 2, \cdots, J \\ &0 \leqslant \theta_k \leqslant 1, \; \sum_{k=1}^{K} z_k = 1 \\ &z_k \geqslant 0, \; k = 1, 2, \cdots, K \; \} \end{aligned} \tag{7}$$

We created a numerical example comparing $\hat{\mathbf{T}}_{\text{W}}$, $\hat{\mathbf{T}}_{\text{S}}$, $\hat{\mathbf{T}}_{\text{K}}$, and $\hat{\mathbf{T}}_{\text{New}}$ by output sets $\hat{\mathbf{P}}_{\text{W}}(\mathbf{x})$, $\hat{\mathbf{P}}_{\text{S}}(\mathbf{x})$, $\hat{\mathbf{P}}_{\text{K}}(\mathbf{x})$ and $\hat{\mathbf{P}}_{\text{New}}(\mathbf{x})$. Considering the production activity with one input, one desirable output, and one undesirable output, four observed DMUs and constructed output sets were analyzed as presented in Table 1 and Figure 1.

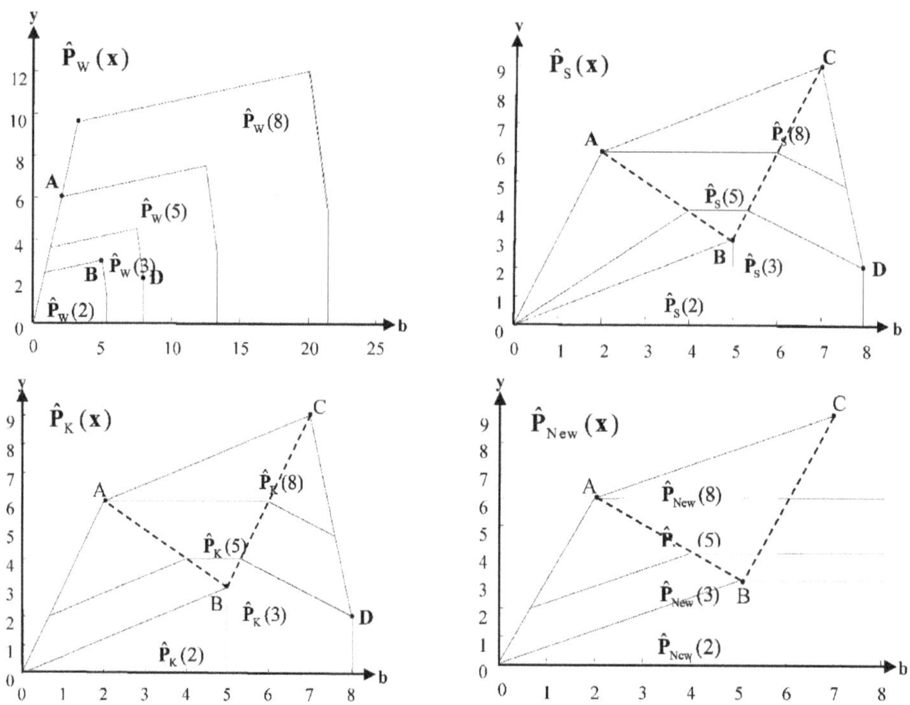

Figure 1. $\hat{\mathbf{P}}_{\text{W}}(\mathbf{x})$, $\hat{\mathbf{P}}_{\text{S}}(\mathbf{x})$, $\hat{\mathbf{P}}_{\text{K}}(\mathbf{x})$ and $\hat{\mathbf{P}}_{\text{New}}(\mathbf{x})$ induced by observed DMUs.

Table 1. Comparison between $\hat{\mathbf{P}}_{\mathrm{W}}(\mathbf{x})$, $\hat{\mathbf{P}}_{\mathrm{S}}(\mathbf{x})$, $\hat{\mathbf{P}}_{\mathrm{K}}(\mathbf{x})$ and $\hat{\mathbf{P}}_{\mathrm{New}}(\mathbf{x})$.

DMU	A	B	C	D
b	2	5	7	8
x	5	2	8	3
y	6	3	9	2

As shown in Figure 1, all the output sets satisfy Assumptions (1) and (2). Meanwhile, all the output sets above are the compact sets, but $\hat{\mathbf{P}}_{\mathrm{New}}(\mathbf{x})$ is not a bounded and closed set due to the strong disposability set on input. Because $\hat{\mathbf{P}}_{\mathrm{K}}(\mathbf{x})$ possesses multi abatement factors and $\hat{\mathbf{P}}_{\mathrm{S}}(\mathbf{x})$ possesses a single abatement factor, which makes $\hat{\mathbf{P}}_{\mathrm{K}}(\mathbf{x})$ larger than $\hat{\mathbf{P}}_{\mathrm{S}}(\mathbf{x})$ (see $\hat{\mathbf{P}}_{\mathrm{K}}(3)$ is larger than $\hat{\mathbf{P}}_{\mathrm{K}}(3)$). Therefore, the DMU in $\hat{\mathbf{P}}_{\mathrm{K}}(\mathbf{x})$ has more choices to conduct undesirable output abatement. However, we often observe a negative slope in the frontier of $\hat{\mathbf{P}}_{\mathrm{K}}(\mathbf{x})$ (see piecewise lines are located on the right side of line BC in $\hat{\mathbf{P}}_{\mathrm{K}}(\mathbf{x})$), on which the shadow price of undesirable output is positive (e.g., $-\partial y/\partial c > 0$)and the marginal abatement cost is negative (e.g., $\partial y/\partial c < 0$). Hence, this anti-fact situation must be improved. We improved $\hat{\mathbf{P}}_{\mathrm{K}}(\mathbf{x})$ to $\hat{\mathbf{P}}_{\mathrm{New}}(\mathbf{x})$ according to a suggestion made by Leleu [55], and set strong disposability on undesirable output, making $\hat{\mathbf{P}}_{\mathrm{New}}(\mathbf{x})$ satisfies two necessary assumptions. The success of doing so is evidenced by piecewise horizontal lines on the right side of line BC in $\hat{\mathbf{P}}_{\mathrm{New}}(\mathbf{x})$, which can avoid the positive shadow price emerging on undesirable output.

3.2. Centralized Model for New Environmental DEA Technology

The centralized model based on new environmental DEA technology was designed to reallocate undesirable outputs to allow the substitution of CO_2 emissions between DMUs in order to maximize total desirable outputs. After centralized optimization, all DMU activities are as efficient as possible under the potential desirable outputs and allocated undesirable outputs set by the model. All DMU activities in the time period t are marked with a subscript t. There is an emissions control coefficient δ for period t, which implies that the total emissions permitted is equal to δ multiplied by the aggregate baseline emissions for all the DMUs. Then, the centralized DEA model for reallocation of CO_2 emissions can be formulated as follows:

Total control strategy (Primal):

$$
\begin{aligned}
&\max \sum_{t=1}^{T}\sum_{s=1}^{S}\sum_{l=1}^{K} \hat{y}_{s,l}^{t} \\
\text{s.t.} \quad & \sum_{k=1}^{K} (z_{s,k,l}^{t} + \lambda_{s,k,l}^{t}) x_{s,m,k}^{t} \leqslant x_{s,m,l}^{t}, m = 1, \cdots, M \\
& \sum_{k=1}^{K} z_{s,k,l}^{t} y_{s,k}^{t} \geqslant \hat{y}_{s,l}^{t} \\
& \sum_{k=1}^{K} z_{s,k,l}^{t} b_{s,k}^{t} \leqslant \hat{b}_{s,l}^{t} \\
& \sum_{k=1}^{K} (z_{s,k,l}^{t} + \lambda_{s,k,l}^{t}) = 1 \\
& \sum_{t=1}^{T}\sum_{s=1}^{S}\sum_{l=1}^{K} \hat{b}_{s,l}^{t} = \delta \cdot \sum_{t=1}^{T}\sum_{s=1}^{S}\sum_{l=1}^{K} b_{s,l}^{t} \\
& z_{s,k,l}^{t} \geqslant 0, \quad \lambda_{s,k,l}^{t} \geqslant 0, \hat{y}_{s,l}^{t} \geqslant \varepsilon, \quad \hat{b}_{s,l}^{t} \geqslant \varepsilon \\
& t = 1, \cdots, T; \quad s = 1, \cdots, S; \quad l = 1, \cdots, K;
\end{aligned}
\tag{8}
$$

Sectoral control strategy (Primal):

$$
\begin{aligned}
&\max \sum_{t=1}^{T}\sum_{s=1}^{S}\sum_{l=1}^{K} \hat{y}_{s,l}^{t} \\
\text{s.t.} \quad & \sum_{k=1}^{K} (z_{s,k,l}^{t} + \lambda_{s,k,l}^{t}) x_{s,m,k}^{t} \leqslant x_{s,m,l}^{t}, m = 1, \cdots, M \\
& \sum_{k=1}^{K} z_{s,k,l}^{t} y_{s,k}^{t} \geqslant \hat{y}_{s,l}^{t} \\
& \sum_{k=1}^{K} z_{s,k,l}^{t} b_{s,k}^{t} \leqslant \hat{b}_{s,l}^{t} \\
& \sum_{k=1}^{K} (z_{s,k,l}^{t} + \lambda_{s,k,l}^{t}) = 1 \\
& \sum_{t=1}^{T}\sum_{l=1}^{K} \hat{b}_{s,l}^{t} = \delta_{s} \cdot \sum_{t=1}^{T}\sum_{l=1}^{K} b_{s,l}^{t} \\
& z_{s,k,l}^{t} \geqslant 0, \quad \lambda_{s,k,l}^{t} \geqslant 0, \hat{y}_{s,l}^{t} \geqslant \varepsilon, \quad \hat{b}_{s,l}^{t} \geqslant \varepsilon \\
& t = 1, \cdots, T; \quad s = 1, \cdots, S; \quad l = 1, \cdots, K;
\end{aligned}
\tag{9}
$$

In Models (8) and (9), the number of time periods, environmental production technology and observed activity are T, S and K. We used the piecewise linear programming constraints to form Kuosmanen environmental production technology. The first and second constraints represent the strong disposability of M inputs and one desirable output, respectively. The third constraint represents the strong disposability of one undesirable output. $z_{s,k,l}^{t}$ and $\lambda_{s,k,l}^{t}$ are the intensity variables valued according to Kuosmanen environmental production technology. The fourth constraint denotes variable returns to scale. All the possible combinations of desirable and undesirable outputs $\left(\hat{y}^t_{s,l}, \hat{b}^t_{s,l}\right)^T$ with fixed input $\mathbf{x}^t_{s,l}$ will be optimized by maximizing the aggregated $T \times S \times K$ activities' desirable outputs. $\hat{y}^t_{s,l} \geqslant \varepsilon$ and $\hat{b}^t_{s,l} \geqslant \varepsilon$ ensure that the optimized desirable output and undesirable output take the value greater than zero. In order to control the undesirable outputs, we designed two models to denote sectoral and total undesirable outputs control strategies in Models (8) and (9), respectively:

(1) $\sum_{t=1}^{T}\sum_{s=1}^{S}\sum_{l=1}^{K}\hat{b}^t_{s,l} = \delta \cdot \sum_{t=1}^{T}\sum_{s=1}^{S}\sum_{l=1}^{K} b^t_{s,l}$, can be defined as total control strategy, which implies that the total undesirable outputs permitted is equal to δ multiplied by the aggregate baseline undesirable outputs from all DMU.

(2) $\sum_{t=1}^{T}\sum_{l=1}^{K}\hat{b}^t_{s,l} = \delta_s \cdot \sum_{t=1}^{T}\sum_{l=1}^{K} b^t_{s,l}$, we define it as sectoral control strategy, which denotes the optimized sum of aggregated undesirable outputs produced by $T \times K$ activities refer to environmental production technology s, which will be equal to δ_s times from observed aggregated undesirable outputs produced by $T \times K$ activities.

We can deduce the dual forms of Models (8) and (9) with the aim to explore the shadow price of inputs, desirable and undesirable output, and the formulations can be induced as follows:

Total control strategy (Dual):

$$
\begin{aligned}
\min \quad & \sum_{s=1}^{S}\sum_{t=1}^{T}\sum_{l=1}^{K}\sum_{m=1}^{M} v^t_{s,m,l}x^t_{s,m,l} + \delta \cdot \sum_{s=1}^{S}\sum_{t=1}^{T}\sum_{l=1}^{K} b^t_{s,l} \cdot \sigma + \sum_{s=1}^{S}\sum_{t=1}^{T}\sum_{l=1}^{K} \varphi^t_{s,l} \\
\text{s.t.} \quad & \sum_{m=1}^{M} v^t_{s,m,l}x^t_{s,m,l} - \mu^t_{s,l}y^t_{s,l} + \omega^t_{s,l}b^t_{s,l} + \varphi^t_{s,l} \geqslant 0 \\
& \sum_{m=1}^{M} v^t_{s,m,l}x^t_{s,m,l} + \varphi^t_{s,l} \geqslant 0 \\
& \mu^t_{s,l} \geqslant 1, \\
& \sigma + \omega^t_{s,l} \geqslant 0, \\
& v^t_{s,m,l} \geqslant 0,\ \ \mu^t_{s,l} \geqslant 0,\ \ \omega^t_{s,l} \geqslant 0,\ \ \sigma : URS,\ \ \varphi^t_{s,l} : URS \\
& m = 1,\cdots,M;\ t = 1,\cdots,T;\ \ s = 1,\cdots,S;\ \ l = 1,\cdots,K
\end{aligned}
\tag{10}
$$

Sectoral control strategy (Dual):

$$
\begin{array}{ll}
\min & \sum_{s=1}^{S}\sum_{t=1}^{T}\sum_{l=1}^{K}\sum_{m=1}^{M} v_{s,m,l}^{t} x_{s,m,l}^{t} + \sum_{s=1}^{S} (\delta_s \cdot \sum_{t=1}^{T}\sum_{l=1}^{K} b_{s,l}^{t} \cdot \sigma_s) + \sum_{s=1}^{S}\sum_{t=1}^{T}\sum_{l=1}^{K} \varphi_{s,l}^{t} \\
\text{s.t.} & \sum_{m=1}^{M} v_{s,m,l}^{t} x_{s,m,l}^{t} - \mu_{s,l}^{t} y_{s,l}^{t} + \omega_{s,l}^{t} b_{s,l}^{t} + \varphi_{s,l}^{t} \geqslant 0 \\
& \sum_{m=1}^{M} v_{s,m,l}^{t} x_{s,m,l}^{t} + \varphi_{s,l}^{t} \geqslant 0 \\
& \mu_{s,l}^{t} \geqslant 1, \\
& \sigma_s + \omega_{s,l}^{t} \geqslant 0, \\
& v_{s,m,l}^{t} \geqslant 0, \ \ \mu_{s,l}^{t} \geqslant 0, \ \ \omega_{s,l}^{t} \geqslant 0, \ \ \sigma_s : URS, \ \ \varphi_{s,l}^{t} : URS \\
& m = 1, \cdots, M; \ t = 1, \cdots, T; \ \ s = 1, \cdots, S; \ l = 1, \cdots, K
\end{array} \quad (11)
$$

In Models (10) and (11), $v_{s,m,l}^{t}$, $\mu_{s,l}^{t}$ and $\omega_{s,l}^{t}$ are the dual variables of first three constraints of primal model. They interpret the marginal contributions for aggregated $T \times S \times K$ desirable outputs from inputs, desirable output and undesirable output, so we name them shadow prices. $\varphi_{s,l}^{t}$ is the dual variable of the convexity constraint of primal model.

4. Empirical Study

4.1. Data

The mathematical model provided above was instantiated using Chinese production data at the sectoral level for several provinces. We set the time period t from 1995 to 2011 (In general, reference base year sets are 1990, 1995, 2000, 2005. The Kyoto Protocol set the reduction target for industrial countries, in which average CO_2 emission from 2008 to 2012 reached 95% by 1990. Because China established its market-based economic institution in 1992, and its economy is yet transitioning, we set the base year as 1995 [56]. Chinese CO_2 emissions from fuel combustion in 1995–2010 accounted for 64.3% of the total emissions from 1971–2010 [57]. We collected as much recent data as possible.), constructed environmental DEA technologies S = {agriculture, manufacturing, construction, transportation, service} at provincial, multi-sectoral level, and the observed DMUs K = {Beijing, Tianjin, Hebei, Shanxi, Inner Mongolia, Liaoning, Jilin, Heilongjiang, Shanghai, Jiangsu, Zhejiang, Anhui, Fujian, Jiangxi, Shandong, Henan, Hubei, Hunan, Guangdong, Guangxi, Hainan, Chongqing+Sichuan (since Chongqing was a part of Sichuan before 1997, the data for Chongqing in 1995–1996 is inseparable from Sichuan; therefore, we combined them together from 1997–2011), Guizhou, Yunnan, Shaanxi, Gansu, Qinghai, Ningxia, Xinjiang}. The number of activities collected was $T(17) \times S(5) \times K(29) = 2465$.

We described activities as capital stock (K) and labor force (L) as inputs, and sectoral value added (Y) as desirable outputs and CO_2 emissions (C) as undesirable outputs. We collected energy data for all provinces in each sector, but did not use them as inputs, because CO_2 emission is the transformation form of energy inputs considering the energy mix weighted with CO_2 emission coefficients. As shown in Model (8), we gave strong disposability to the undesirable output, making it similar to the common input; that way, when we added energy and CO_2 emission simultaneously as inputs, a substitutional relationship formed. However, there is a positive correlation between CO_2 and energy. Given this, we leave out energy input. The similar processing can be referred to [4,58].

While estimating capital stock at sectoral level, it is easy to obtain the provincial-level capital stock according to the method developed by Zhang [59], but there is no support specifically for obtaining sectoral capital stock data by province. We implemented an approach suggested by Guo [60], Gan and Zheng [61] and Lv and Zhou [62]: first, we collected the new fixed assets at province/sector level from 1981 to 2011. We then took the five-year moving average of the province/sector time series (the new fixed asset data for years before 1985 took the moving average from 1981, and our methods were developed in effort to eliminate interference between these data, price indices, and investment depreciation rate), then used it to aggregate the new fixed assets from 1981 to each year province/sector level. We then computed the sectoral new fixed asset proportions province-by-province, then used the proportions to allocate the provincial-level capital stock to the province/sectoral level. The new fixed asset data 1981–1985, 1996–1998, and 2002–2011 we used came from the Statistical Yearbooks of China's Investment in Fixed Assets [63]. The data from1986–1995 and 1999–2001 came from the China Statistical Yearbook [64], and the provincial-level capital stock was estimated by perpetual inventory approach taking 1952 as the base year. In order to estimate provincial capital stock, we took gross fixed capital formation as the annual investment data, then converted it to a 2000 constant price using the investment price index. We estimated the labor force at province/sector level according to 1995–2011 employment data by sector and region from the 1996–2012 China Labor Statistical Yearbook [65].

We estimated value added at province/sector level according to 1995–2011 data from the 1996–2012 China Statistical Yearbook, and converted the data to 2000 constant price according to the value added index [66]. Energy consumption at province/sector level was estimated according to end-use energy consumption, with basic data collected from the energy balance table in the 1996–2012 China Energy Statistical Yearbook [67]. We used the standard coal conversion coefficient to convert these data to standard coal equivalent, and the portion of energy consumption of raw material in the manufacturing sector was removed from the energy balance table.

Table 2. Descriptive statistics of inputs and outputs for province/sector level over 1995–2011.

Index	Sector	Unit	Dimension	Quantity	Mean	Standard Deviation	Minimum	Maximum
Capital Stock (2000 constant price)	Agriculture	100 million RMB	provincial sector	29 × 17	379.46	417.68	25.61	2843.24
	Manufacturing	100 million RMB	provincial sector	29 × 17	6182.51	6558.73	274.76	48,577.60
	Construction	100 million RMB	provincial sector	29 × 17	186.14	219.64	9.30	2134.61
	Transportation	100 million RMB	provincial sector	29 × 17	1905.75	1687.42	61.25	9526.33
	Service	100 million RMB	provincial sector	29 × 17	6370.96	7042.74	145.72	39,632.95
Labor Force	Agriculture	10 thousand persons	provincial sector	29 × 17	1099.34	853.02	37.09	3996.00
	Manufacturing	10 thousand persons	provincial sector	29 × 17	426.05	399.76	19.60	2283.46
	Construction	10 thousand persons	provincial sector	29 × 17	146.53	141.98	9.80	715.10
	Transportation	10 thousand persons	provincial sector	29 × 17	87.86	84.37	8.40	792.02
	Service	10 thousand persons	provincial sector	29 × 17	594.30	396.37	40.70	1915.42
Energy (standard coal equivalent)	Agriculture	10 thousand tons	provincial sector	29 × 17	179.18	122.51	7.96	691.64
	Manufacturing	10 thousand tons	provincial sector	29 × 17	3702.02	3135.19	76.72	17,649.98
	Construction	10 thousand tons	provincial sector	29 × 17	74.77	82.01	1.12	605.01
	Transportation	10 thousand tons	provincial sector	29 × 17	493.44	477.83	13.73	2716.30
	Service	10 thousand tons	provincial sector	29 × 17	318.02	297.73	3.13	1969.78
Value Added (2000 constant price)	Agriculture	100 million RMB	provincial sector	29 × 17	600.36	450.00	33.51	2065.31
	Manufacturing	100 million RMB	provincial sector	29 × 17	2567.06	3115.48	42.99	19,950.01
	Construction	100 million RMB	provincial sector	29 × 17	343.27	299.55	14.45	1760.92
	Transportation	100 million RMB	provincial sector	29 × 17	399.12	409.84	8.44	2916.87
	Service	100 million RMB	provincial sector	29 × 17	1671.10	1735.37	50.31	10,590.03
CO_2	Agriculture	10 thousand tons	provincial sector	29 × 17	565.88	374.80	22.43	1948.69
	Manufacturing	10 thousand tons	provincial sector	29 × 17	12,408.36	10,493.94	265.52	58,033.89
	Construction	10 thousand tons	provincial sector	29 × 17	220.54	220.70	8.03	1598.10
	Transportation	10 thousand tons	provincial sector	29 × 17	1148.49	1064.41	33.30	6123.26
	Service	10 thousand tons	provincial sector	29 × 17	1082.89	1004.35	17.84	6034.86

To estimate CO_2 emissions at province/sector level, we considered 20 distinct types of end-use energies as the sources of carbon emission. The emission coefficients were taken from 2006 IPCC guidelines for national greenhouse gas inventories. To calculate the emission coefficients of electricity and heat generation, we computed the total CO_2 emissions from energy mix inputs for generating electricity and heat nationally, then converted these energy mix inputs to standard coal equivalent as the denominator, and plugged in the total CO_2 emissions to be divided by this denominator to obtain the emission coefficients of electricity and heat.

To obtain enough activities to construct the environmental DEA technology, we instantiated the proposed model using the province/sector level data gathered as discussed above (as shown in Table 2). We further aggregated the sector-level data by province into sectoral data by region, which provided more convenient analysis (with less influence due to administrative divisions). We followed the technique proposed by the China State Council Development Research Center and divided the 29 provinces into four areas: east, central, west and northeast. The four areas were then further divided into eight economic regions, as shown in Table 3.

Table 3. Compositions of four areas and eight economic regions in China.

Area	Economic Region	Provinces
East	Northern Coastal	Beijing, Tianjin, Hebei, Shandong
	Eastern Coastal	Shanghai, Jiangsu, Zhejiang
	Southern Coastal	Fujian, Guangdong, Hainan
Central	Middle Yellow River	Shanxi, Inner Mongolia, Henan, Shaanxi
	Middle Yangtze River	Anhui, Jiangxi, Hubei, Hunan
West	Southwest	Guangxi, Chongqing+Sichuan, Guizhou, Yunan
	Northwest	Gansu, Qinghai, Ningxia, Xinjiang
Northeast	Northeast	Liaoning, Jilin, Heilongjiang

4.2. Main Results

We take CO_2 emissions at province/sector level as the research objective, using Models (8) and (9) with the data in Table 2 to research optimal CO_2 emissions allocations in different sectors within different provinces between 1995 and 2011. To appropriately set the emission control coefficient δ, we employed two patterns of restriction. First, we set $\delta = 1$ corresponding to Model (8), made the aggregated CO_2 emissions for five sectors from all provinces mixed equally to their gross emissions from 1995–2011. Second, we set $\delta_s = 1$ following Model (9), controlled the aggregated sectoral CO_2 emissions including all provincial sub-sectors equal to its actual emissions from 1995–2011. Then, the annually optimal allocation of CO_2 emissions was solvable using above models, for all eight economic regions.

In the regional agriculture sectors, as shown in Figure 2, under sectoral emissions control strategy, compared to actual emissions, middle Yangtze River and southwest regions were allocated more emission quotas than middle Yellow River, northwest, or northeast regions. More CO_2 emissions allocated to middle Yangtze River and southwest regions would then produce more value-added yield than the other regions. This can be attributed to different agricultural mechanization levels and water resources endowment in northern and southern China. Mechanized production processes in southern China are more difficult than in northern China due to the abundance of mountains and hills in the south. Mechanization is carbon-intensive. At the same time, however, the natural supply of water in southern China saves costs that would otherwise accrue for irrigation, also saving energy input. As far as grain production overall, rice in southern China can be harvested 2–4 times more often than the wheat in northern China. To this effect, southern China has overall low energy input and high grain output compared to northern China. Additionally, under total emissions control strategy, the agriculture sector showed low emissions overall and as such should accept emission quotas from other sectors.

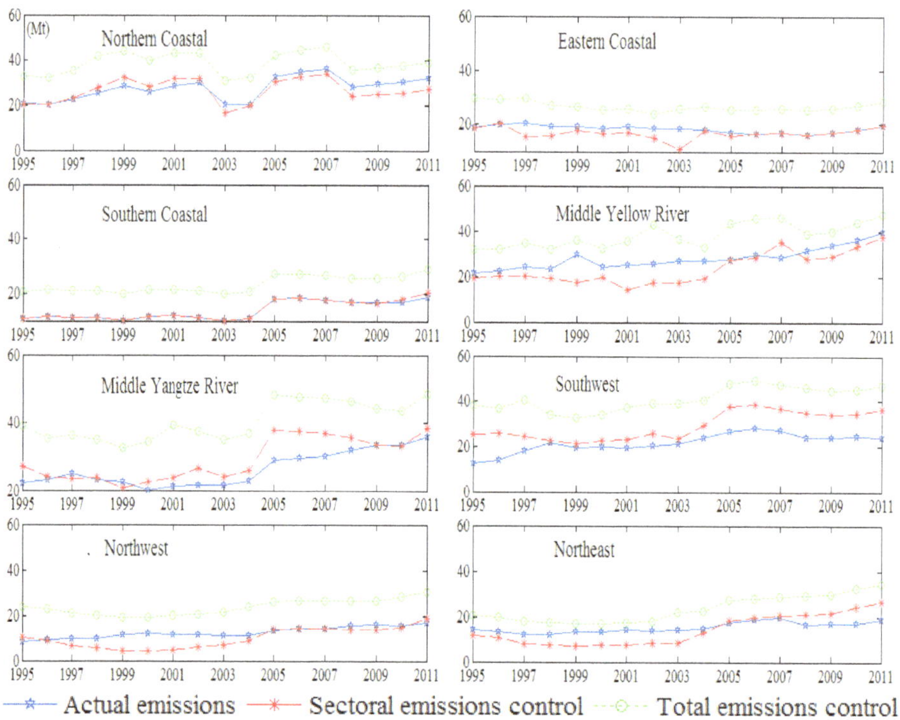

Figure 2. Optimal emission paths for regional agriculture sectors.

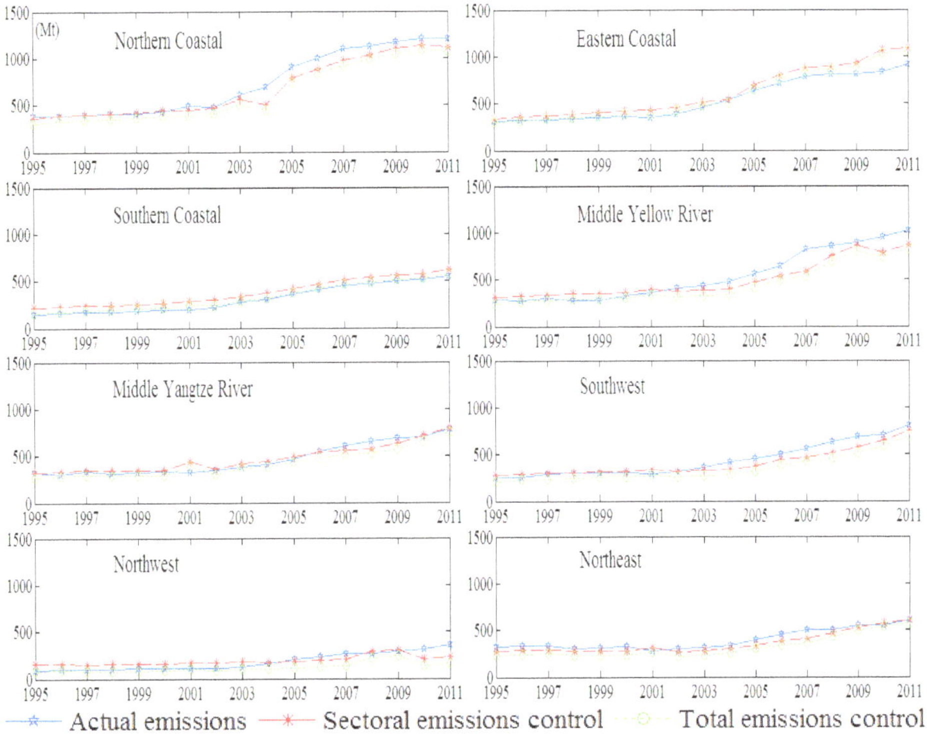

Figure 3. Optimal emission paths for regional manufacturing sectors.

Figure 3 shows the optimal emission paths for regional manufacturing sectors under sectoral emissions control and total emissions control strategies with their actual emissions. Under sectoral emissions control strategies, eastern coastal and southern coastal regions (*i.e.*, more developed parts of the country) showed high energy use efficiency and the potential to produce more value-added yield. To this effect, eastern and southern coastal regions were given more emission quotas in 1995–2011. Other regions, conversely, were limited by emissions reduction regulations of varying degree. Northern coastal, middle Yellow River, and southwest regions began dramatically overproducing CO_2 emissions, so those regions must improve energy use efficiency as soon as possible. Under total emissions control strategy, the manufacturing sector in each region obtained lower emission quotas than the control sectoral emissions allowance, suggesting that policies should be implemented to ensure that manufacturing emission quotas are extended to other sectors.

Given that the aggregated emissions from 29 provinces is equal to the actual emissions aggregated at the province level for regional construction sectors. As marked by red lines in Figure 4, construction in the northern coastal region must be strongly regulated. Northern coastal region inputs excessive energy into building infrastructure and housing, and absolutely must significantly improve energy use efficiency. Northwest and northeast regions obtained emission quotas greater than their actual emissions from 2003 to stay financially viable. The eastern coastal region faced similar emissions reduction regulations as the northern coastal region. The middle Yellow River, middle Yangtze River, and southwest regions should reduce their emissions in early periods and increase emissions in latter periods, indicating that the energy efficiency of the construction sector in these regions improved rapidly. Suppose that emission quotas are transferable across sectors and regions, as marked by green lines in Figure 4, the construction sector in each region obtained more emission quotas than their actual emissions, suggesting that construction should accept emission quotas from other sectors and does not require any major emissions reduction policies be enacted.

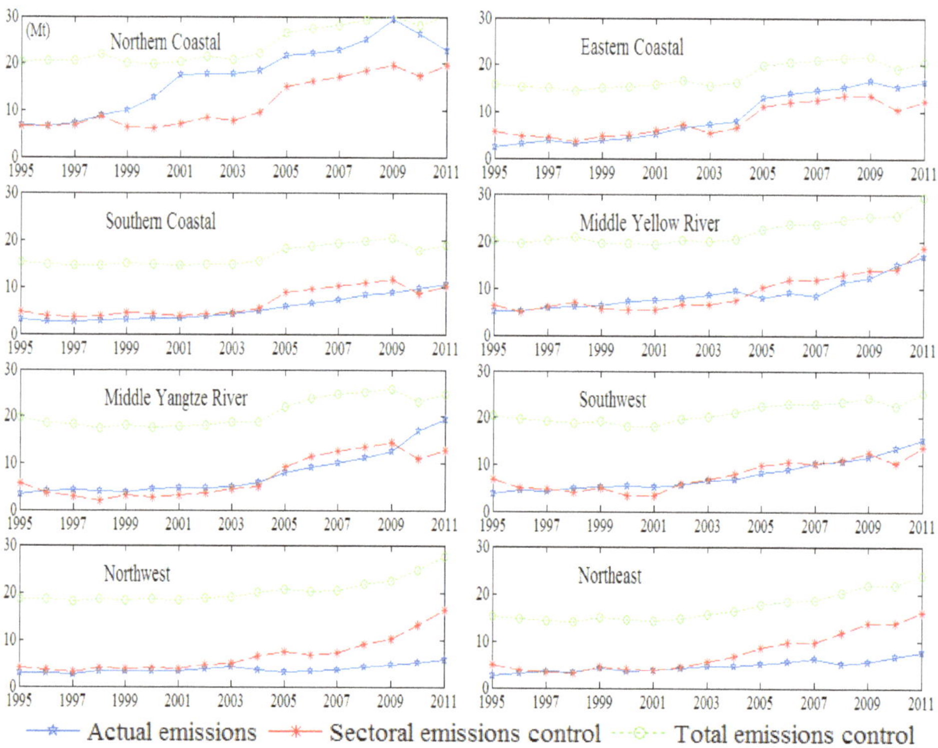

Figure 4. Optimal emission paths for regional construction sectors.

As shown by red lines in Figure 5, for regional transportation sector, goods transported most commonly in northwest and northeast regions are coal, industrial equipment, and production materials, which have low value-added yield compared to agriculture products and tourism. To this effect, northwest and northeast regions received emissions reduction regulations in 1995–2011. Southwest region was allowed to increase its emissions from 2003 to 2011 even though its actual emissions dramatically increased, which can be attributed to energy efficiency for the transportation sector in southwest region improved rapidly in 2003–2011. The energy efficiency of the transportation sector was higher in southern coastal region than other regions, so it obtained emission quotas from other regions in 2005–2010. Northern coastal, eastern coastal, middle Yellow River, and middle Yangtze River regions were required to reduce emissions in 1995–2003, but still received more emission quotas in 2003–2011. These results altogether suggest that strategies where emissions are reduced first and increased later are suitable for these regions, evidenced by their enhanced energy efficiency over time. Results also suggested that the transportation sector should accept emission quotas from other sectors, because it is not a major producer of emissions.

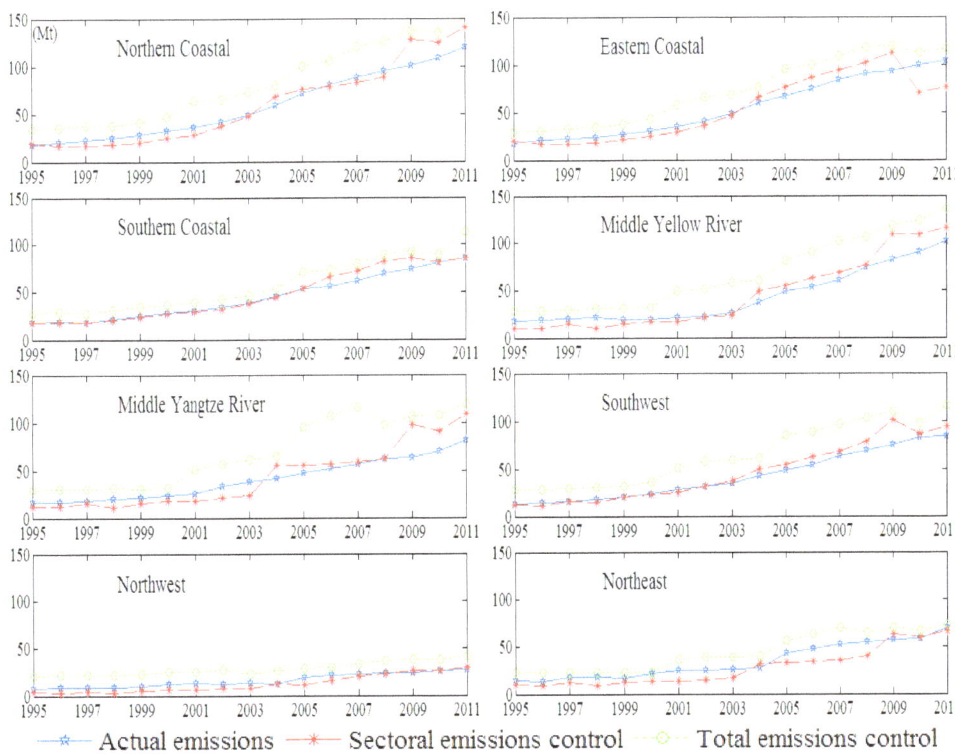

Figure 5. Optimal emission paths for regional transportation sector.

As marked by red lines in Figure 6, under sectoral emissions control strategy, northern coastal regions received strict emissions reduction regulations. The energy input for the service sector in the northern coastal region was excessive, and efficiency of energy utilization should be improved immediately. The northeast region also received emissions reduction regulation, as it employs extensive heating equipment (which is carbon-intensive) to cope with its cold climate. Middle Yangtze River and southwest regions obtained their emission quotas continuously in 1995–2011, suggesting that these regions should increase emissions to release value-added potential in observed time series. Eastern coastal and southern coastal regions, as mentioned above, are China's most developed areas and showed high energy efficiency and value-added yield in the service sector, so they do not particularly require emissions reduction regulations. Under total emissions control strategy, as marked by green lines in Figure 6, the service sector in almost all regions received more emissions than actual—the only exception was the northern coastal region, which received strict emissions reduction regulations in 2006–2011. Northern coastal region, to this effect, is urgently tasked with improving its service sector's energy efficiency. In general, the service sector should obtain emission quotas from other sectors as it does not require any major reduction in emissions.

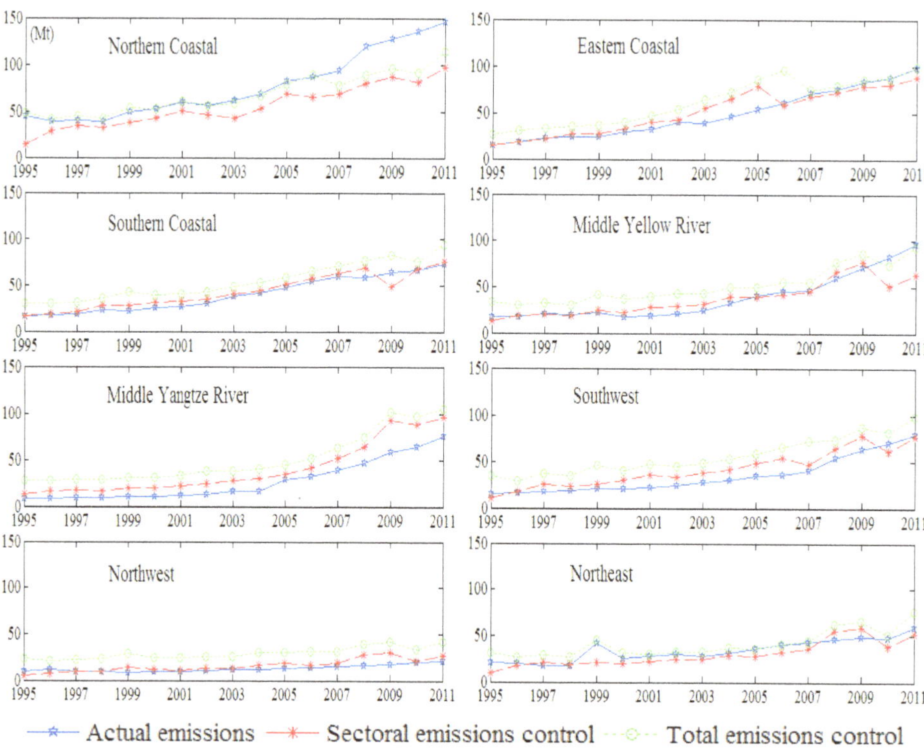

Figure 6. Optimal emission paths for regional service sectors.

In order to compare the differences between optimal emission paths and actual emission paths according to controlled sectoral and total emissions, we defined the degree of cumulative deviation as follows:

$$d(s)_l = \sum_{t=1}^{T} \left| \frac{\bar{b}(s)_l^t - b_l^t}{b_l^t} \right| / T \tag{12}$$

Similarly, we defined the following absolute quantity of cumulative deviation for optimal emission path and actual emission path:

$$d(s)_l^t = \left| \bar{b}(s)_l^t - b_l^t \right| \tag{13}$$

In Models (12) and (13), $s \in$ {sectoral emissions control, total emissions control}, $t = \{1995 - 2011\}$, $l \in$ {agriculture, manufacturing, constructing, transportation, service}. $\bar{b}(s)_l^t$ denotes optimal emissions from emissions control strategy s in sector l at period t, and b_l^t denotes actual emissions from sector l at period t.

As shown in Table 4, under sectoral emissions control strategy, the degree of deviation was largest in the construction sector and smallest in manufacturing. In effect, there were significant differences in construction emissions efficiency between provinces, and the manufacturing sector needs tightly controlled energy inputs overall. Under total emissions control strategy, the degree of deviation was largest in the transportation sector (attributed to receiving the largest emissions quota, 259.38 million tons, from the manufacturing sector, as transportation networks expanded).The second largest accepter of emission quotas was the agriculture sector, which required mechanization (and related energy input) to substitute for labor force input. The service sector also has rigid energy demands to maintain operation, but its energy input is not the important production factor—instead, most of its energy consumption was due to mechanical heating and refrigeration.

As shown in Table 5, western and northeastern areas had the largest degree of deviation under sectoral emissions control, indicating that western and northeastern China utilize energy very differently between their respective provinces and sectors. Eastern area had the smallest degree of deviation under sectoral emissions control, indicating that the gaps in emissions efficiency between provinces in eastern China are fairly narrow. Under total emissions control strategy, the degrees of deviation are similar to those under sectoral emissions control strategy. Central, western, and northeastern areas show low emission efficiency overall, so these areas should provide emission quotas to eastern areas, which have higher emission efficiency.

We used Models (10) and (11) to measure the shadow price on CO_2 emissions under sectoral and total emissions control strategies. We also measured the shadow price under variable returns to scale, based on the directional distance function.

The directional distance function serves to increase desirable outputs and reducing undesirable outputs with the directional vector $(g^y, g^c) = (1, -1)$ and the same scaling factor β. The specific expression can be found in a previous study [68].

The average shadow price for all province/sectors over 1995–2011 from strategies of sectoral emissions control (SEC), total emissions control (TEC), and directional distance function (DDF) can be calculated.

Table 4. Cumulative deviations between optimal and actual emission paths at sector level.

Sector	Deviation Under Sectoral Emissions Control		Deviation Under Total Emissions Control	
	Degree	Quantity (Unit: 10 Thousand Tons)	Degree	Quantity (Unit: 10 Thousand Tons)
Agriculture	0.635	0	0.091	173655
Manufacturing	0.128	0	0.072	-784174
Construction	1.984	0	0.141	173426
Transportation	0.539	0	0.183	259380
Service	0.439	0	0.081	177712

Table 5. Cumulative deviations between optimal and actual emission paths at area level.

Area	Deviation Under Sectoral Emissions Control		Deviation Under Total Emissions control	
	Degree	Quantity (Unit: 10 Thousand Tons)	Degree	Quantity (Unit: 10 Thousand Tons)
East	0.951	117,672	0.908	123,870.9
Central	1.146	−47,519.2	1.123	−28,893.7
West	1.758	1932.418	1.939	−28,708.4
Northeast	1.794	−72,085.3	1.905	−66,268.8

As shown in Table 6, under sectoral emissions control strategy, the shadow price of manufacturing and construction sectors were both zero. To this effect, these sectors can obtain maximum value-added yield by allocating CO_2 emission quotas appropriately within each sector. Agriculture, transportation, and service sectors showed negative values after optimization, indicating that these sectors should receive more emission quotas from other sectors to increase their value-added yield. The transportation sector had the largest negative value, suggesting that it needs more emission quotas to reach its value-added potential. The shadow prices for all sectors were zero under total emissions control strategy, indicating that all productive activities obtained maximum value-added yield by allocating CO_2 emission quotas across all sectors.

Table 6. Average values of provincial CO_2 emission shadow prices for each sector.

Year (1995–2011)	Agriculture			Manufacturing			Construction			Transportation			Service		
	DDF	SEC	TEC	DDF	SEC	TEC	DDF	SEC	TEC	DDF	SEC	TEC	DDF	SEC	TEC
Average	−0.504	−0.050	0	−0.090	0	0	−1.471	0	0	−0.152	−0.109	0	−0.921	−0.057	0

The directional distance function optimization model aims to establish policies that target CO_2 emissions while increasing value-added yield for all productive activities. Our proposed model aims to optimize allocation of fixed, undesirable output quantity to maximize the desirable output. The shadow price generated through our model is smaller than that of the directional distance function model due to the lack of necessity for severe regulations to decrease undesirable output when increasing desirable output simultaneously.

5. Conclusions

The optimal path for controlling CO_2 emissions, classified by both sector and region, represents a cross-analytical foundation for sectoral and regional emissions reduction policies. Many previous studies have explored the DEA approach to allocate emissions allowances at different levels, but in this study, we developed a new environmental DEA technology based on Kuosmanen environmental DEA technology which changes weak disposability to strong disposability on undesirable outputs, and ensures non-positive shadow price on undesirable outputs. We proposed a centralized DEA model based on multi-sector and multi-region data over time to explore efficiency maximization in China. The model was applied to study the optimal allocation of CO_2 emissions for each province and its sectors in China from 1995 to 2011, specifically, and the optimal paths for controlling CO_2 emissions at sectoral and regional levels under sectoral emissions control and total emissions control strategies were derived. We also designed a dual model for our centralized DEA model, and used it to analyze the shadow price on CO_2 emissions.

We conducted an empirical analysis on a sample of province/sector level data from 1995–2011. The results indicate that the manufacturing sector showed the most dramatic reduction in emissions, and emissions in the agriculture, construction, transportation, and service sectors all increased. The transportation sector was the largest accepter of emission quotas. Under sectoral emission control regulation, the construction sector must significantly increase its energy use efficiency, as its allowable emissions must undergo the most dramatic adjustment. Under total emission control regulation, the emission quotas should be removed from the manufacturing sector.

Agriculture, construction, transportation, and service sectors should receive more emission quotas to release their value-added potential. At the region level, western and northeastern areas require the most drastic adjustment to allowable emissions, which can be attributed to differences in emission efficiency among regions (as described in detail above). Eastern area has high energy use efficiency, so it can obtain more emission quotas from other areas. Based on our shadow price analysis, total emission control regulation policies can indeed enhance value-added potentials for all sectors. If optimization of CO_2 emissions is centralized, emissions increase or decrease bidirectionally, ensuring that the shadow price on CO_2 emissions is lower than the directional distance function regulation, decreasing CO_2 emissions and increasing value-added yield.

Despite the merits of this study, it did encounter limitations. First, we only considered agriculture, manufacturing, construction, transportation, and service sectors. The manufacturing sector, in particular, is divisible into more detailed sub-sectors. Second, our centralized model aims to realize the optimal allocation of CO_2 emission quotas while maximizing desirable output, which keeps capital and labor force input fixed, and fails to account for the price level of other production factors. In future, we plan to include the price of production factors into the dual model of our centralized model and research their influence on the allocation of CO_2 emission quotas. Third, our model mainly depicts environmental production technology and focuses on desirable output potentials while allocating undesirable outputs, but selecting detailed technology to improve emission efficiency is the next necessary step. We plan to discover how to integrate our centralized DEA model into energy technology system optimization platforms, such as TIMES (The Integrated MARKAL-EFOM System) (http://www.iea-etsap.org/web/Times.asp), LEAP (Long range Energy Alternatives Planning System) (http://www.energycommunity.org/default.asp?action=47), and AIM (The Asian Pacific Integrated Model) (http://www-iam.nies.go.jp/aim/datalibrary.htm) in future studies.

Acknowledgments: This work was supported by the Shandong Provincial Natural Science Foundation, China (Grant No. 2014ZRE27593), the China Postdoctoral Science Foundation Grant (Grant No. 2014M561895), and the Postdoctoral Science Foundation Grant of Shandong Province, China (Grant No. 201403015).

Author Contributions: This work was designed, analyzed, and written by the first author. The data were gathered by the coauthor and corresponding author.

Conflicts of Interest: The authors declare no conflict of interest.

References

1. Stern, N. *The Economics of Climate Change: The Stern Review*; Cambridge University Press: Cambrige, UK, 2007.
2. Liu, Z. *China's Carbon Emissions Report 2015*; Harvard Kennedy School: Cambridge, UK, 2015.
3. Färe, R.; Grosskopf, S.; Kokkelenberg, E.C. Measuring plant capacity, utilization and technical change: A nonparametric approach. *Int. Econ. Rev.* **1989**, *41*, 655–666.
4. Färe, R.; Grosskopf, S.; Margaritis, D.; Weber, W. Technological change and timing reductions in greenhouse gas emissions. *J. Product. Anal.* **2012**, *37*, 205–216.
5. Färe, R.; Grosskopf, S.; Pasurka, C.A., Jr. Potential gains from trading bad outputs: The case of US electric power plants. *Resour. Energy Econ.* **2014**, *36*, 99–112.

6. Zhou, P.; Sun, Z.R.; Zhou, D.Q. Optimal path for controlling CO_2 emissions in China: A perspective of efficiency analysis. *Energy. Econ.* **2014**, *45*, 99–110.
7. Aparicio, J.; Pastor, J.T.; Zofio, J.L. On the inconsistency of the Malmquist–Luenberger index. *Eur. J. Oper. Res.* **2013**, *229*, 738–742.
8. Färe, R.; Grosskopf, S.; Pasurka, C. Technical change and pollution abatement costs. *Eur. J. Oper. Res.* **2016**, *248*, 715–724.
9. Charnes, A.; Cooper, W.W.; Rhodes, E. Measuring the efficiency of decision making units. *Eur. J. Oper. Res.* **1978**, *2*, 429–444.
10. Kuosmanen, T.; Cherchye, L.; Sipiläinen, T. The law of one price in data envelopment analysis: Restricting weight flexibility across firms. *Eur. J. Oper. Res.* **2006**, *170*, 735–757.
11. Dantzig, G.B. Programming of interdependent activities: II mathematical model. *Econometrica* **1949**, *17*, 200–211.
12. Debreu, G. The coefficient of resource utilization. *Econometrica* **1951**, *3*, 273–292.
13. Koopmans, T.C. Analysis of production as an efficient combination of activities. *Act. Anal. Prod. All.* **1951**, *13*, 33–37.
14. Koopmans, T.C. Efficient allocation of resources. *Econometrica* **1951**, *19*, 455–465.
15. Koopmans, T.C. *Three Essays on the State of Economic Science*; McGraw-Hill: New York, NY, USA, 1957.
16. Afriat, S.N. Efficiency estimation of production functions. *Int. Econ. Rev.* **1972**, *13*, 568–598.
17. Banker, R.D.; Maindiratta, A. Nonparametric analysis of technical and allocative efficiencies in production. *Econometrica* **1988**, *56*, 1315–1332.
18. Färe, R.; Grosskopf, S. Nonparametric tests of regularity, Farrell efficiency, and goodness-of-fit. *J. Econom.* **1995**, *69*, 415–425.
19. Hanoch, G.; Rothschild, M. Testing the assumptions of production theory: A nonparametric approach. *J. Polit. Econ.* **1972**, *80*, 256–275.
20. Varian, H.R. The nonparametric approach to production analysis. *Econometrica* **1984**, *52*, 579–598.
21. Golany, B.; Tamir, E. Evaluating efficiency-effectiveness-equality trade-offs: A data envelopment analysis approach. *Manag. Sci.* **1995**, *41*, 1172–1184.
22. Lozano, S.; Villa, G. Centralised target setting for regional recycling operations using DEA. *Omega* **2004**, *32*, 101–110.
23. Li, S.K.; Ng, Y.C. Measuring the productive efficiency of a group of firms. *Int. Adv. Econ. Res.* **1995**, *1*, 377–390.
24. Asmild, M.; Paradi, J.C.; Pastor, J.T. Centralized resource allocation BCC models. *Omega* **2009**, *37*, 40–49.
25. Mar-Molinero, C.; Prior, D.; Segovia, M.-M.; Portillo, F. On centralized resource utilization and its reallocation by using DEA. *Ann. Oper. Res.* **2014**, *221*, 273–283.
26. Fang, L. A generalized DEA model for centralized resource allocation. *Eur. J. Oper. Res.* **2013**, *228*, 405–412.
27. Korhonen, P.; Syrjänen, M. Resource allocation based on efficiency analysis. *Manag. Sci.* **2004**, *50*, 1134–1144.

28. Hadi-Vencheh, A.; Foroughi, A.A.; Soleimani-damaneh, M. A DEA model for resource allocation. *Econ. Model.* **2008**, *25*, 983–993.
29. Amirteimoori, A.; Tabar, M.M. Resource allocation and target setting in data envelopment analysis. *Expert. Syst. Appl.* **2010**, *37*, 3036–3039.
30. Athanassopoulos, A.D. Goal programming & data envelopment analysis (GoDEA) for target-based multi-level planning: Allocating central grants to the Greek local authorities. *Eur. J. Oper. Res.* **1995**, *87*, 535–550.
31. Gomes, E.G. Modelling undesirable outputs with zero sum gains data envelopment analysis models. *J. Oper. Res. Soc.* **2007**, *59*, 616–623.
32. Serro, A. Reallocating Agricultural Greenhouse Gas Emission in EU 15 Countries. In Proceeding of the Agricultural & Applied Economics Association 2010 AAEA, CAES, & WAEA Joint Annual Meeting, Denver, CO, USA, 25–27 July 2010.
33. Wang, K.; Zhang, X.; Wei, Y.-M.; Yu, S. Regional allocation of CO_2 emissions allowance over provinces in China by 2020. *Energy Policy* **2012**, *54*, 214–229.
34. Pang, R.-Z.; Deng, Z.-Q.; Chiu, Y.-H. Pareto improvement through a reallocation of carbon emission quotas. *Renew. Sustain. Energy Rev.* **2015**, *50*, 419–430.
35. Chiu, Y.-H.; Lin, J.-C.; Su, W.-N.; Liu, J.-K. An efficiency evaluation of the EU's allocation of carbon emission allowances. *Energy Sour. Part B Econ. Plan. Policy* **2015**, *10*, 192–200.
36. Lozano, S.; Villa, G.; Brännlund, R. Centralised reallocation of emission permits using DEA. *Eur. J. Oper. Res.* **2009**, *193*, 752–760.
37. Wu, H.; Du, S.; Liang, L.; Zhou, Y. A DEA-based approach for fair reduction and reallocation of emission permits. *Math. Comput. Model.* **2013**, *58*, 1095–1101.
38. Singh, S.; Majumdar, S.S. Efficiency improvement strategy under constant sum of inputs. *J. Math. Model. Algorithms Oper. Res.* **2014**, *13*, 579–596.
39. Sun, J.; Wu, J.; Liang, L.; Zhong, R.Y.; Huang, G.Q. Allocation of emission permits using DEA: Centralised and individual points of view. *Int. J. Prod. Res.* **2014**, *52*, 419–435.
40. Homayounfar, M.; Amirteimoori, A.; Toloie-Eshlaghy, A. Production planning considering undesirable outputs-A DEA based approach. *Int. J. Appl.Oper.Res.* **2014**, *4*, 1–11.
41. Feng, C.; Chu, F.; Ding, J.; Bi, G.; Liang, L. Carbon emissions abatement (CEA) allocation and compensation schemes based on DEA. *Omega* **2015**, *53*, 78–89.
42. Zhu, J. Data envelopment analysis with preference structure. *J. Oper. Res. Soc.* **1996**, *47*, 136–150.
43. Halme, M.; Joro, T.; Korhonen, P.; Salo, S.; Wallenius, J. A value efficiency approach to incorporating preference information in data envelopment analysis. *Manag. Sci.* **1999**, *45*, 103–115.
44. Zhou, P.; Fan, L. An extension to data envelopment analysis with preference structure for estimating overall inefficiency. *Appl. Math. Comput.* **2010**, *216*, 812–818.
45. Allen, R.; Athanassopoulos, A.; Dyson, R.G.; Thanassoulis, E. Weights restrictions and value judgements in data envelopment analysis: Evolution, development and future directions. *Ann. Oper. Res.* **1997**, *73*, 13–34.

46. Roll, Y.; Cook, W.D.; Golany, B. Controlling factor weights in data envelopment analysis. *IIE Trans.* **1991**, *23*, 2–9.
47. Podinovski, V. Production trade-offs and weight restrictions in data envelopment analysis. *J. Oper. Res. Soc.* **2004**, *55*, 1311–1322.
48. Førsund, F.R. Weight restrictions in DEA: Misplaced emphasis? *J. Prod. Anal.* **2013**, *40*, 271–283.
49. Färe, R.; Grosskopf, S.; Noh, D.-W.; Weber, W. Characteristics of a polluting technology: Theory and practice. *J. Econom.* **2005**, *126*, 469–492.
50. Zhou, P.; Ang, B.W.; Poh, K.L. Measuring environmental performance under different environmental DEA technologies. *Energy Econ.* **2008**, *30*, 1–14.
51. Färe, R.; Grosskopf, S. Nonparametric productivity analysis with undesirable outputs: Comment. *Am. J. Agric. Econ.* **2003**, *85*, 1070–1074.
52. Kuosmanen, T. Weak disposability in nonparametric production analysis with undesirable outputs. *Am. J. Agric. Econ.* **2005**, *87*, 1077–1082.
53. Kuosmanen, T.; Podinovski, V. Weak disposability in nonparametric production analysis: Reply to Färe and Grosskopf. *Am. J. Agric. Econ.* **2009**, *91*, 539–545.
54. Kuosmanen, T.; Kazemi Matin, R. Duality of weakly disposable technology. *Omega* **2011**, *39*, 504–512.
55. Leleu, H. Shadow pricing of undesirable outputs in nonparametric analysis. *Eur. J. Oper. Res.* **2013**, *231*, 474–480.
56. United Nations Framework Convention on Climate Change (UNFCCC). *Kyoto Protocol to the United Nations Framework Convention on Climate Change*; Kyoto Protocol: Kyoto, Japan, 1997.
57. International Energy Agency (IEA). *CO_2 emissions from fuel combustion (2012 Edition)*; OECD Publishing: Paris, France, 2012.
58. Färe, R.; Grosskopf, S.; Pasurka, C.A., Jr. Accounting for air pollution emissions in measures of state manufacturing productivity growth. *J. Reg. Sci.* **2001**, *41*, 381–409.
59. Zhang, J. Estimation of China's provincial capital stock (1952–2004) with applications. *J. Chin. Econ. Bus. Stud.* **2008**, *6*, 177–196.
60. Guo, K.S. The driving factors to value added growth and the characteristics of changing for three industries. *Jingji Yanjiu (Econ.Res. J.)* **1992**, *27*, 51–61. (In Chinese).
61. Gan, C.; Zheng, R. An empirical study on change of industrial structure and productivity growth since the reform and opening-up—A test for the structure-bonus hypotheses from 1978 to 2007 in China. *Zhongguo Gongye Jingji (China Ind.Econ.)* **2009**, *251*, 55–65. (In Chinese).
62. Lv, T.; Zhou, S. Upgrading of an industrial structure and transformation of economic growth pattern in China. *Guanli Shijie (Manag. World)* **1999**, *1999*, 113–125. (In Chinese).
63. National Bureau of Statistics of China. *Statistical Yearbook of China's Fixed Asset Investment 1982–1986, 1997–1999, and 2003–2012*; China Planning Press: Beijing, China, 2012.
64. National Bureau of Statistics of China. *China Statistical Yearbook 1987-1996 and 2000–2002*; China Statistics Press: Beijing, China, 2002.

65. National Bureau of Statistics of China. *China Labor Statistical Yearbook 1996–2012*; China Statistics Press: Beijing, China, 2012.
66. National Bureau of Statistics of China. *China Statistical Yearbook 1996–2012*; China Statistics Press: Beijing, China, 2012.
67. National Bureau of Statistics of China. *China Energy Statistical Yearbook 1996–2012*; China Statistics Press: Beijing, China, 2012.
68. Chung, Y.H.; Färe, R.; Grosskopf, S. Productivity and undesirable outputs: A directional distance function approach. *J. Environ. Manag.* **1997**, *51*, 229–240.

Did the Establishment of Poyang Lake Eco-Economic Zone Increase Agricultural Labor Productivity in Jiangxi Province, China?

Tao Wu and Yuelong Wang

Abstract: In this paper, we take the establishment of Poyang Lake Eco-Economic Zone in 2009 as a quasi-natural experiment, to evaluate its influence on the agricultural labor productivity in Jiangxi Province, China. The estimation results of the DID method show that the establishment of the zone reduced agricultural labor productivity by 3.1%, lowering farmers' net income by 2.5% and reducing the agricultural GDP by 3.6%. Furthermore, this negative effect has increased year after year since 2009. However, the heterogeneity analysis implies that the agricultural labor productivities of all cities in Jiangxi Province will ultimately converge. We find that the lack of agricultural R&D activities and the abuse of chemical fertilizers may be the main reasons behind the negative influence of the policy, by examining two possible transmission channels—the R&D investment and technological substitution. Corresponding policy implications are also provided.

Reprinted from *Sustainability*. Cite as: Wu, T.; Wang, Y. Did the Establishment of Poyang Lake Eco-Economic Zone Increase Agricultural Labor Productivity in Jiangxi Province, China?. *Sustainability* **2016**, *8*, 8.

1. Introduction

Poyang Lake, which is located in the northern part of Jiangxi Province and connects to the lower Yangtze River, is the largest freshwater lake in China. It provides the residential and industrial water use for almost half of the regions of Jiangxi Province, especially as a water source for the production of agriculture industry. On 11 September 2008, The Standing Committee of Jiangxi Provincial Party reviewed "the Planning of Poyang Lake Eco-Economic Zone" and then submitted it to the State Council of China. According to the planning document, the Poyang Lake Eco-Economic Zone includes the three cities of Nanchang, Jiujiang, Jingdezhen, and a total of 38 counties, including some counties from Yingtan, Xinyu, Fuzhou, Yichun, Shangrao and Ji'an. The objective of the zone is to maintain sustainable environmental, social, and economic development. In December 2009, the State Council gave an official statement to support the establishment of Poyang Lake Eco-Economic Zone. After that, the central government of China provided more subsidies to the farmers in the zone and allocated more fiscal funds to Jiangxi Province.

Due to its exogeneity, we can take the establishment of Poyang Lake Eco-Economic Zone in 2009 as a quasi-natural experiment. In this paper, we use the DID method to evaluate its influence on agricultural labor productivity. The estimation results of the DID method show that the establishment of the zone reduced agricultural labor productivity by 3.1%. In terms of agricultural development, the establishment of the zone caused farmers' net income to be reduced by 2.5% and agricultural GDP by 3.6%.

Then we give the analysis of heterogeneity. We identify the agricultural labor productivities of food crops and other multiple crops, and study the dynamic changes in policy over time. The results show that the labor productivities of food crops and multiple crops did not reduce significantly and the negative effect of the zone on the agricultural labor productivity has increased year after year since 2009. However, the agricultural labor productivities of all cities in Jiangxi Province will ultimately converge during the progress of economic development.

To explain the reasons for the negative influence of the policy, we analyze two different transmission channels of the influence: the R&D investment and technological substitution. We find that, due to the low efficiency in allocating fiscal funds, the significant growth of fiscal funds into Jiangxi Province does not promote the improvement of the agricultural labor productivity. On the other hand, farmers increase the use of new technology (e.g., fertilizer) to substitute for the labor force, thereby inhibiting the progress of agricultural labor productivity. Hence, the lack of agricultural R&D activities and the abuse of chemical fertilizers may be the main reasons behind the negative influence.

The paper is organized as follows. A literature review is given in Section 2. Section 3 introduces the estimation methodology and data. Section 4 gives the main results of an empirical analysis. Section 5 provides the policy implications. Finally, Section 6 concludes.

2. Literature Review

One can find many scholars contributing to the studies of determinant factors of agricultural labor productivity. Hayami [1] and Hayami *et al.* [2] are the pioneers who initiated studies of agricultural labor productivity differences among countries. They identify the sources of the agriculture productivity gap using international data. Kawagoe *et al.* [3] give a closer look at the difference in agricultural labor productivity between developed countries and less developed countries. Instead of labor productivity, researchers also provide the analysis of total factor productivity growth in agriculture [4,5]. Another important branch of literature focuses on the relationship between R&D activities and the growth of agricultural labor productivity. The long-run effects of agricultural research on the productivity growth are estimated [6–8]. On the other hand, some scholars study the relationship between farm size and agricultural labor productivity [9,10].

In relation to the empirical research of agricultural labor productivity in China, Rozelle *et al.* [11] and Bhattacharyya *et al.* [12] believe that migration plays an important role in the determination of agricultural labor productivity, because there is a huge labor force migrating out of agriculture during the process of urbanization in China. Other empirical studies focus on the effects of economic reform after the 1980s [13,14] and regional disparity [15,16] on China's agricultural labor productivity. Among all of those studies, few of them focus on the determination of agricultural labor productivity in Jiangxi Province, especially after the establishment of Poyang Lake Eco-Economic Zone in 2009.

The economic and social effects of Poyang Lake Eco-Economic Zone are analyzed by some scholars. Xie *et al.* [17,18] discuss the ecological land use in Poyang Lake Eco-Economic Zone. Chen *et al.* [19] give a case study for Poyang Lake to discuss the sustainable land use and economic development. Other related research focuses on either biochar utilization [20,21] or forest protection [22] in Poyang Lake Eco-Economic Zone. However, empirical study on the economic influence of Poyang Lake Eco-Economic Zone on agricultural productivity is still lacking. This study aims to fill this gap by using the DID method to estimate the influence of the policy on agricultural labor productivity, which will provide policy implications for agricultural development in Jiangxi Province.

3. Estimation Methodology and Data

3.1. Estimation Methodology

The main scope of Poyang Lake Eco-Economic Zone is the five nearby cities of Nanchang, Jiujiang, Yingtan, Shangrao, and Jingdezhen, that locate closely to the center of the Poyang Lake; therefore, we chose these five cities as the component parts of Poyang Lake Eco-Economic Zone. Another reason to make this classification is that we use city-level data to estimate the influence of policy. We have to separate the 11 cities of Jiangxi Province into either the treatment group or reference group to match with the data. Hence, the five cities (*i.e.*, Nanchang, Jiujiang, Yingtan, Shangrao, and Jingdezhen) are chosen for the treatment group of our evaluation, and the other six cities in Jiangxi Province (*i.e.*, Fuzhou, Yichun, Ji'an, Pingxiang, Ganzhou, and Xinyu) are chosen as the reference group. By using DID estimation, the econometric model is set as follows:

$$y_{it} = \alpha_i + \gamma_t + \beta_1 \times du + \beta_2 \times dt + \beta_3 \times du \times dt + X\theta \qquad (1)$$

We used the panel data of 11 cities in Jiangxi Province from 2005 to 2013 to estimate Equation (1). In Equation (1), y_{it} is the agricultural labor productivity, which equals agricultural GDP divided by agricultural labor force (denoted as *productivity*). α_i and γ_t represent the city-fixed effects and the time-fixed effects, respectively. *du* is the group dummy variable; $du = 1$ for the cities of Nanchang, Jiujiang, Yingtan, Shangrao, and Jingdezhen in the treatment group, and $du = 0$ for the cities of Fuzhou, Yichun, Ji'an, Pingxiang, Ganzhou, and Xinyu in the reference group. *dt* is the time dummy variable; $dt = 0$ for 2005–2008, before the implementation of the policy, and $dt = 1$ for 2009–2013, after the implementation of the policy. The coefficient β_3 of the interaction term $du \times dt$ is our main focus of the policy effect. X is a set of control variables: we control the proportion of primary industry in GDP (denoted as *ratio_agriculture*), trade openness (*i.e.*, the total volume of foreign trade divided by total GDP, denoted as *trade*), the share of infrastructural investment in GDP (denoted as *invest_infrastructure*), the share of agricultural fixed capital investment in GDP (denoted as *invest_agriculture*), and foreign direct investment (denoted as *fdi*).

3.2. Data

The panel data for 11 cities in Jiangxi Province from 2005 to 2013 were collected from the Jiangxi Statistical Yearbook [23]. To avoid the problem of heteroscedasticity, we take the natural logarithms of all data except proportions. All the data related to prices have been deflated, by choosing year 2005 as the reference base year. The descriptive statistics of the main variables in the paper are shown in Table 1.

Table 1. Descriptive statistics of variables.

Variables (units)	Mean	ST.D.	Median	Min	Max	Samples
productivity (RMB/capita)	9266	2136	8872	5505	14,646	99
ratio_agriculture (%)	14	6.4	12.78	4.71	27.5	99
invest_agriculture (%)	0.9	1.28	0.37	0.00	7.05	93
invest_infrastructure (%)	2.43	1.99	1.69	0.36	10.62	99
expenditure_agriculture (10,000 RMB)	1.6×10^5	1.5×10^5	1.1×10^5	6563	6.1×10^5	99
expenditure_r&d (10,000 RMB)	5159	13,504	929	31	70,754	99
trade (%)	15.18	16.77	9.8	1.88	77.8	99
fdi (10,000 RMB)	3×10^5	2.7×10^5	2.1×10^5	41,785	1.3×10^6	99
fertilizer (ton/10,000 RMB)	0.38	0.12	0.36	0.17	0.64	99
pesticide (ton/10,000 RMB)	0.01	0.00	0.01	0.00	0.02	99
income_agriculture (RMB)	5768	2133	5077	2760	11,173	99
gdp_agriculture (10,000 RMB)	9.8×10^5	6.2×10^5	8.8×10^5	1.7×10^5	2.7×10^6	99
productivity_food (RMB/capita)	3902	1214	3902	1122	6411	98
productivity_multi (RMB/capita)	11,080	2840	10,745	4958	19,654	98

The time trends of agricultural labor productivity for 2005–2013 inside and outside the Poyang Lake Eco-Economic Zone are shown in Figure 1. The solid curve, which represents the agricultural labor productivity of the treatment group, lies below the dotted curve, which represents the agricultural labor productivity of the reference group.

In Figure 1, we can see that the two curves are almost parallel before 2009, which implies that our samples satisfy the assumption of parallel trend that is required by the DID method. After 2009, one can find that growth rate of agricultural labor productivity in the Poyang Lake Eco-Economic Zone has slowed down compared to that of the reference group. This implies that the establishment of Poyang Lake Eco-Economic Zone may have a negative influence on agricultural labor productivity, which will be revealed by the following econometric analysis.

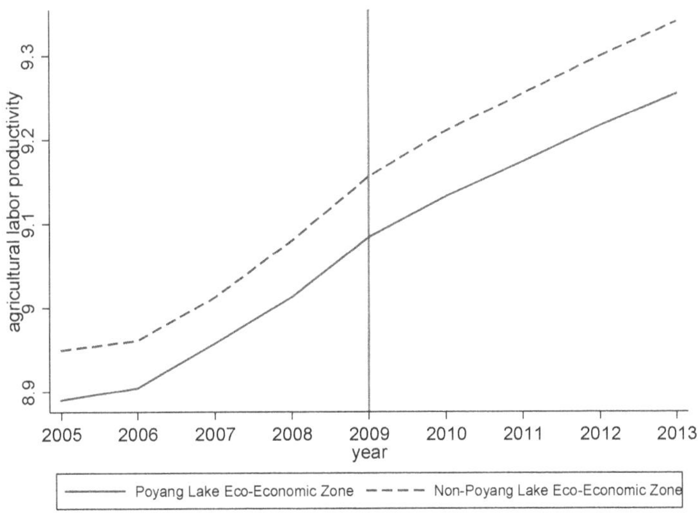

Figure 1. Time trends of agricultural labor productivity.

4. Results

4.1. Benchmark

In Table 2, the results of model 1 show that the establishment of Poyang Lake Eco-Economic Zone reduced agricultural labor productivity by 2%. Furthermore, by adding some control variables, we can derive model 2. It shows that the negative effect increased up to 3.1%, and is still significant at the 1% level. Out of all the control variables, only the coefficient of the proportion of primary industry (denoted as *ratio_agriculture*) is significantly positive, which implies that the higher the proportion of primary industry and the better the local agricultural basic conditions are, the higher the labor productivity is. We also give a placebo test to verify that the establishment of Poyang Lake Eco-Economic Zone in 2009, instead of any other policies, lowered the agricultural labor productivity. By assuming that the policy occurred in 2007 (*i.e.*, $dt0 = 1$ for 2007 and 2008; $dt0 = 0$ for 2005 and 2006), we find that the coefficient of interaction term (β_3) representing the policy influence in model 3 is not significant. Hence, we can conclude that agricultural labor productivity decreased after 2009, due to the establishment of the Poyang Lake Eco-Economic Zone.

Table 2. Benchmark model.

Variables	Model 1	Model 2	Model 3
	Lnproductivity	*Lnproductivity*	*Lnproductivity*
du	0.306 *** (37.15)	0.390 *** (8.04)	0.398 *** (4.85)
dt	0.388 *** (35.44)	0.428 *** (21.32)	
$du \times dt$	−0.0203 *** (−2.76)	−0.0311 *** (−3.82)	
invest_agriculture		0.000821 (0.23)	−0.0201 (−0.98)
ratio_agriculture		0.00619 *** (2.87)	0.00569 (1.65)
invest_infrastructure		−0.000906 (−0.64)	0.00532 (0.78)
trade		1.58×10^{-5} (0.04)	−0.000302 (−1.52)
lnfdi		0.00816 (0.45)	0.00476 (0.33)
dt0			0.159 *** (9.36)
$du \times dt0$			−0.0139 (−1.56)
Constant	8.870 *** (791.18)	8.624 *** (40.42)	8.654 *** (54.41)
Individual fixed effects	Yes	Yes	Yes
Time fixed affects	Yes	Yes	Yes
Observations	99	93	40
R-squared	0.995	0.995	0.998

Note: (1) ***, **, and * represent the significance at the 1%, 5%, and 10% levels; (2) the *t*-statistics in parentheses are given by three heteroscedastic robustness standard errors.

4.2. Heterogeneity Analysis

Did the establishment of Poyang Lake Eco-Economic Zone causes reduction of agricultural labor productivity in terms of all agricultural products? To answer this question, models 4 and 5 in Table 3 study the influences of the establishment on the labor productivity of food crops (denoted as *productivity_food*) and that of multiple crops (denoted as *productivity_multi*). Both results are negative but not significant, which implies that the negative effects do not exist in the production process of food crops and multiple crops.

Table 3. Heterogeneity analysis.

Variables	Model 4	Model 5	Model 6	Model 7	Model 8
	Lnfood	*Lnmulti*	*Lnproductivity*	*Lnproductivity*	*Lnproductivity*
du	0.231 (0.93)	0.392 (1.58)	0.389 *** (8.12)	0.392 *** (8.08)	0.387 *** (7.84)
dt	0.656 *** (6.15)	0.656 *** (6.15)	0.427 *** (21.57)	0.430 *** (21.21)	0.436 *** (20.80)
policy2009					−0.0212 ** (−2.61)
policy2010					−0.0312 *** (−2.93)
policy2011					−0.0327 ** (−2.26)
policy2012					−0.0369 ** (−2.54)
policy2013					−0.0433 *** (−3.29)
invest_agriculture	−0.0378 *** (−3.20)	−0.0378 *** (−3.20)	0.000726 (0.20)	0.000577 (0.16)	−0.000679 (−0.17)
ratio_agriculture	0.0164 (1.39)	0.0164 (1.39)	0.00616 *** (2.87)	0.00633 *** (2.94)	0.00667 *** (3.16)
invest_infrastructure	0.00105 (0.19)	0.00105 (0.19)	−0.000871 (−0.62)	−0.000838 (−0.59)	−0.000518 (−0.33)
trade	−0.000231 (−0.20)	−0.000231 (−0.20)	1.28×10^{-5} (0.03)	9.91×10^{-6} (0.03)	-5.34×10^{-5} (−0.13)
lnfdi	0.0269 (0.33)	0.0269 (0.33)	0.00858 (0.47)	0.00841 (0.47)	0.0122 (0.64)
policy	−0.0508 (−1.10)	−0.0508 (−1.10)			
policy_gdp			−0.00231 *** (−3.90)		

Table 3. *Cont.*

Variables	Model 4	Model 5	Model 6	Model 7	Model 8
	Lnfood	*Lnmulti*	*Lnproductivity*	*Lnproductivity*	*Lnproductivity*
policy_income				−0.00358 *** (−3.84)	
Constant	7.530 *** (7.73)	8.405 *** (8.63)	8.620 *** (40.34)	8.618 *** (40.49)	8.568 *** (38.22)
Individual fixed effects	Yes	Yes	Yes	Yes	Yes
Time fixed effects	Yes	Yes	Yes	Yes	Yes
Observations	93	93	93	93	93
R-squared	0.970	0.938	0.995	0.995	0.995

Note: (1) ***, **, and * represent the significance at the 1%, 5%, and 10% levels; (2) the *t*-statistics in parentheses are given by heteroscedastic robustness standard errors.

Then, we introduce the interaction term of policy and agricultural GDP (denoted as *policy_gdp*), and the interaction term of policy and farmers' net income (denoted as *policy_income*) in models 6 and 7, respectively. We can see that both coefficients of the interaction terms are significantly negative, which implies that in the regions of higher agricultural GDP and farmers' net income (*i.e.*, regions of higher agricultural development), the reduction of the agricultural labor productivity caused by the policy is higher. Taking these effects into consideration, we can predict that the agricultural labor productivities of all cities in Jiangxi Province will ultimately converge during the progress of agricultural development. By adding in dummy variables for the interaction terms of the policy as well as each year from 2009 to 2013, model 8 examines the effect of the policy on dynamic heterogeneity over time. The results show that after the establishment of Poyang Lake Eco-Economic Zone, the negative effect increased year after year.

4.3. Analysis of Transmission Channels

Because labor productivity is mutually influenced by the allocation of the government's funds and the substitution of labor by new technologies (or new materials), we analyze two possible transmission channels of the policy influence: R&D investment and technological substitution.

Models 9 and 11 in Table 4 give the DID estimations of fiscal expenditure on agriculture (denoted as *expenditure_agriculture*) and R&D expenditure (denoted as *expenditure_r&d*), respectively. The results show that after the establishment of Poyang Lake Eco-Economic Zone in 2009, the government's fiscal expenditure on agriculture in Jiangxi Provence increased by 6.9%, but there is no obvious change in the R&D expenditure of Jiangxi Provence. Then, on the basis of model 2, by adding two possible channel variables (*i.e.*, *expenditure_agriculture* and *expenditure_r&d*), we got models 10 and 12. The estimated coefficient of the policy interaction term in model 10 does not change significantly, compared to that of model 2. However, the estimated coefficient of the policy interaction term in model 12 has changed significantly; meanwhile, the coefficient of the R&D expenditure is significantly negative. As a conclusion, the establishment of Poyang Lake Eco-Economic Zone does not promote agricultural labor productivity through government fiscal expenditure. There is no corresponding increase in R&D expenditure because of the inappropriate use of fiscal funds, which results in a lack of R&D activities on agriculture and a reduction of agricultural labor productivity.

Table 4. Transmission channel I: R&D investment.

Variables	Model 9	Model 10	Model 11	Model 12
	Lnexpenditure_Agriculture	*Lnproductivity*	*Lnexpenditure_R&D*	*Lnproductivity*
du	−0.408 ** (−2.09)	0.397 *** (8.18)	5.526 *** (8.37)	0.447 *** (10.18)
dt	2.512 *** (21.15)	0.345 *** (5.15)	1.592 *** (4.29)	0.434 *** (24.56)
du × *dt*	0.0692 * (1.71)	−0.0309 *** (−3.71)	−0.000990 (−0.01)	−0.0270 *** (−3.36)
lnexpenditure_agriculture		0.0310 (1.40)		
invest_agriculture		0.00182 (0.50)		0.00361 (1.16)
ratio_agriculture	−0.0211 (−1.62)	0.00683 *** (3.04)	0.0895 ** (2.34)	0.00683 *** (3.34)
invest_infrastructure		−0.00105 (−0.74)		−0.000774 (−0.58)
trade		-1.66×10^{-5} (−0.04)		−0.000240 (−0.51)
lnfdi		0.0108 (0.60)		0.0128 (0.73)
lnexpenditure_r&d				−0.0128 *** (−2.67)
Constant	10.66 *** (31.18)	8.265 *** (23.08)	3.661 *** (3.64)	8.632 *** (43.39)
Individual fixed effects	Yes	Yes	Yes	Yes
Time fixed effects	Yes	Yes	Yes	Yes
Observations	99	93	99	93
R-squared	0.994	0.995	0.944	0.995

Note: (1) ***, **, and * represent significance at the 1%, 5%, and 10% levels; (2) the *t*-statistics in parentheses are given by heteroscedastic robustness standard errors.

The output of agricultural products depends on the input of production factors such as labor and land, and also depends on the use of new production technologies. Due to the availability of data, we use the chemical fertilizer usage per unit of agricultural GDP (denoted as *fertilizer*) and the pesticide usage per unit of agricultural

GDP (denoted as *pesticide*) to study the transmission channels of technological substitution. Models 13 and 15 in Table 5 show that after Poyang Lake Eco-Economic Zone was established in 2009, the chemical fertilizer usage per unit of agricultural GDP significantly increased by 3.1%; meanwhile, there was no obvious change in the use of pesticides. Based on model 2, we get models 14 and 16 by adding two possible channel variables (*i.e., fertilizer* and *pesticide*). The estimated coefficient of the policy interaction term in model 14 is significantly different from that in model 2, and the use of chemical fertilizers has an insignificantly negative effect on the agricultural labor productivity. Meanwhile, there is no obvious change in the coefficient of the policy interaction term in model 16, compared to model 2. We can explain the estimation results in the following way. After the establishment of Poyang Lake Eco-Economic Zone in 2009, farmers used more new production technology, such as chemical fertilizers, to substitute for labor input. In the short run, the utilization of chemical fertilizer increased the output of agricultural products. However, it causes pollution of waterway, acidification and mineral depletion of the soil, therefore damages the value of agricultural products. The utilization of chemical fertilizer will reduce the growth rate of agricultural GDP, thus resulting in a loss of agricultural labor productivity in the long run.

4.4. *Effects on Agriculture Development*

The reduction of agricultural labor productivity will directly slow the growth rate of farmers' income and agricultural development. Hence, we provide a further study of the effects of agricultural labor productivity on agricultural development. We use farmers' net income (denoted as *income_agriculture*) and agricultural GDP (denoted as *gdp_agriculture*) as indicators of agricultural development. The estimation results of models 17 and 18 in Table 6 show that as agricultural labor productivity increases by 1%, farmers' net income increases by 0.81% and agricultural GDP increases by 1.1%. Recall that in model 2, we have shown that the establishment of Poyang Lake Eco-Economic Zone reduced agricultural labor productivity by 3.1%. Hence, the establishment of the zone reduced farmers' net income by 2.5% and agricultural GDP by 3.6%.

Table 5. Transmission channel II: technological substitution.

Variables	Model 13	Model 14	Model 15	Model 16
	Fertilizer	*Lnproductivity*	*Pesticide*	*Lnproductivity*
du	−0.186 *** (−13.53)	0.373 *** (7.16)	−0.00571 *** (−6.06)	0.391 *** (7.73)
dt	−0.295 *** (−24.14)	0.400 *** (13.54)	−0.00539 *** (−7.24)	0.428 *** (19.20)
du × *dt*	0.0310 ** (2.40)	−0.0274 *** (−3.42)	-3.17×10^{-5} (−0.04)	−0.0311 *** (−3.80)
fertilizer		−0.116 (−1.43)		
invest_agriculture		−0.000293 (−0.08)		0.000833 (0.23)
ratio_agriculture		0.00629 *** (3.02)		0.00620 *** (2.80)
invest_infrastructure		−0.000991 (−0.73)		−0.000912 (−0.63)
trade		-9.87×10^{-5} (−0.27)		1.37×10^{-5} (0.04)
lnfdi		0.00799 (0.42)		0.00806 (0.45)
pesticide				0.0740 (0.05)
Constant	1.292 *** (4.99)	8.699 *** (36.17)	0.0596 *** (4.76)	8.624 *** (40.00)
Individual effect	Yes	Yes	Yes	Yes
Time trends	Yes	Yes	Yes	Yes
Observation	98	93	98	93
R-squared	0.963	0.995	0.798	0.995

Note: (1) ***, **, and * represent the significance at the 1%, 5%, and 10% levels; (2) the *t*-statistics in parentheses are given by heteroscedastic robustness standard errors.

Table 6. The effects on agricultural development.

Variables	Model 17	Model 18
	Lnincome_Agriculture	*Lngdp_Agriculture*
lnpgdp	0.130 *** (3.15)	
ratio_employ	0.00169 (1.20)	
lnproductivity	0.808 *** (3.66)	1.100 *** (3.24)
lnemploy_agriculture		0.153 (1.37)
lngdp		0.139 (1.43)
Constant	−0.445 (−0.21)	0.503 (0.14)
Individual effect	Yes	Yes
Time trends	Yes	Yes
Observations	98	98
R-squared	0.994	0.997

Note: (1) ***, **, and * represent the significance at the 1%, 5%, and 10% levels; (2) the *t*-statistics in parentheses are given by heteroscedastic robustness standard errors.

5. Policy Implications

Based on the results of our analysis, the government should focus on reversing the negative impact of Poyang Lake Eco-Economic Zone on agricultural labor productivity to improve social welfare and agricultural development. The policy implications are as follows: 1) increase the proportion of R&D expenditure in Jiangxi Province, especially to encourage the R&D activities related to agricultural technologies and the primary industry in Poyang Lake Eco-Economic Zone; 2) issue a series of incentive-compatible subsidy policies to encourage farmers to reduce the excess usage of chemical fertilizers, and increase the values of agricultural products in the long run; 3) to shift the industry focus from manufacture to tourism by regulation, which will cause a positive externality on agricultural development. The combination of these policies will increase farmers' income and encourage agricultural development by improving agricultural labor productivity.

6. Conclusions

In this paper, we use the establishment of the Poyang Lake Eco-Economic Zone in 2009 as a quasi-natural experiment to evaluate its influence on the agricultural labor productivity in Jiangxi Province, China. The estimation results of the DID method show that the establishment of the zone reduced agricultural labor productivity by 3.1%. Hence, the establishment of the zone caused a negative influence on agriculture development, reducing the farmers' net income by 2.5% and the agricultural GDP by 3.6%. The analysis of heterogeneity implies that this negative effect has increased year after year since2009, and the agricultural labor productivities of all cities in Jiangxi Province will ultimately converge. We also examine two possible transmission channels (*i.e.*, R&D investment and technological substitution) to show that the main reasons behind the negative policy influence may be the lack of agricultural R&D activities and the abuse of chemical fertilizers.

Because the establishment of Poyang Lake Eco-Economic Zone in 2009 can be taken as a quasi-natural experiment, we can also apply the similar DID method to analyze the influence of policy on total factor productivity in Jiangxi Province. Another interesting further extension would be the relationship between agricultural labor productivity and water resources of Poyang Lake, which play an important role in agricultural production and are also greatly influenced by the planning of Poyang Lake Eco-Economic Zone.

Acknowledgments: Yuelong Wang would like to acknowledge the financial support from the University Youth Project 2014 of Humanities and Social Science Research in Jiangxi Province (#JJ1444), the Key Base Project 2014 of Humanities and Social Science in Jiangxi Province (#JD1454), and the Science and Technology Project 2016 of Jiangxi Provincial Department of Education, "Public Investment and Urban Development".

Author Contributions: Tao Wu conceived and designed the study, also devoted to the writing of the paper; Yuelong Wang collected and analyzed the data.

Conflicts of Interest: The authors declare no conflict of interest.

References

1. Hayami, Y. Sources of agricultural productivity gap among selected countries. *Am. J. Agric. Econ.* **1969**, *51*, 564–575.
2. Hayami, Y.; Ruttan, V.W. Agricultural productivity differences among countries. *Am. Econ. Rev.* **1970**, *60*, 895–911.
3. Kawagoe, T.; Hayami, Y.; Ruttan, V.W. The intercountry agricultural production function and productivity differences among countries. *J. Dev. Econ.* **1985**, *19*, 113–132.
4. Coelli, T.J.; Rao, D.S. Total factor productivity growth in agriculture: A Malmquist index analysis of 93 countries, 1980–2000. *Agric. Econ.* **2005**, *32*, 115–134.
5. Restuccia, D.; Yang, D.T.; Zhu, X. Agriculture and aggregate productivity: A quantitative cross-country analysis. *J. Monetary Econ.* **2008**, *55*, 234–250.

6. Alston, J.M.; Beddow, J.M.; Pardey, P.G. Agricultural research, productivity, and food prices in the long run. *Science* **2009**, *325*, 1209–1210.
7. Thirtle, C.; Lin, L.; Piesse, J. The impact of research-led agricultural productivity growth on poverty reduction in Africa, Asia and Latin America. *World Dev.* **2003**, *31*, 1959–1975.
8. Sumberg, J. Systems of innovation theory and the changing architecture of agricultural research in Africa. *Food policy* **2005**, *30*, 21–41.
9. Cornia, G.A. Farm size, land yields and the agricultural production function: An analysis for fifteen developing countries. *World Dev.* **1985**, *13*, 513–534.
10. Feder, G. The relation between farm size and farm productivity: The role of family labor, supervision and credit constraints. *J. Dev. Econ.* **1985**, *18*, 297–313.
11. Bhattacharyya, A.; Parker, E. Labor productivity and migration in Chinese agriculture: A stochastic frontier approach. *China Econ. Rev.* **1999**, *10*, 59–74.
12. Rozelle, S. J.; Taylor, J.E.; de Brauw, A. Migration, remittances, and agricultural productivity in China. *Am. Econ. Rev.* **1999**, *89*, 287–291.
13. McMillan, J.; Whalley, J.; Zhu, L. The impact of China's economic reforms on agricultural productivity growth. *J. Political Econ.* **1989**, *97*, 781–807.
14. Lin, J.Y. Rural reforms and agricultural growth in China. *Am. Econ. Rev.* **1992**, *82*, 34–51.
15. Cai, F.; Wang, D.; Du, Y. Regional disparity and economic growth in China: The impact of labor market distortions. *China Econ. Rev.* **2002**, *13*, 197–212.
16. Kanbur, R.; Zhang, X. Fifty years of regional inequality in China: A journey through central planning, reform, and openness. *Rev. Dev. Econ.* **2005**, *9*, 87–106.
17. Xie, H.; Wang, P; Huang, H. Ecological risk assessment of land use change in the Poyang Lake eco-economic zone, China. *Int. J. Environ. Res. Public Health* **2003**, *10*, 328–346.
18. Xie, H.; Liu, Z.; Wang, P.; Liu, G.; Lu, F. Exploring the mechanisms of ecological land change based on the spatial autoregressive model: A case study of the Poyang lake eco-economic zone, China. *Int. J. Environ. Resear. Public Health* **2003**, *11*, 583–599.
19. Chen, W.; Carsjens, G. J.; Zhao, L.; Li, H. A spatial optimization model for sustainable land use at regional level in China: A case study for Poyang Lake region. *Sustainability* **2014**, *7*, 35–55.
20. Chang, M.S.; Kung, C.C. Nonparametric forecasting for biochar utilization in Poyang Lake Eco-Economic Zone in China. *Sustainability* **2014**, *6*, 267–282.
21. Kung, C.C.; Kong, F.; Choi, Y. Pyrolysis and biochar potential using crop residues and agricultural wastes in China. *Ecol. Indic.* **2015**, *51*, 139–145.
22. Liu, R.Z.; Song, P.; Sheng, Q.Y.; Gong, W.J. Legal system for China's forest resource protection: A case study of Poyang Lake Eco-economic Zone. *Asian Agric. Resear.* **2012**, *4*, 65–68.
23. Statistic Bureau of Jiangxi. *Jiangxi Statistical Yearbook, 2006–2014*; China Statistics Press: Beijing, China, 2006–2014.

MDPI AG
St. Alban-Anlage 66
4052 Basel, Switzerland
Tel. +41 61 683 77 34
Fax +41 61 302 89 18
http://www.mdpi.com

Sustainability Editorial Office
E-mail: sustainability@mdpi.com
http://www.mdpi.com/journal/sustainability

www.ingramcontent.com/pod-product-compliance
Lightning Source LLC
Chambersburg PA
CBHW041939260926
51850CB00018B/296
9783038422747